W9-AGX-852

# The Social Impact
# of the Telephone

**MIT Bicentennial Studies**

# The Social Impact of the Telephone

Ithiel de Sola Pool, Editor

301.161
S678

The MIT Press
Cambridge, Massachusetts,
and London, England

First MIT Press paperback edition, 1981
Second printing, 1978
Copyright © 1977 by
The Massachusetts Institute of Technology

All rights reserved. No part of this book may be reproduced in any form or by any means, electronic or mechanical, including photocopying, recording, or by any information storage and retrieval system, without permission in writing from the publisher.

This book was set in V-I-P Palatino by The MIT Press Media Department Computer Composition Group, and printed and bound by Murray Printing Company in the United States of America.

Library of Congress Cataloging in Publication Data
Main entry under title:

The Social impact of the telephone.

   (MIT Bicentennial studies ; 1)
   Includes index.
   1. Telephone—Social aspects—Addresses, essays, lectures. I. Pool, Ithiel de Sola.
HE8735.S65   301.16′1   77–4110
ISBN 0-262-16066-8 (hard)
ISBN 0-262-66048-2 (paper)

N. C. Wesleyan College Library
Rocky Mount, North Carolina

69708

Contents

Contents

# Series Foreword

As part of its contribution to the celebration of the U.S. Bicentennial, MIT has carried out studies of several social and intellectual aspects of the world we inhabit at the beginning of our third century. Our objective has been to inquire how human beings might deal more intelligently and humanely with these factors, most of which are closely linked to developments in science and technology.

The papers prepared for these inquiries are being published in a Bicentennial Studies Series of which this volume is a part. Other studies in the series deal with the future of computing and information processing, linguistics and cognitive psychology, the economics of the new international economic order, administrative and economic factors in air pollution, and world change and world security.

It is our hope that these volumes will be of interest and value to those concerned now with these questions and, additionally, will provide useful historical perspective to those concerned with the same or similar questions on the occasion of the U.S. Tricentennial.

Jerome B. Wiesner

# Acknowledgments

The papers in this volume were originally presented at a series of seminars held at MIT in celebration of the centennial of the telephone. This centennial series was supported by AT&T; their help, without strings, made this study possible. The telephone centennial seminars were part of a continuing MIT seminar series on communications policy that has been supported by the Markle Foundation. Work on the social impact of the telephone at MIT, part of which is reported in this volume, has also been supported by the National Science Foundation.

The editor of a collective volume of this sort owes much to many people. The authors, who are busy people, agreed to turn their attention to the subject of our inquiry and to edit their papers for publication in a very short period of time. Norman Dahl, the coordinator of both this series of centennial seminars, and of those represented in the companion volumes as well, worked long and hard to make those occasions the memorable events that they were. And particular thanks go to two individuals who struggled with every detail of the arrangements from the first invitations to the last error in the proofs: Suzanne Planchon and Doris Lannoy.

# Introduction

How much difference has a device invented 100 years ago
by Alexander Graham Bell, enabling people to converse across
distance, made in our lives?

Like other day-to-day appurtenances of life—water, air, met-
als, plastics, streets, and electricity—we take the phone for
granted; we use it without a thought. Historians or sociologists
rarely mention it in their books; when they do, it is usually to
assert that it has had a profound impact on society. Has it, and
if so, how? That is what this book is about.

In this century of the telephone, life has changed more rapidly than ever before. In 1870, three-fourths of the American population was rural; in 1970 one-fourth was; and the population
itself had grown five times. In 1870, only one person in fifty
had finished high school; in 1970, one in two had. In 1870, no
electric light illuminated the night, no automobiles traversed the
streets; no buildings rose above walk-up height; no freezers or
tin cans made foods available outside their season. Doctors, it is
generally agreed, saved no more people than would be saved by
chance. Great music or entertainment was available only to the
few who could sit in a handful of theatres on a given night;
many went from cradle to grave without ever experiencing such
things.

Men never flew, either into the air or outer space. Radio
waves, X-rays, viruses, chromosomes, genes, quanta, black
holes, plastics, and electronics were unheard of. There were far
fewer things dreamed of in their philosophy than in our experience. People lived close to their relatives, and if not—if for example they immigrated to a foreign country or moved West—it

was a break approaching unto death; they might never speak to their kin again. Locally, when people wanted to go someplace, they usually walked; they shopped, worked, and visited mostly in their own neighborhoods.

Religion played a large role in community life. Respectable ladies did not work among men; vulgar words were not used in their presence or in print; obedience to elders was a cardinal virtue of the young.

In some complex causal way, all the changes of the century may have stemmed from science and technology. But how important was the telephone's specific technology in these processes of change? It is extraordinary how little has been written exploring this question.

The same thing, Keller notes below, could be said of most inventions. The impact of a few technologies on society has been traced closely by historians and sociologists: among them are the steam engine, the railroad, and the cotton gin—but not many more.

Social scientists have neglected the telephone not only along with, but also relative to, other technologies. As a cause of social change, transportation, Chapter 6 tells us, has been much more studied than communication. And among communication media, TV, radio, movies, even the telegraph have been studied more than the telephone.

The degree of this neglect, however, has varied over time. Earlier, the phone was an exciting novelty and received more attention. John Brooks, in Chapter 9, finds that the device's use as a symbol declined as it became more commonplace. During the first quarter of the twentieth century, magazine editors would occasionally run a piece about the telephone in the countryside, the "hello girls" (see Maddox), or the growth of the nation's largest corporation and its competitors. By the time the sociology of invention emerged as a field in the 1930s, however, the telephone was already becoming old hat and the new excitement was broadcasting.

The sociology of science can be traced back into the nineteenth century. Like so much else in the social sciences, the paradigm was the industrial revolution in which power machinery fostered factory production and thereby urbanism, which in turn shaped industrial civilization; theoretical interpretation was provided by Marx's thesis that productive technology shapes social relations and men's ideas. By the twentieth century, that insight had found its way, with appropriate restriction and partial incorporation, into the accepted wisdom of academic social science. Max Weber's counterstatement demolished the overly simple notion of a one-way causal process from the material base to social organization and then to the ideological superstructure, but it preserved the insight of the close causal linkage among them.

In the first quarter of the twentieth century, sociology flourished most in Germany; it was fed by the debate over Marx, and it transmuted Social Democratic impulses into social theory. American sociologists in the 1910s and 1920s studied in and borrowed from Germany.

Thus stimulated, there emerged in the 1930s a lively American literature on technology and social change—what today would be called "technology assessment." William F. Ogburn was the leading exponent, while monographs by such authors as S. C. Giffillan, Lewis Mumford, and Hans Zilsel analyzed how various technologies (from the pump, sailing ship, and chronometer, to the automobile, airplane, and TV) changed social mores and relations. Perhaps the largest public impact was the study *Recent Social Trends*, commissioned by Herbert Hoover. In all the literature on technology and society, however, the telephone received little attention, much less than the cinema and radio; it had already passed its peak of popular attention by the 1930s. As a result, no sociologist of science in that first wave of technology assessment seriously examined the ways in which the telephone had changed society.

That question has been evaded for other reasons, too. All literature on how technology affects society, from the Marx-Weber debate through the sociology of science and to technology assessment today, is permeated by the subtle issue of what is meant by an effect. Causality, for all its importance, is an obscure concept in the social sciences. The problem recurs in the chapters in this book, for it is particularly well highlighted by the case of the telephone.

The telephone is distinctive (so several argue) in that it is a facilitating rather than a constraining device. Some other inventions have consequences that seem obvious and linear: Pasteur's discovery of vaccination reduced smallpox, and thereby suffering, and increased life expectancy. It had other more obscure and unanticipated second-order consequences, too, such as increasing the incidence of diseases of aging and complicating border regulations for travelers. Yet the primary effects are fairly straightforward. Some inventions such as the telephone seem to defy definition of even the primary effects; these seem polymorphous though indubitably large.

Wherever we look, the telephone seems to have effects in diametrically opposite directions. It saves physicians from making house calls, but physicians initially believed it increased them, for patients could summon the doctor to them rather than travel to him.[1] The phone invades our privacy with its ring, but it protects our privacy by allowing us to transact affairs from the fastness of our homes. It allows dispersal of centers of authority, but it also allows tight continuous supervision of field offices from the center. It makes information available, but reduces or eliminates written records that document facts.

The telephone's inherently dual effects are one reason for the paucity of literature on its social impact. Its impacts are puzzling, evasive, and hard to pin down. No matter what hypothesis one begins with, reverse tendencies also appear.

The phone, in short, adds to human freedom, but those who gain freedom can use it however they choose. Rather than constraining action in any one direction, the telephone is an agent of effective action in many directions.

That conclusion does not imply that the phone has no impact or that there is nothing to study. On the contrary, it implies that the study of the telephone's social impact belongs to the important and subtle class of problems in the social sciences which demands a logic more complex than that of simple causality—a logic that allows for purposive behavior as an element in the analysis. The theory of games and economic behavior is an example; economists do not predict a uniform consequence from a single input into the economic process; the specific consequences of the new input depend on purposive human calculations.

That, too, is how to look at the entry into man's experience of a device that permits instantaneous communication between persons at a distance. It is hard to imagine a more revolutionary change in man's relation to his universe; it transformed what previously seemed an eternal aspect of space and time. But other social variables shaped the use of that newfound power.

Because simple generalizations about the telephone's consequences are, therefore, likely to prove false, the chapters that follow are mostly empirical testing of what did happen. They report how the telephone was used in different societies, times, and institutions during its first 100 years. They compare the exceptional telephone experience in the United States in the private market to that in France (see Attali and Stourdze) and England (see Perry) where governmental concerns played a more controlling role.

The authors below find many surprising reversals; telephones had different impacts when they were scarce and limited to centers of commerce and power than when they became ubiquitous. For instance, several chapters (Gottmann's, Abler's, Keller's, Moyer's, and Thorngren's) consider the telephone's impact on patterns of urban life and settlement. The consequences are far from obvious; the telephone contributed considerably to urban sprawl and the mass migration to suburbia. It also helped create the congested downtowns from which people are now fleeing. The telephone's role in skyscraper development is largely unknown and fascinating, as the reader will discover. It also turns out that the phone had an important effect on the location of offices and factories, profoundly affecting the character of the American business elite. After reading Section III (on the telephone and the city), the reader may want to ask himself how captains of industry who spent their time at factory sites differed in perspective from those in Manhattan.

So the telephone's relation to the city has been complex and changing; it has been a catalyst in a series of changes from the 1880s until today. Neighborhoods decomposed as the city grew from a walking city to a vehicled one, and the telephone promoted that process. Neighborhoods were frozen by zoning and the telephone played a part in that, too. Cities grew and then exploded into megalopolitan regions also with the telephone's help. The story of changing urban ecology cannot be understood without examining its interaction with the phone system. Yet that relationship has not been examined in depth before.

Rural life has also been profoundly affected by the telephone,

but in quite different ways before and since the advent of automobile and radio. A telephone system with operators and party
lines had a much different impact in the country from that of
the modern automatic telephone. The party-line system created
a community that proved a favorite subject for short stories and
essays.[2] In many places, farmers gathered every evening by the
phone to exchange news and gossip; today they sit by their TV
sets or ride into town. While the modern automatically-switched
telephone no longer performs the party line's functions, the new
efficient instrument for private conversation serves purposes
that the community channel never could.

Since these and other profound and complex impacts of telephony were largely unstudied, it seemed appropriate to celebrate the telephone's centenary by a symposium on these social
impacts; that is the origin of this book. The papers in it (with
certain exceptions) were originally presented in a symposium
series funded by AT&T and held at MIT as an activity of the
MIT Research Program on Communications Policy. The centennial of a device invented in Boston, which in its unobtrusive
way has so changed our lives, seemed a natural occasion for a
serious reassessment of its significance.

The authors have varied backgrounds and points of view.
Pierce and Boettinger have made their careers in the Bell System. Aronson is almost unique as a social scientist who has tak-
en the telephone as his special subject. Others, indeed most of
the scholars whose work is included in this volume, had not
been particularly concerned with telephone research until some
tangential involvement recently drew them to the subject. Perry
had been working on the history of the British Post Office,
Briggs on the history of radio and telegraphy; each had come
upon data about telephony as a related technology. Wurtzel and
Turner took note when a catastrophe occurred affecting telephone service and siezed the opportunity for a study. Attali was
working on the economics of information; Schegloff's specialty
is the psychology of interpersonal recognition; and Keller is
studying the sociology of new towns.

The authors confronted the telephone as an element of their
particular study. In most instances, the editor asked them to
distill from their work whatever they had learned about the telephone in the process.

Our authors also vary in disciplines and approaches. Brooks, Maddox, and Mayer are writers; Pierce, Cherry, and Rao are engineers; Reid is an architect and planner; Attali and Thorngren are economists; Briggs, Perry, and Moyer are historians, and Gottmann and Abler are geographers. Most of the other authors are social scientists (sociologists, psychologists, political scientists). It takes a variety of approaches to understand a technology that has permeated all aspects of life over a 100-year history.

Some differences in view appear among the authors, though clearly they are fewer than if our subject were policy alternatives or future forecasts. This book concerns what has actually happened. Yet even on that there are differences of interpretation. One issue arising in several papers is how far the physical possibility (created by the phone) of developing communities without contiguity is, in fact, being realized. How far and under what circumstances do human relations develop among people who do not meet face-to-face? Keller, Reid, Gottmann, Abler, and Thorngren all address this issue and marshall substantial evidence.

The answer from the past suggests that social relations are rarely initiated in that way; one might question, however, whether that is an eternal prediction. How far may purely telephone-based relationships develop if the long-distance/local cost differential continues to decline in the trend that Abler documents, if teleconferencing becomes common, or if video at some point enters into two-way use? We shall not try to predict; this is not a book of predictions, though clearly forecasting is a goal underlying anyone's interest in technology assessment. In this book we seek to understand the telephone's impact upon society in the past 100 years. Yet, because that analysis is a prelude to forecasts, it is interesting to ask how good our predecessors' forecasts were on the telephone's social impact. What did they expect the phone to do? Among those who forecast, who saw clearly and who did not, and how good were they as prophets? Those questions are addressed in several chapters, and most extensively in Chapters 1, 2, and 6.

The first two chapters deal with a particularly interesting set of anticipations. It was not initially obvious just how the new device, the telephone, would be used. Perhaps it could be used for central authorities to broadcast public messages; or it could be used, as it turned out to be, for individual conversations among pairs of citizens. Aronson and Briggs raise the question of why the broadcast concept, though often tried, had to wait fifty years for implementation by radio. Cherry notes that society would have been very different and less democratic if it had not come out that way.

Bell, Hubbard, Vail, and other telephone system founders had a remarkable record of prescience about the phone system's future. From the start, Bell anticipated a network of private phones serving not only the rich and powerful (who then could afford it) but also the poor. We raise the question below of how the system's founders saw its future with such clarity. There are a number of explanations, and several may be partially true. In this introduction, however, one of them deserves particular attention: to some degree, there may be an inherent determinism to the technology. Despite everything said before about the telephone's polymorphous character, its many uses, and the variation in its chosen use at different times and places, some direct consequences for society are also traceable from the very technology of the telephone.

Choices about how to use the phone are not all equally cost-effective. When governments chose to protect older modes of communication by restricting telephone growth (as happened in Europe), there was a price to pay. If a telephone entrepreneur at the turn of the century sought to use the telephone in broadcast mode, he had to find a market for a service that required substantial electric current and cope with serious interference problems; the Bell system had adjudged the market to lie rather in conversation given the service that the technology then permitted. The technical and economic constraints were real.

No law of nature, however, says that societies will always choose the solutions which make the most cost-effective use of the available technologies. In telephony, as Pierce argues, there are many examples of degraded systems arising from poor policy choices. If technology is in command, if there is any determinism to the technology itself, it is only under those circumstances where some Darwinian or rational process rewards the good choices and penalizes the bad.

In the past, sovereign authorities protected themselves from such processes of comparison and selection. In many countries, their attitude seemed to be that if their national communication systems did not work very well, so be it; there was nothing to force them to change. Today, however, national isolation becomes increasingly difficult because the telecommunication network is now a global one. New technologies of telecommunication are burgeoning. The successful systems that make good use of technology's growing capabilities are interconnected with, and in obvious contrast with, systems that protect obsolete practices. Those who do not run hard in the race may pay a large penalty in standard of living, national power, and level of development.

In some circumstances and respects, therefore, technology is in command. The policy decisions adopted must reflect the requirements of the technology, or there will be a heavy price to pay.

The telephone's social impact is liberating, as we have already noted. It offers options to serve society's many varied and even contradictory needs. But those who fashion the use of the system need to take account of the specific technology and economics of this very complex machine. Indeed, to gain its liberating potential requires understanding of its technology.

The telecommunications network of the second century will undoubtedly do many great and wondrous things, only some of which we can now guess. Yet we can say with confidence that the institutions and societies that press forward most vigorously in the use of this technology will tend to shape the way of life of that century.

## NOTES

1. *Telephony*, vol. 9, no. 6, p. 492.
2. Cf. Dalton Trumbo, *Johnny Got His Gun* (New York: Bantam Paperback, 1970), p. 12.

# I

**Alternative Paths
of Development:
The Early Years**

# Editor's Comment

A century ago, in 1876, the first intelligible words transmitted by electricity to a distant hearer were Graham Bell's simple plea: "Mr. Watson, come here, I want you." Thus the telephone was born, but it had already been conceived and thought about for perhaps a century.

To understand the telephone system's shape as it matured with subscribers, exchanges, and public lines, one must look not only at the transmitting and receiving instruments invented and patented in 1876 but also at the conceptions, and indeed misconceptions, about how to use such a device—which had already occurred to many.

Several authors in Part I of this volume discuss early notions of a telephone system, as they existed in the 1870s, as they developed from then until the turn of the century, and also as they emerged during the phone's prehistory during the nineteenth-century age of telegraphy. Asa Briggs, Sidney Aronson, Charles Perry, Colin Cherry, Jacques Attali and Yves Stourdze, and Ithiel Pool all deal in their papers with the early formation of the character of the telephone system.

The Gutenberg mode of presentation is linear, so we cannot link as we go along those various authors' discussions on all points of tangency. It reminds me of Rousseau's plea to his readers that he cannot discuss all points at once. The Editor's Comments, here and later, may help to flag some common topics that appear in various chapters—in this instance chapters dealing with the telephone system's gestation and birth.

Surprisingly, the idea of the telephone and indeed the word itself existed fairly widely before the thing itself. It is probably irrelevant that in 1683 a French play was entitled *Telephonte* or that a Dutch version, *Telephon: Konig van Messene*, appeared in 1707, for the etymology of the word is different. It is not irrelevant that in 1854 Charles Bourseul wrote a piece on "Electric Telephony" in *Illustration de Paris* (August 26), nor that Alexander Graham Bell's paper to the American Academy of Arts and Sciences on May 10, 1876, gave an extensive historical treatment. He traced the telephone idea from galvanic music, described by C. G. Page in 1837, through a variety of works including Matteucci's in 1845 and Reis' 1863 article entitled "Telephonie."

Once the notion got around that electrical impulses could be transmitted invisibly over a distance—whether over wire, through the ground, or in the air—the idea of using variations in the current for communicating messages was not far behind. Once telegraphers realized that current sent at various frequencies or intensities could give output mimicking the input, the idea of sending sounds or pictures electrically was also imminent. One must look at the discussions of those early ideas before their successful fruition to understand the ways the inventions came to be used when they finally arrived.

Attali and Stourdze begin their story of the telephone in France and how it was used—or not used—with C. Chappe's visual telegraph in 1793. By the mid-nineteenth century, France was dotted with semaphore towers between which visible signals were sent, primarily for national defense. The precedents and interests formed in that experience shaped electric telegraphy in the following half century (as Attali and Stourdze describe it); that experience, in turn, shaped the French government's approach to telephony in the next half century.

Asa Briggs, the historian of broadcasting in Britain, and Charles Perry, also a historian, trace the development of the telephone idea in Britain and describe the incubus imposed by the post office's telegraph monopoly, an investment the government felt it had to protect.

Briggs presents extraordinary quotations from *Punch* in 1848 and the 1850s, pleading for the extension of telegraphy from an office-to-office service to one direct to homes and also misreporting that in the fabulous United States songs were being sent by telegraph between New York and Boston. Little distinguishes such tales of the mid century from ecstatic reports of

Bell's feats when he presented the actual telephone a quarter
century later, except that one was fantasy and the other reality.
Indeed, satirical remarks about the endless chit-chat of telephone gossip can be found not only in the 1920s accounts of
sociologists and fiction writers about housewives' use of the
phone (see the chapters by Brooks, Keller, Pierce, Maddox,
Mayer, etc.), but also in Bell's early speeches about his new invention long before it existed anywhere except in a few business installations (see Aronson) and in *Punch* in 1858 when it
was discussed as a yet nonexistent "house telegraph."

The first two chapters of our book (Aronson's and Briggs'),
discuss two alternative nineteenth-century conceptions of what
telecommunications might become; both conceptions have been
realized since the second quarter of the twentieth century. The
alternatives are the *switched point-to-point* system and *broadcasting*. Since the 1920s the former has served us over wires and the
latter by wireless; each meets a long recognized social need. But
for the first years of the telephone, there was only one technology available and two conceptions of what society needed it for.

In the first chapter, Aronson describes the evolution of Bell's
own thinking about the use of the telephone in conversational
versus broadcast mode. In the second chapter, Briggs discusses
various attempts to provide much the same thing as a modern
broadcasting service by means of what he calls "The Pleasure
Telephone."

**Bell's Electrical Toy:**
**What's the Use?**
**The Sociology of Early**
**Telephone Usage**

Sidney H. Aronson

The telephone came to America—and to the rest of the world—
on March 10, 1876; on that day, as far as is known, Alexander
Graham Bell became the first person to transmit speech electrically. The American response to that event included a mixture
of wonder, confusion, and sheer disbelief. That spoken words
could be converted into electrical waves, transmitted along wire,
and then reconverted into sound at the other end of the line
could not easily be comprehended even after the telephone had
been widely described. Perhaps the best way to understand one
of Graham Bell's incredulous contemporaries is to imagine how
we would feel if we were told that a way had been devised to
make extrasensory perception a means of communication.

Even after the telephone had been widely discussed and its
principle had begun to be understood, for many the telephone—as remarkable as the idea seemed—had no obvious use.
There is even evidence that some who should have known better did not immediately appreciate the possibilities of the amazing new device. Sometime in the late fall or winter of 1876–1877,
Gardiner Greene Hubbard, Bell's future father-in-law and a
partner in the telephone venture, offered to sell all the rights to
the telephone patent to the Western Union Telegraph Company
for $100,000.[1] Western Union's president, William Orton, turned
down that extraordinary opportunity to monopolize all electrical
communications at a time when it had become customary for
big business leaders to seek to dominate their entire industry.
Orton apparently did not even think it was worth buying the
patent just to keep the thing off the market. According to one of
Bell's biographers, Orton dismissed the offer with the words,

"What use could this company make of an electrical toy?"[2]
Thomas A. Watson, Bell's assistant, was disappointed at Western Union's rejection because he lost the chance to receive
$10,000 for his share of ownership of the patent. This suggests
that the future of the telephone was not evident from the mere
fact of its invention, as $10,000 was a considerable sum in those
days.[3] Furthermore, the officials of the British Post Office Department, who controlled the telegraph industry in England,
also turned down the chance to purchase Bell's English patent.[4]

Matthew Josephson, chronicler of the activities of the first
generation of avaricious industrialists, attributed Western
Union's blunder to "customary bureaucratic caution" and the
reluctance to bear the considerable expense of substituting telephone equipment for the telegraph.[5] Those reasons indeed suggest that Orton did not fully appreciate the telephone and that
his legendary retort to the offer was shortsighted.[6] He and his
British counterparts were probably no different from most of
those who grew up in the telegraph age and witnessed or read
about one improvement in telegraphy after another. By 1876, it
seemed that future developments would take place exponentially, as inventors who seemed to be congregated in Boston but
were, of course, dispersed throughout the Western world were
trying to strike it rich by adding to the techniques for instantaneously transmitting more and more messages. At the electrical
shop of Charles Williams, Jr., on Court Street in Boston, a group
of especially talented inventors, including Joseph B. Stearns,
Thomas Edison, and Graham Bell himself, all vied at one time
or another for the honor of making still another breakthrough in
telegraphy that would bring a fortune.[7]

What these and other inventors had devised by 1876 may help
explain this initial indifference to Bell's invention by the telegraphists. By that year telegraphy had so developed that people
were still marveling over a means of communication that had already been in operation for over forty years. Over 200 submarine cables had been laid and succeeded in making virtually
every corner of the earth a link in a worldwide chain of communications. Direct service could be conducted by telegraph between points several thousand miles apart. Printing telegraphs
recorded messages in Roman letters as early as 1841. The advantage of the telegraph over the telephone, as those who ran the

telegraph industry saw it, was that telegraphy left a permanent record.[8]

In 1876, a number of specialized telegraphs catered either to the public at large or to special interests. For example, telegraphs for financial and commercial offices provided investment houses and brokerage firms with the latest prices in stocks, gold, and merchandise printed on what came to be known as ticker tape.[9] By that time, the fire alarm telegraph—first invented in 1851—was in use in seventy-five major cities and towns in the United States. Systems consisting of a series of locked signal boxes were placed at convenient intervals throughout a locality—New York City had 600—each containing a mechanism for transmitting a numerical signal along connecting telegraph lines that indicated which box had initiated the alarm. That device made it possible to dispatch fire-fighting equipment within a minute or two after the alarm. By 1871, W. B. Watkins improved the fire alarm telegraph so that the fire itself triggered the signal automatically.[10] By 1870, district telegraphs, or signal boxes, were being placed in homes in virtually every major city in the United States and were connected telegraphically with a central station. By simply turning a crank at any hour of the day or night, a messenger or policeman could be instantly summoned or a fire alarm transmitted.[11]

The speed of telegraph transmission was accelerating. Automatic telegraphy held the promise that the telegraph operator using a Morse instrument and painstakingly tapping out 25 words a minute would soon be eliminated by a set of devices that could send 800 words a minute. Automatic telegraphy also made it possible to send messages to many places simultaneously and was especially useful in sending large quantities of news in duplicate to various parts of the country.[12]

One of the chief disadvantages of telegraphy was that a skilled operator had to decipher all messages; perhaps it does not sound too complicated a task to "read" Morse code, but it was generally regarded then as a highly skilled occupation. By 1874, however, autographic telegraphy—a process devised as early as 1848—allowed Americans to write a telegram on an ordinary piece of paper that could then be transferred to a metal plate for transmission. Some entrepreneurs expected that such a device would someday be placed in every American home and business.[13]

Another disadvantage of the older means of communication was the limited capacity of the line. At first, telegraph wire could carry only one message in one direction at any one time. That limitation often meant telegrams were piled on tall mounds in telegraph offices, waiting hours to be sent and ultimately delivered. But the capacity of telegraph technology to transmit messages was continually improving as multiplexing allowed two or more messages to be transmitted simultaneously. The duplex telegraph, perfected around 1870, allowed operators at each end of a line to transmit simultaneously. Between 1871 and the time Alexander Graham Bell succeeded in transmitting speech over wire, Western Union used these methods to increase the working capacity of its lines by 25 percent, the equivalent of nearly 50,000 miles of wire at a fraction of the expense.[14]

By 1876, the United States was criss-crossed by 214,000 miles of telegraph wire delivering 31,703,181 telegrams through 8,500 telegraph offices.[15] Even these figures do not complete the story of the predominance of the telegraph, since they do not include telegraph companies organized for general commercial and business purposes, telegraph systems owned and operated by the various branches of the federal government, and municipal electric and police patrol systems.[16] If all that were not enough, one of the most promising developments was the creation in 1877 of the Social Telegraph Association in Bridgeport, Connecticut. It installed instruments in subscribers' homes that could be connected, through a central switchboard, to one another so that subscribers could "speak" to one another through the Morse Code once they had been taught how.[17] Those unwilling or unable to learn the Morse Code would simply have to wait for the autographic telegraph.

Bell himself, as well as his backers, for a time showed signs of being members in good standing of the society of telegraphists. This is not surprising because he discovered the telephone while working on the harmonic or musical telegraph. This device would have greatly increased the efficiency of telegraphy, since it would have permitted "thirty or forty messages . . . [to be] sent simultaneously" over the same wire.[18]

Given the hold of telegraphy over communications in the

America of 1876, it is not surprising that many could not see any immediate use for Bell's invention. The inventor and his backers thus faced the formidable task of inventing uses for the telephone and impressing them on others. What complicated that task is that many members of the public did not distinguish between the telephone and the telegraph, and when the terms were separated, the word telephone was often associated with the transmission of music. What further compounded the difficulty of finding uses for the telephone was the quality and character of the instrument in the early days of its existence.

In the experimental telephones used on March 10, 1876, when Bell first succeeded in talking to Watson over the telephone, the same mechanism was used alternately as a transmitter and as a receiver. Two-way conversation, however, could be done only on very short circuits. Furthermore, transmission and reception tended to be poor—in subsequent demonstrations it often did not work at all—and Bell decided to concentrate on developing the "membrane" telephone as a transmitter and the "iron-box" telephone as a receiver, with the result that sound was transmitted in one direction only.[19] That limitation posed a serious problem for Bell. What possible use could there be for that kind of telephone—as extraordinary as it was—which would enable him and his backers to recoup their investment, let alone make money? How could such an instrument be made to pay off? [20]

Bell's first attempts to deal with this problem can be seen in his early public lectures and demonstrations of the telephone. On May 10, 1876, at a meeting of the American Academy of Arts and Sciences in Boston, Bell disclosed his theory of the telephone and related something of its practice. Although he was not sufficiently confident to attempt the transmission of speech on his still primitive apparatus, he hinted about a possible use of the telephone. At that lecture he sent a telegraphic signal to a confederate who waited at a "telegraphic organ" in Bell's office at Boston University, a few doors away from the lecture hall. "To the astonishment of the audience" at the Academy, a box on the table emitted the melody "Old Hundred."[21] At the Centennial Celebration of American Independence in Philadelphia in the summer of 1876, Bell left the judges and went to the other side of the exhibition hall where he had placed the transmitter. He proceeded to recite Hamlet's soliloquy verbatim.[22] In his first telephonic transmission from Boston to New York in July 1876, Bell played "Yankee Doodle" on an organ and then asked the telegraph operator on the other end to name that tune. [23]

Later that summer at the family home in Brantford, Ontario, Bell devised the "triple mouthpiece" to see if three persons could speak simultaneously over a single transmitter. Since the triple mouthpiece was relegated to the museum of phone fossils not long after it was devised, one may wonder what Bell had in mind. The answer seems to lie in the first use of the triple mouthpiece. Bell arranged for several performers in Brantford to sing into that curious device and to transmit their words and music a distance of several miles. Similarly, at a demonstration in Salem, Massachusetts, on February 12, 1877, Bell had Watson in Boston read the latest news from Washington.[24]

The telephone's use to transmit music, drama, and news— what can now be called a radio concept of telephony—resulted from two sources. First, Bell and Watson had not yet sufficiently improved the telephone so that the same instrument could both transmit and receive effectively. That prevented Bell from putting the telephone into immediate commercial use doing what he intended it to do: to make conversation possible for people who were at a distance from one another. The demonstrations described—with the exception of the one in Salem—were done before the problem of two-way communication over long circuits was satisfactorily resolved. Yet Bell was under pressure to produce profits for his impatient backers, and he may have seen entertainment and enlightenment before paying assembled audiences as the immediate practical use of the telephone until such time as "long-distance" talks could be perfected.

Although Bell and Watson succeeded in improving the telephone so that they could hold a satisfactory two-way conversation by October 9, 1876, Bell continued to give radio-like performances. The reason was that Bell's years as an inventor had left him without a comfortable income. Demonstrating the telephone before large audiences would serve both to popularize the telephone and raise badly needed cash.[25] Bell and Watson thus perfected the art of transmitting music, drama, and enlightenment, as Rosario Tosiello has amply documented.[26]

The "lecture performances" began early in 1877; they lasted

between two and two-and-a-half hours. Introduced by a prominent member of the community, Bell lectured on electricity and
demonstrated his several instruments. Then he signalled for
music—usually by telegraph in the early lectures because he did
not yet trust the telephone as the sole means of intercommunication—and Watson, located at a distance from the lecture hall,
would play the organ. His repertoire included popular, religious, and patriotic favorites such as "America," "Hail Columbia," and "Killarney." Then Bell and Watson would attempt a
conversation. Watson would shout sentences such as "How do
you do?", "What Do You Think of the Telephone?", and "Good
Evening." For the more discriminating New York audiences,
Bell engaged a professional singer, but the singer, who was reluctant to put his lips close to the transmitter, did not project
his voice as well as Watson. In May of 1877, the opera singer Si-
gnor Brignoli sang selections over the telephone from the stage
of a theater in Providence for transmission to Boston. Occasionally there was a brass band or a cornetist of renown. Seminar
discussions, the state of the weather, the time of the day, and
newspaper dispatches were also carried over the telephone. In
April of 1877, another radio-type feature was added to the
format of Bell's performances when Watson, situated in Middle-
town, Connecticut, transmitted his shouts, songs, and conversations to Skiff's Opera House in New Haven and to Robert's
Opera House in Hartford—shades of the radio network.[27]

It would be interesting to know what sense audiences made
of these demonstrations; beyond the descriptions of their
amazement and wonder, there is little information about the
possible uses they attributed to the telephone. That Elisha Gray,
who later claimed priority in the invention of the telephone,
also took to the lecture circuit in 1877 to demonstrate his device
may have added to any lingering confusion about the telephone, for Gray's telephone could not transmit articulate speech
and could only be used to carry music over telegraph wire.[28]

During his performances, Alexander Graham Bell never lost
sight of the primary use of the telephone; he always set aside
some time for the discussion of its future. He told his audience
that he envisioned a central office system wherein telephones in
various locations could be connected by means of a "switch."[29]
Soon rapid improvements in the quality of telephone transmis-

sion and reception, as well as the appropriation of the switchboard from telegraphy, made that prediction seem absolutely
prophetic. These advances allowed Bell to be even more discern-
ing. In a letter he wrote on March 25, 1878, to the "Capitalists of
the Electric Telephone Company," a group organized to develop
his invention in Britain, it was obvious that he had already conceived the true character of the telephone system:

It is conceivable that cables of telephone wires could be laid underground, or suspended overhead, communicating by branch
wires with private dwellings, country houses, shops, manufactories, etc., etc., uniting them through the main cable with a
central office where the wire could be connected as desired, establishing direct communication between any two places in the
city. Such a plan as this, though, impracticable at the present
moment, will, I firmly believe, be the outcome of the introduction of the telephone to the public. Not only so, but I believe in
the future wires will unite the head offices of the Telephone
Company in different cities, and a man in one part of the
country may communicate by word of mouth with another in a
different place.[30]

If Bell's vision of the future was prophetic, magazine and
newspaper writers (surely among the better informed members
of the public) were still not sure of its purpose and did not always distinguish between the telegraph and the telephone.
Thus, *Frank Leslie's Illustrated Newspaper* in the summer of 1876
referred to Bell's device as an "electric speaking trumpet." A
*Puck* cartoon of April 1877 placed the telephone at the front door
suggesting that the magazine had confused the new device with
a speaking tube. In a subsequent issue, *Puck* pictured a telephone connection consisting of a speaking tube extending
beneath the Atlantic Ocean from Europe to New York. By December 1, 1877, *Puck* was improving its conception of the telephone although it was still drawn as a speaking tube. Moreover,
the radio concept of telephony persisted. The *Boston Transcript*
of July 18, 1876, speculated about the "wonderful results which
are sure to follow" the improvements in "telegraphy" brought
about by Bell's invention: "But if the human voice can now be
sent over the wire, and so distinctly that when two or three
known parties are telegraphing, the voice of each can be recognized, we may soon have distinguished men delivering speeches in Washington, New York or London, and audiences assembled in Music Hall or Faneuil Hall to listen."[31]

*Nature*, on August 24, 1876, also associated Bell's telephone with what later became radio. Remarking on Bell's "notable improvement" on the electric telegraph, the article predicted that at some future date it would be possible to

at a distance, repeat on one or more pianos the air played by a similar instrument at the point of departure. There is a possibility here . . . of a curious use of electricity. When we are going to have a dancing party, there will be no need to provide a musician. By paying a subscription to an enterprising individual who will, no doubt, come forward to work this vein, we can have from him a waltz, a quadrille, or a galop, just as we desire. Simply turn a bell handle, as we do the cock of a water or gas pipe and we shall be supplied with what we want. Perhaps our children may find the thing simple enough.[32]

Gradually the concept of the "speaking" or "talking" telephone took hold, and notions about usage changed accordingly. The first telephones were installed between two places such as home and office and were known as "private lines." Under that arrangement a telephone subscriber could call only one other "party." By June 30, 1877, when the telephone was little more than a year old, 230 telephones were in use; by July that figure rose to 750 and by the end of August to 1,300. These were mainly substitutes for telegraph instruments used to communicate between two points. The telephone was becoming more popular, doubtlessly because it eliminated the necessity for the skilled telegraphist on private lines, but it was not too different from the previously existing system of private telegraph lines.[33]

Even the introduction of the central office, which paved the way for modern telephony by breaking through the severe limitations of private lines, did not at first change usage. The first experimental telephone exchange was in operation in Boston on May 17, 1877, and connected Brewster, Bassett and Company, bankers, the Shoe and Leather Bank, and the Charles Williams Company. These "stations" were repeatedly interconnected, and many conversations were held. [34]

Despite the operational success of that system, E. T. Holmes, who held the license to operate the telephone business in Boston, was not immediately aware of the most profitable use of the central exchange. Holmes did proceed to build a central office at which subscribers' lines terminated, but instead of providing customers with the opportunity to talk to one another, he used the system to receive orders which he retransmitted to a general express agency. Making money that way seemed to be assured, and it was along that line that the telephone was first developed in Boston.[35]

Holmes was not the only one who used the telephone this way; it also seemed ideally suited for district telegraph systems, which were connected telegraphically to subscribers' homes and were used by the latter to summon a messenger. Although some district telegraph companies resisted substituting a telephone for a telegraph, others took to the new instrument. For example, in March 1878, the American District Telegraph Company of Missouri discontinued renting telegraphic signal boxes at $12 per year and instead rented telephones at $20 a year. The subscriber of a district *telephone* company found that he could not only call a central office for a messenger or other related service but he could also speak to other subscribers on the same line.[36]

Early in 1878, the Bell Telephone Company directed its agents to urge district telegraph companies to switch to the telephone on grounds that they were logical places for establishing a central exchange system. By September, however, Hubbard was encouraging telephonic central exchange systems instead of district systems, for he had come to see the telephone system as distinct from the district telegraph and feared that the latter would retard the achievement of his goal of interconnecting all telephone subscribers.[37]

If any single event deserves credit for demonstrating the value of exchange service and the utility of the telephone as a means of prompt communication, it was a railway accident at Tariffville, Connecticut, in January 1878. The first news of the disaster was wired to the Western Union office in Hartford, urging that as many surgeons as possible be sent. Fortunately, Isaac D. Smith, the proprietor of the Capitol Avenue Drug Store in Hartford, had obtained an agency of the New England Telephone Company in July 1877 and used it to organize a central telephone system whose subscription list was made up primarily of doctors and a livery stable. By November 1877, Smith had twenty-one subscribers who could all call one another on request.

On the night of the accident, the Western Union operator in Hartford immediately telegraphed Isaac Smith's drug store with news of the accident and relayed the plea for the doctors. The clerk, in turn, telephoned the twenty-one physicians, called the livery stable for an express wagon, and packed it with bandages, morphine, chloroform, and other first-aid equipment. Meanwhile the doctors were rushed to Tariffville by special train. This was the first example of the benefits that the telephone and the exchange system could provide in a catastrophe, and press coverage aroused public interest in the invention and its applications.[38]

But the events of the Tariffville accident did not necessarily suggest to ordinary Americans to what use *they* could put Bell's marvel. It may be obvious now when the telephone often seems to be an appendage to the mouth and ear how many different uses (and misuses) it has, but those living in the telegraph age had to learn them. To convince Americans that they needed the telephone, they first had to be taught *how* to use the telephone and what to use it for. Although many of the lessons were taught informally by a flood of favorable articles in newspapers and magazines, it should come as no surprise that American entrepreneurs who took control of the business decided to advertise the new medium's possibilities.

The campaign began modestly and concentrated more on the "how" than on the "what for." In May 1877, Bell and his partners distributed a circular assuring the reader that using a telephone was not difficult: "Conversation can be easily carried on after slight practice and with occasional repetition of a word or sentence." Since the owners decided they would sell service only and retain title to the instruments, the circular referred to the uses of the phone only indirectly in connection with the costs for renting the telephones. "The terms for leasing two Telephones," the circular read, "for social purposes connecting a dwelling house with any other building will be $20 a year, for business purposes $40 a year, payable semiannually in advance." Thus, two broad categories of telephone usage were set forth. In a circular issued April 26, 1879, by the National Bell Telephone Company—a predecessor of the American Telephone and Telegraph Company—"social or club purposes" were defined as being "for use between residences . . . for social purposes exclusively" designed to serve neighborhood groups. Just how the telephone was to serve neighborhood groups was left to the imagination.[39]

N. C. Wesleyan College Library
Rocky Mount, North Carolina

On October 27, 1877, a circular issued by a Bell licensee, Ponton's Telephone Central Service of Titusville, Pennsylvania, added a new aspect to the teaching of telephone usage. By that time, the exchange system was operating, and it was time to get people to shift from the private line to central service. Thus, Ponton's circular read:

The system is extremely simple. All parties who wish to adopt to it must have a separate wire from their house, office, factory, hotel, store, bank or restaurant to a central switch room, where any one wire can instantaneously be connected with any other wire. Supposing that one hundred persons adopt this system, and that the average length of each wire is half a mile, it would give each person the privilege of using fifty miles of wire at less cost than it could be done with only one mile in the private line.[40]

Then the circular described explicitly what a telephone connected to a central office could do for the subscriber. "In domestic life," it stated, "the telephone can put the user in instant communication with the grocer, butcher, baker." The circular went on to list 176 other occupations "and other places and persons too numerous to mention." [41]

The American public did not always find these advertising appeals irresistible; in fact, the Ponton Company was out of business by January 1878. But the exchange did better in the larger cities and towns, perhaps because of the greater familiarity in such localities with the latest in telegraph technology. The district telegraph or messenger services had been used by many businessmen and had, moreover, educated others in the concept of the central office. Furthermore, as early as August 1867, a telegraph exchange was installed in Philadelphia and connected private telegraph wires through a central office. The subscribers to this system of intercommunication were the largest banking houses in that city and numbered fifty by 1872. The business of such central telegraph offices—New York had its first in 1869—was chiefly to report to each bank (by means of printing telegraphs) its daily debit and credit balance and to repeat to any bank messages received for it by wire from other banks. The wires of any two banks desiring communication, however, could also be connected together at a switch. Thus, the idea of intercommunication in telegraphy had begun to replace the concept of the telegraphed message.[42]

American businessmen were beginning to learn the value of exchange *telephone* service; the directors of the *telegraphic* exchange companies began to replace the Morse instruments with telephones as early as 1878, and their subscribers started speaking to one another. Telephone exchanges were widely introduced in 1878—the same year Alexander Graham Bell told the British that they were not yet practicable—and quickly became an aid to commerce. They enabled conversation between points over distances "not exceeding 20 miles," a distance within which many, if not most, of one's business associates were located. The telegraph, with all its improvements, became a cumbrous means of communication compared to the telephone; confidence replaced the incredulity with which the business community had first regarded the telephone.[43]

In its first decades, the telephone tended to be the monopoly of businessmen not only because they often simply substituted it for a telegraph, but also because they could afford AT&T's high rates better than most other Americans. At that time the Bell Company concentrated on the business community probably because company officials were still somewhat influenced by concepts that governed the telegraph industry; these saw that medium primarily as an aid to commerce and had given little priority to social telegraphy.[44] Bell's bias toward businesses can be seen, for example, in the sixteen advertising appeals in its *National Telephone Directory* of 1897, which listed all subscribers connected to the long-distance lines. Half the advertisements were directed at the business community; the remainder could appeal to any user of the telephone.[45]

The extent to which businessmen filled the ranks of telephone subscribers is also seen in the six-page Pittsburgh telephone directory of 1879; of the 300 telephones it listed, all but six were installed in business concerns. Even the six residential telephones were used for business rather than social purposes, since they belonged to anxious entrepreneurs who felt the need to keep in constant touch with their establishments.[46]

In many other places practically all subscribers were businessmen. Tullahoma, Tennessee, had nine telephones in 1897; one was a public station, two were residential phones, two were connected to banks, one to a hotel, one to a flour mill, one to a factory, and one to a Western Union office. In that same year,

Pawtucket, Rhode Island, had a total of 161 telephones in service; 3 were public phones, 25 were connected to municipal offices including police, fire, and school departments, 115 were business and professional phones, and only 16 were in residences. Two additional phones were connected to the homes of the chief of police and the commissioner of public works.[47] Thus, only 11 percent of the phones in Pawtucket were residential, a pattern characteristic of the rest of the country.

This distribution suggests that most calls at that time were not retail orders from the public to retail establishments but were calls made from one businessman to another. Most communities had public telephones, however, and practically every exchange had one or more offices at which nonsubscribers could make calls for ten to fifteen cents. Nonsubscribers could also use the telephone at a local drugstore where telephone use was free. In May 1879, D. M. Finley, a businessman in Providence, Rhode Island, advertised that he was the "first plumber and coal dealer whose establishment had been connected with the new telephone wires and that householders wanting coal the same day as ordered might telephone to the dealer from Mr. Leith's drug store."[48]

The early history of telephone usage, then, is largely the story of how commercial and professional communities adopted the new means of communication. That topic in itself could fill volumes, and only a few highlights of that usage can be suggested. At a time when raising money quickly and quietly was the key to corporate empire building, the telephone was ideally suited to the needs of the "robber barons"; it was simple to operate and left no written record, a decided advantage when the message often involved the violation of laws and values against monopolies. Banker George W. Perkins, an associate of J. P. Morgan, originated the "Perkins Plan" of "rapid transit telephony" in which a list of ten to thirty men were phoned in succession as quickly as the operator could reach them.[49]

In the organization of industrial and manufacturing activities, the telephone's uses seemed unlimited. Factories installed internal telephone systems enabling the main office to keep in touch with shop foremen. Once long-distance service was available (development was underway by 1881), the main office could learn almost instantly the condition of any order in a main branch or factory. In the raising of skyscrapers, temporary wire was commonly strung vertically along the elevator shaft so that the architect on the ground could confer with a foreman sitting astride a girder hundreds of feet in the air, a system obviously superior to the whistles and messengers that had been previously used.[50]

One of the first historians of the telephone, Herbert Casson, wrote: "To give New York the seven million electric lights that have abolished night in that city requires twelve private exchanges and 512 telephones." Casson was referring to the way those who directed the flow of traffic for power companies supervised the distribution of electricity.[51] In 1879, the Pennsylvania Railroad became the first railroad to install a telephone; by 1910, its own private telephone system had 175 exchanges, 400 operators, 13,000 phones, and 20,000 miles of wire. By the beginning of the twentieth century, the telephone was an intercommunications system for a number of large-scale construction projects involving earthwork over extensive areas. In the construction of the Panama Canal and in reclamation projects for watering the arid regions of the West, an appropriate local telephone service was found to facilitate the handling of men and materials.[52] The infant telephone industry benefited from an 1877 Pennsylvania law that required the several hundred anthracite coal mines to have some means of oral communication between the interior and the surface.[53]

In areas such as those surrounding developed water power, where the operation of factories, street car lines, and lights depended on electric power transmitted from a distance, the necessity of avoiding interruption was great. Defective conditions could be promptly reported and remedied if a telephone line was installed parallel to the power lines.[54]

The telephone was also regarded as a boon for both the vacation industry and the tired executive who would have been reluctant to take a vacation in the pretelephonic age. Thus, the Special Census of Telephones of 1907 reported: "The last few years have seen such an extension of telephone lines through the various summer resort districts of the country that it has become practicable for businessmen to leave their offices for several days at a time and yet keep in close touch with their offices."[55]

The role of the telephone in the hotel industry is especially interesting. Before its invention, the hotel depended exclusively on the messenger to satisfy the requests of its guests and to deliver instructions and receive information from employees stationed on each floor. When a guest wanted service he could signal a messenger who would have to make two trips, one to learn what the guest wanted and one to provide it. Under these circumstances the elevators and stairways were crowded with bell hops, and hotels had to employ many messengers. By 1900, practically every large and medium-sized hotel and many small hotels had telephones in every guest room. These provided for the interior needs of the hotel and also gave access to the local and long-distance lines. Hotels were able to make money on the charges received from customers.

By 1909, the hundred largest hotels in New York City had 21,000 telephones—nearly as many as the continent of Africa and more than Spain—which were averaging six million calls annually. The telephone system of the Waldorf-Astoria with 1,120 telephones and 500,000 calls a year (1904) constituted the largest concentration of telephones under one roof in the world. The guests of this American hotel used the telephone service freely but not for free; during the first month of the Waldorf-Astoria's installation, the business amounted to 30,000 calls exclusive of those made within the hotel.[56]

The telephone was so important to business, commerce, and industry then that the fortunes of any Bell agency depended on proximity to concerns that could use the new medium. In the first few years, the Pacific Coast territory was a smaller field for telephonic development than the East Coast, because there were fewer manufacturing establishments or other concerns requiring communications between an office and the place where the work was actually being conducted.[57] But if that meant little business for Bell agents in areas with few industries, businessmen's demands for the telephone in other places were greater than Bell's capacity to manufacture telephones; the company had a problem supplying instruments as early as 1877 and 1878.[58]

The rapid inroads of the telephone into the industrial and commercial worlds reduced the cost of service and started the trend toward a telephone in every American home. Further, many new "independent" companies entered the telephone industry when Bell patents expired in 1894. As a consequence, the telephone became widely available to women and adolescents—two groups who, according to telephone folklore, distinguished themselves as talkers. Women could use the home phone for shopping, but the housewife did not always have to take the initiative in the new practice of shopping by telephone. In some localities, grocers would call their customers every morning to take the day's orders. Big-city department stores placed telephones at counters and advertised the telephone service that allowed a patron to call the salesman, who could generally perform the same services involved in a sale conducted across the counter.[59]

Even small business owners, such as cobblers, cleaners, and laundresses, considered the telephone service necessary for keeping in touch with customers and reducing the amount of idle time. With early introduction of the private branch exchange in 1879, a separate line did not have to be run from the telephone company's exchange to every department in the store. Instead, a small switchboard, built and operated much like a central-office switchboard and attended by an operator, received all incoming calls and fed them to the appropriate department.[60]

Although the telephone company early distinguished between business and social uses of the phone, it was vague about the nature of the purely social. Even shopping by phone does not clearly belong to that category since that always involved someone in business. Interestingly enough, Bell himself had a clear conception of an important social use of the telephone; he suggested during his lectures of 1877 that his invention would enable people to socialize at a distance. The time would come, he predicted, when Mrs. Smith would spend an hour with Mrs. Brown "very enjoyably in cutting up Mrs. Robinson" over the telephone; connection between the two women would be made possible by means of a central switching office.[61] With the widespread diffusion of the telephone after 1894, such a use of the phone became commonplace.

Even with the advent of the independent telephone companies, the high cost of urban commercial service restricted it to the well-to-do. As early as 1882, a residential phone cost $150 a year in New York and $100 in Chicago, Philadelphia, and Boston.[62]

These initially high charges were to become even higher in the 1890s as improvements in telephone technology and the growth of the number of subscribers drove costs up. In some urban areas, the high rates frequently brought complaints.[63] They were beyond the means of the average factory worker as well as the white collar employee.

But the *use* of a telephone was certainly not limited to the affluent. In its early days, as we have noted, some establishments such as banks and drugstores offered free telephone use to their customers. By the end of the century, the pay telephone was well established. In 1902, of the 2,315,000 telephones in the United States 81,000 (3½ percent) were "nickel-in-the-slot" (more accurately three nickels) phones. Most inhabited places in the United States had at least one public station, and in the larger cities and towns these might be found in every conceivable retail establishment from liquor stores to fish markets, from real estate offices to undertakers' parlors. Above all, the public phone was found in drugstores; although it would be impossible to tell whether every drugstore had a pay phone, it would not be far from the truth.[64] Finally, those who had no phone might have a neighbor who would permit them to use it.[65]

In discussing early patterns of telephone usage, we should take further note of what has been called the radio concept of telephony. Alexander Graham Bell gave up his idea of the telephone as a commercial medium of entertainment and enlightenment when he solved the problem of reciprocal communication over distances, but the concept developed just the same and in a number of ways. In many communities the first to transmit news were not professional reporters or broadcasters but the telephone operators themselves. It is likely that in the role of informal broadcasters, operators illustrated the possibilities of the telephone. By the time increased telephone traffic made it impossible for them to continue that service, it was relatively easy for the enterprising to see the direction that a new medium might take.

Thus, in the telephone's early days, the operators were accustomed to giving news about crises like fires and floods, missing persons reports, man-wanted bulletins, crimes, and so forth. Some of the information was of interest only to certain subscribers. For example, operators often took calls requesting the services of particular doctors who were out making home visits. These doctors would then "call in" for their messages.[66]

The nature of telephone technology in its early days allowed the operator to eavesdrop. That obviously made it easier for her to cover her beat as a newscaster, but it also gave telephony one of its enduring characteristics—the absence of privacy.[67]

Once it had become customary for operators to inform individual subscribers about local events, the latter came to feel that they had the right to receive such information when they called Central. The newscasting services of the operator started in city exchanges in the early days of telephony and then spread to the farm after 1894 when the independents and farmers' cooperative lines brought the telephone to rural America. The country operator continued to perform these services even after the introduction of radio, though it relieved much of the pressure to do so.

The radio concept of telephony ultimately took the form of the "telephone-newspaper," introduced in Budapest, Hungary, in 1898, and described in the next chapter. The telephone newspaper broadcast news, music, and theatrical performances. Similar telephone-newspapers were established in the United States. In Philadelphia, for example, the Bell Telephone Company arranged with the newspaper the *North American* for operators to give callers news summaries at any hour of the day and night.[68]

Before long, in the larger cities and towns the large numbers of subscribers and the small number of operators made it virtually impossible to provide the personal service that continued to be rendered in rural regions. Company officials were aware of the changing nature of urban telephony as it grew, but they found it impractical to continue the previous kind of service. Increasing complexity, the size of the traffic, the nature of the equipment, as well as operating arrangements for handling the load demanded more formality in managing the calls. Furthermore, customers gradually came to regard telephone employees as impersonal representatives of the industry and to judge the service on a less personal basis than was customary in smaller localities. In cities most calls involved little of the unusual, and the main consideration was prompt, accurate handling. There was no need for the operator to comment or to raise questions. The customer's major requirement of the operator was an attentive and pleasing attitude. At times, of course, a more individual approach to the urban subscriber was required, but these were only a small percentage of all calls and were well within the capacity of an obliging operator. These involved requests for charges on toll calls, reports of service troubles, and emergency calls such as fires, accidents, and burglaries. They also included long-distance calls, especially those in which there was difficulty in locating the called parties.[69]

The process of urban telephony's development thus depersonalized the service from the standpoint of the operator's acquaintance with the individual subscriber. This made it possible for her to sense urgency on any call while it also discouraged customers from seeking preferential treatment from an operator who had come to know them better. Nor was it easy for some customers to get acquainted with the operators and thus to make greater claims than less friendly subscribers. Thus, all calls and callers had to be treated alike, and it is reasonable to expect that since the operators could not, in most cases, know the importance of a particular call, they tended to handle each call as though it were urgent. This businesslike sense of urgency for each call contrasted with the situation in rural places where the operator not only could tell how pressing each call was—and therefore placed some before others—but also knew which callers merited personal consideration.[70]

The tremendous growth in the number of subscribers connected to urban switchboards—the Cortlandt exchange in New York City had 5,000 lines in use in 1890 and 9,000 by 1900—obviously transformed the character of a job that had once provided much satisfaction for genial and helpful young women. In 1912, Caroline Crawford studied telephone operators in Boston and concluded that making connections at the switchboard was difficult. That operation required "constant employment of the muscles of the eye in different directions [and] constant use of the optic nerve." The ear, too, she asserted, "is obliged . . . to distinguish between a number of different voices, to ascertain at once, so as to avoid repetition, the number asked for, no matter how indistinctly or ill-pronounced the number may be; this necessitates constant alertness of the auditory nerve; while the vocal organs are scarcely less constantly in use in answering calls and such conversations as may be necessary." Under such circumstances and the pressure to complete 200 calls per hour,

Crawford concluded that work in an exchange involved a tremendous strain upon the "mental constitution as well as upon the nervous system."[71]

Crawford found particularly objectionable the job of the "all-night" operators. They went on duty at 7 P.M. and worked until midnight; they would then sleep on a cot in a rest room of the exchange until 6 A.M. when they returned to the switchboard for an hour. That really was not a six-hour day, according to Crawford, because "it is understood that their sleep may be interrupted whenever the number of calls becomes abnormal." Also reprehensible was the fact that the two girls—who might be no older than seventeen—who paired up at night in the exchange were locked up alone. "Think a little," Crawford wrote, "about what might happen while they are there." Also offensive were the "tests" or "listening in" of supervisors—often officious college graduates—regarded by the company as necessary to see that the operators attended strictly and efficiently to their duties. It had the opposite effect, according to Crawford, since it deadened the connection and subjected the operator to much extra strain. Furthermore, the tests overlooked the need of the operator to use "phrases not in the 'ritual,'" especially when talking to foreigners. Yet they were reproved for the slightest deviation from the set of words prescribed for them.[72] Thorstein Veblen, the sociologist-critic of large-scale organizations such as the telephone company, later referred to such strict and self-defeating adherence to bureaucratic rules as "trained incapacity."

Crawford completes her critique of the working conditions of the operators by noting the frequent confrontation with subscribers who accused the operator of not working diligently to complete calls, of supervisors' incessant pressures to "hurry, hurry," of the difficulty of getting relief to go to the lavatory or to get a drink of water, and of bearing heavy apparatus on her chest. But perhaps the unkindest cut of all was the "split trick" which divided the operators' work time between two parts of the day and which involved either an extra carfare or the alternative of loafing around the city. Crawford cited the opinion of social workers who "say that a very real danger for girls lies in 'killing time' at the moving picture shows."[73] Parts of the indictment were later repeated, and the frequent turnover of operators was later a major problem for the telephone company.

Other patterns of telephone usage that warrant interest can only be alluded to in the absence of space or research. Hardly mentioned has been the spread of the long-distance service which, by the turn of the century, resulted in the substitution of telephone calls for traveling and which was not confined to short distances or the beaten paths of commercial traffic.[74] The widespread use of the party line not only on the farm but in cities and towns, too,   :ses questions about how Americans learned to share lines, how much conflict there was when some tried to monopolize the line, and how conversations were influenced when callers knew they might be overheard. Patterns of usage among professionals—whether lawyers or doctors or professional criminals—have hardly begun to be examined.[75]

Although this chapter has focused on usage, the more complicated and intriguing question is that of the impact of the telephone. Are we justified in assuming that the telephone actually changed the character of American society, that we are different because of it, and that the differences between a society that has an effective telephone system and one that does not are as great as those between literate and preliterate societies? But those are different questions that no doubt will receive their deserved attention now that social scientists have discovered the telephone, almost 100 years after Alexander Graham Bell did.

## NOTES

1. Robert V. Bruce, *Bell: Alexander Graham Bell and the Conquest of Space* (Boston, 1973), p. 229.
2. Catherine Mackenzie, *Alexander Graham Bell* (Boston and New York, 1928), pp. 157–158.
3. Bruce, *Bell*, p. 229.
4. John E. Kingsbury, *The Telephone and Telephone Exchanges* (New York, 1972), p. 209.
5. Rosario J. Tosiello, *The Birth and Early Years of the Bell Telephone System, 1876–1880* (unpublished doctoral dissertation, Boston University, 1971), p. 82.
6. Orton was well aware, however, that a powerful corporation like Western Union had "many ways of getting control of a new product once it was proved practicable; by imitation; by evasion; or by the erosive effect on the original inventor of lawsuits conducted at enormous

costs, at the end of which . . . the invention might fall to the corporation." At the same time, Orton had employed Thomas Alva Edison
who, at the time of the offer of Bell's patent rights, was already at work
improving the telephone, and by January of 1877 the latter had "succeeded in conveying over wires many articulated sentences." Matthew
Josephson, *Edison* (New York, 1959), p. 142; Bruce, *Bell*, p. 229.

7. Josephson, *Edison*, p. 62.

8. *Johnson's New Universal Cyclopedia* (New York, 1877), pp. 762, 759.

9. Ibid., p. 760; E. A. Marland, *Early Electrical Communication* (London, 1964), p. 137.

10. *Johnson's Cyclopedia*, p. 760.

11. Kingsbury, *The Telephone and Telephone Exchanges*, p. 92.

12. *Johnson's Cyclopedia*, p. 761.

13. Marland, *Early Electrical Communication*, p. 185.

14. *Johnson's Cyclopedia*, p. 761.

15. The number of telegrams is for the year 1880.

16. *Johnson's Cyclopedia*, p. 762; Department of Commerce and Labor, Bureau of the Census, *Telegraph Systems: 1907* (Washington, 1909), pp. 10, 22–39.

17. Frederick L. Rhodes, *Beginnings of Telephony* (New York, 1929), p. 149.

18. Ibid., p. 4; *The Bell Telephone: The Deposition of Alexander Graham Bell* (Boston, 1908), p. 302.

19. *The Deposition of Alexander Graham Bell*, p. 437; Tosiello, *The Birth and Early Years*, pp. 62–63; Rhodes, *Beginnings of Telephony*, p. 35.

20. Rhodes stated Bell's dilemma as follows: "But this [the Brantford experiments in the summer of 1876] was a demonstration of telephonic transmission in one direction only. When it became necessary to con-vey information from the receiving end of the wire to the sending end, this had to be accomplished by telegraphing over another wire. Until a sustained conversation, in both directions, alternately, had been car-ried on, the complete practicability of the telephone as a means of oral communication between distant points remained to be proved." *Begin-nings of Telephony*, p. 36. See also *The Deposition of Alexander Graham Bell*, p. 379; M. D. Fagan, ed., *A History of Engineering and Science in the Bell System* (1975), pp. 13–14.

21. Bruce, *Bell*, p. 189.

22. Rhodes, *Beginnings of Telephony*, pp. 30–32.

23. *The Deposition of Alexander Graham Bell*, p. 114.

24. Ibid., p. 379; Tosiello, *The Birth and Early Years*, pp. 125–126.

25. Tosiello, *The Birth and Early Years*, pp. 44–46.

26. Ibid., pp. 45–62.

27. Ibid.

28. Ibid., p. 59.

29. Ibid., p. 57.

30. Kingsbury, *The Telephone and Telephone Exchanges*, p. 90.

31. *Some Findings To Date in Study of "Telephone History,"* Unpublished

Manuscript, Library, American Telephone and Telegraph Company, pp. 1–7.

32. Ibid., p. 7.

33. Kingsbury, *The Telephone and Telephone Exchanges*, pp. 68–71; Rhodes, *Beginnings of Telephony*, pp. 147–148.

34. Kingsbury, *The Telephone and Telephone Exchanges*, p. 70.

35. Ibid., p. 71.

36. Tosiello, *The Birth and Early Years*, pp. 129–130.

37. Ibid., pp. 131, 135.

38. Kingsbury, *The Telephone and Telephone Exchanges*, pp. 72–73; Rhodes, *Beginnings of Telephony*, p. 148; *Semi-Weekly Advertiser*, Boston, January 25, 1878.

39. Kingsbury, *The Telephone and Telephone Exchanges*, p. 67; Department of Commerce and Labor, Bureau of the Census, *Special Reports, Telephones: 1907* (Washington, 1910), pp. 75–76.

40. Kingsbury, *The Telephone and Telephone Exchanges*, p. 74.

41. Ibid.

42. Ibid., pp. 84–85.

43. Ibid., pp. 86, 67, 75–76; American Telephone and Telegraph Company, *Events in Telephone History* (New York, 1958), pp. 6–7.

44. *Special Reports, Telephones: 1907*, p. 76; Tosiello, *The Birth and Early Years*, p. 110.

45. American Telephone and Telegraph Company, *National Telephone Directory* (New York, 1897).

46. Salvin Schmidt, "The Telephone Comes to Pittsburgh," *The Western Pennsylvania Historical Magazine*, XXXV (June 1952), p. 105.

47. *National Telephone Directory* (1897), pp. 708–709, 723.

48. *Events in Telephone History*, p. 9; *Special Reports, Telephones: 1907*, p. 80; *Some Findings To Date in Study of "Telephone History,"* p. 13; Fagen, ed., *A History of Engineering and Science in the Bell System*, pp. 155–156.

49. Herbert N. Casson, *The History of the Telephone* (Chicago, 1910), p. 205; Sidney H. Aronson, "The Sociology of the Telephone," *International Journal of Comparative Sociology*, XII (September 1971), pp. 154–158; Bancroft Gerhardi and F. B. Jewett, "Telephone Communications System of the United States," *The Bell System Technical Journal*, IX (January 1930), p. 41.

50. Casson, *History of the Telephone*, pp. 200, 207.

51. Ibid., p. 207.

52. Ibid.; *Special Reports, Telephones: 1907*, pp. 75, 99.

53. Tosiello, *The Birth and Early Years*, p. 113.

54. *Special Reports, Telephones: 1907*, pp. 100–101.

55. Ibid., p. 80.

56. Casson, *History of the Telephone*, p. 200; *Special Reports, Telephones: 1907*, p. 80.

57. Tosiello, *The Birth and Early Years*, pp. 112–113.

58. Ibid., pp. 116–117.

59. *Special Reports, Telephones: 1907*, p. 74.

60. Ibid.; *Events in Telephone History*, p. 9.

61. Tosiello, *The Birth and Early Years*, p. 120.

62. Kingsbury, *The Telephone and Telephone Exchanges*, p. 469.

63. J. Ainsworth and G. Johnston, *A Discussion of Telephone Competition* (Independent Telephone Association, February 1908), pp. 7–8.

64. *Special Reports, Telephones: 1907*, pp.21, 81; Horace Coon, *American Tel & Tel* (New York and Toronto, 1939), p. 72; *National Telephone Directory* (1897).

65. "Save the Carpet" *Michigan State Gazette*, IV (September, 1908), p. 8.

66. Aronson, "The Sociology of the Telephone," pp. 159–160.

67. R. W. Wallace, "The Telephone," *Journal of Education* (March 26, 1908), p. 355; C. H. Petty, "From the Subscriber's Viewpoint," *Western Telephone Journal*, XV (December 1909), p. 76; Sylvester Baxter, 'The Telephone Girl," *The Outlook* (May 26, 1906), p. 231.

68. Aronson, "The Sociology of the Telephone," pp. 160–161.

69. R. T. Barrett, "The Changing Years as Seen from the Switchboard," *Bell Telephone Quarterly* (1935), pp. 285–288.

70. Barrett, "The Changing Years as Seen from the Switchboard," pp. 280, 288.

71. Caroline Crawford, "The Hello-Girls of Boston," *Life and Labor*, II (September, 1912), p. 260.

72. Ibid., pp. 261–263.

73. Ibid., pp. 262–264.

74. *Special Reports, Telephones: 1907*, p. 75.

75. Aronson, "The Sociology of the Telephone," pp. 157–158, 166.

# 2

## The Pleasure Telephone: A Chapter in the Prehistory of the Media

### Asa Briggs

Le téléphone dans la chambre de Papa est notre grand amusement.

The Diary of Victoria Sackville, 9 April 1891*

You stay at home and send your eyes and ears abroad for you. Wherever the electric connection is carried—and there need be no human habitation however remote from social centers, be it the mid-air balloon or mid-ocean float of the weather watchman, or the ice-crusted hut of the polar observer where it may not reach—it is possible in slippers and dressing gown for the dweller to take his choice of the public entertainments given that day in every city of the earth.

Edward Bellamy, *Equality* (1891)

The telephone was neither the first nor the last invention to be considered a toy. Although Thomas Edison objected to coin-in-the-slot phonographs on the grounds that they would give his new invention "the appearance of being nothing more than a mere toy," he himself conceived at first of the kinetoscope as little more.[1]

The sense of pleasure in playing with a new toy—perhaps a necessary part of the inventive process itself—persisted with the users until novelty gave way to routine, with the pleasure usually being private, like that of Queen Victoria playing with telegraphs at the Great Exhibition of 1851, or that of Lady Sackville amusing herself with her father's first telephone forty years later. (Queen Victoria, who found the first telephones "most extraordinary," had installed one as early as 1879.)[2] Guests at Hatfield, the ancestral home of Lord Salisbury, Britain's last nineteenth-century prime minister, were diverted when sitting

---

*I owe this reference to Susan Mary Allsop; my research assistant Carolyn Marvin is responsible for many of the other references.

in their rooms "as they thought alone," to hear their host's "spectral voice" reciting nursery rhymes from "a mysterious instrument on a neighbouring table":

Hey diddle diddle
The Cat and the Fiddle
The Cow jumped over the Moon.[3]

Salisbury was testing his telephone in much the same way that Edison tested the kinetoscope with a performing dog, a trained bear, and a strong man.[4]

Salisbury's guests might have been alarmed rather than diverted, had not many of them already been alarmed enough by his earlier pioneering experiments with electric lighting at Hatfield that used naked uninsulated wires. Indeed, these experiments must have been at least as frightening to spectators as "looping the loops" were in the toy phase of the airplane.[5] Yet Salisbury saw beyond the play: he identified an age of electricity following and contrasting with the age of steam. "You have by the action of the electric telegraph combined together almost at one moment, and acting at one moment upon the agencies which govern mankind, the opinions of the whole intelligent world with respect to everything that is passing at that time upon the face of the globe."[6]

"Intelligence" was one thing, entertainment another. Salisbury's nursery rhymes, like Bell's own *Yankee Doodle*, were not the kind of messages he received by telegraph in his office every day. An element of performance, as in the later case of flying, could turn private diversions into public pleasures; this was to be one of the links between the toy and the medium. Not only instantaneous knowledge could be shared ubiquitously, but instantaneous entertainment programs as well. An element of entertainment was present, indeed, in the first scientific "demonstrations." Reporting a not very exciting occasion of this kind when Sir Oliver Lodge demonstrated "Hertzian waves" at the Royal Institution in 1895, *The Electrician* greatly regretted the absence of mystery or of drama: "A mixed gathering requires its doses of science to be dashed with theatrical effect."[7] The showmanship mattered. Thus, when Sir William Preece of the British Post Office demonstrated Marconi's patents at the Royal Institution in 1897, "the impressive delivery of the lecture" was "in keeping with the wizard-like nature of the experiments: electric bells placed inside galvanised iron dustbins and completely isolated from the outside world, ringing merrily in response to Mr. Marconi's commands from the cellar below."[8]

Orton might dismiss Bell's offer of "an electrical toy" on behalf of Western Union, but Bell's Hamlet soliloquy at the 1876 Centennial Exhibition was not merely incidental to the telephone story.[9] It is interesting, indeed, that some of the first telephone users were said to suffer "stage fright," an anticipation of microphone and television nerves.[10]

Showmanship was even more necessary for the retailers than for the inventors or their sponsors; the Centennial Exhibition was an international event preceded and followed by scores of local events where singing or reciting received more publicity than conversation. In 1878, a Quebec jeweler who was anxious to prove that he had "perfected some new and certainly wonderful improvements in the original telephone of Professor Bell" presented "singers over the wire" to people visiting his shop.[11]

Before long, "brisk young men in hard bowler hats, with handle-bar moustaches and broad check suits" were promoting "telephone concerts" during which young ladies recited or sang "Lord Ullwin's Daughter" or "Kathleen Mavourneen" for the benefit of "audiences" either in private homes or in "central premises," some of them in other cities. The repertoire was limited and conventional, but if "Home Sweet Home" and "Auld Lang Syne" were popular favorites, there was also a place for the two superbly opposite numbers—before there were any *real* telephone numbers—"Thou art so near and yet so far" and "I'll listen to thy voice, thy face I never see."[12]

Appeals to an "audience" widened shared experience of the telephone, pushing it beyond the confined circles of sisters and brothers or hosts and guests. The existence of the telephone was felt therefore, to represent more than a marvel of science; it was an unprecedented extension of ear and voice. It began to raise social and cultural questions that were to become very familiar during the twentieth century; among them were questions concerning "content" and "performance."

Once it became possible to transmit sound along telephone

wires in both directions—a very early achievement of 1876 itself—it might have seemed inevitable that the telephone would
establish itself mainly as an instrument of person-to-person or
organization-to-organization communication rather than broadcast communication.[13] Yet following the brief period before its
two-way capabilities were fully appreciated, it continued to be
publicized as a device to transmit music and news as much as
or more than speech; long after its multiple private and organizational uses had been exploited, it continued to offer the prospect of shared entertainment, information, and instruction.

From London, *Nature* led the way in prediction with its forecast that "by paying a subscription to an enterprising individual
who will, no doubt, come forward to work this vein, we can
have from him a waltz, a quadrille, or a galop just as we
desire."[14]

But similar forecasts were made and "experiments" conducted
in widely scattered contexts, particularly as so-called improved
telephones offered relevant technology. In Paris, for example,
long queues gathered at the International Electrical Exhibition of
1881 to listen to music transmitted by telephone from a mile
away; a few years later "theatrephones" in the Parisian boule-
vards allowed anyone to be "put in communication" with theaters for "five minutes for five pence."[15]

In the United States, during the Bryan/McKinley election of
1896, there was a systematic telephonic relay of election returns,
"a revelation to the public and to the great telegraph and news-
gathering associations." The "elasticity" of the telephone as an
instrument of communication was eloquently pointed out on
this occasion. It was capable "of manipulation by everyone, except they be deaf or dumb." No codes were necessary, no inter-
mediaries: "Thousands transmitted the vote of the country
townships that had never operated a telephone, and thousands
sat with their ear glued to the receiver the whole night long,
hypnotized by the possibilities unfolded to them for the first
time."[16]

If there is an intimation of the excitement of the first televised
election in this contemporary comment, there was more rhetoric
than evaluation in the remarks of Vice-President E. J. Hall of the
American Telephone and Telegraph Company at an 1890 Detroit
convention of the telephone industry:

More wonderful still is a scheme which we now have on foot, which looks to providing music on tap at certain times every day, especially at meal times. The scheme is to have a fine band perform the choicest music, gather up the sound waves, and distribute them to any number of subscribers. Thus a family, club or hotel may be regaled with the choicest airs from their favorite operas while enjoying the evening meal, and the effect will be as real and enjoyable as though the performers were actually present in the apartment.

Then came the qualifications. First, in his own words, his "audience" was limited: he had attracted "over a hundred subscribers, or rather persons who have certified to their anxiety to be subscribers." Second, the proper means of technical transmission were inadequate; the telephone could not successfully distinguish between the notes of a harp and a piano, or among reed, wood, and brass tones.[17] Perhaps the Russians, however, found the answer the same year. The Russian scientist Kildischevski was said to have invented a new form of telephone "of remarkable superiority." It was not necessary to place the ear near the receiver (thereby avoiding the dangers of "telephone ear"),[18] and there was adequate amplification.[19] Speech, songs, and music could be heard in places as distant as Moscow and Rostow on Don. This sounded far more promising than the existing telephone link-ups with Madison Square Garden.

It was in London and Budapest that the most general questions were raised about "pleasure telephone" systems, questions not raised again in such a clear-cut fashion until after the development of radio broadcasting thirty or forty years later. In London what proved to be an unsuccessful "electrophone" service offered relays of music as early as 1883;[20] in Budapest there was a relatively successful regular service, which captured considerable public interest inside and outside Hungary. The form in which the general questions were raised, as well as the answers given to them, reflected contrasting social and cultural features in British and Hungarian society and culture which were to persist. If telephonic fare were to be offered to subscribers, what items should be transmitted? Should there be a balance between entertainment, information, and instruction? How should news, which had already been revolutionized by the telegraph, be handled on the telephone? And to move from the topical to the eternal, what should be the place of religion in

service? What about standards of performance once past the
testing phase?[21] What about the relationship between metropolitan and provincial standards and between metropolitan and
provincial tastes? How much advertising should there be, in
what styles, and when?

Given the very limited use of the "pleasure telephone" and its
many technical inadequacies and limitations, it is remarkable
just how many pivotal questions were asked. As in the later
case of radio broadcasting, technology common to different
countries was being utilized in different ways in different societies. Significantly, perhaps, more emphasis was placed on potential "mass" involvement than on individual choice. In the
United States Edward Bellamy included the key phrase "take his
choice" in the passage quoted at the beginning of this chapter;
but in *Looking Backward* (1888), he talked of "cooperation" in
music as in "everything else," whereby people could hear relayed music in their homes "to suit all tastes and moods" from
central halls of music with twenty-four-hour programs.

Questions from the above list came to the forefront in different countries for many reasons of a nontechnological nature,
particularly during the 1890s: the growth of urban populations,
the rise in consumer incomes, the increase in leisure time, and
the demand for more "entertainment" not just as a "sometime"
thing but as an integrated element in daily life.[22] Moreover,
there was no shortage of "enterprising individuals," as *Nature*
called them, who saw opportunities in "mass audiences" both
for entertainment and for the written word—periodicals in the
form of new-style popular magazines or specialist weeklies,
newspapers, and the "books for the millions" of different kinds,
including novels and works of reference.

The newspapers and periodicals that prophesied what the future of communications would be like were themselves part of
the same complex within which electronic communications were
changing. "We take it that everything which can knit a community together," one of the new specialized periodicals put it in
1884 (and several such periodicals focused on "the age of electricity"), "and which can cause a rapid interchange of sentiment
and ideas, annihilate distance, isolation and prejudice, is of the
greatest happiness to the greatest number."[23]

To consider the history of "the pleasure telephone," therefore,

we must first study it in the light of a far longer history of communications and then relate it to a cluster of other inventions
patented during the last quarter of the nineteenth century. The
first line of investigation means looking backward to the 1840s
and 1850s—to railways, as well as to telegraphs—not least by
reason of the different policies followed by governments toward
them—and the two were very closely associated—and forward
to the 1920s and 1950s, the critical decades in the history of
sound broadcasting and television.

The second line of investigation involves relating "pleasure
telephony" not only to the phonograph and the kinetoscope but
also to radio broadcasting, despite its very long gestation period
of necessary technical development, before Marconi's inventions
of the 1890s became the basis of a "medium." Just as the great
inventions of the late eighteenth century transformed factory
production only in the nineteenth century, so the inventions of
the late nineteenth century transformed home life styles only in
the twentieth century. From the start, the contemporary reading
public of the late nineteenth century, specialized or unsophisti-
cated, tended to think of the different inventions as closely related; of course, they were directly related in the case of Bell's
and Edison's inventions. The *London Times* could write, therefore, in 1878: "Not many weeks have passed since we were star-
tled by the announcement that we could converse audibly with
each other, although hundreds of miles apart, by means of so
many miles of wire wound round a magnet"; another wonder
was "now promised us."[24] This was "an invention purely
mechanical in nature by means of which words spoken by the
human voice can be, so to speak, stored up and reproduced at
will. What shall be said of a machine by means of which the old
familiar voice of one who is no longer with us on earth can be
heard speaking to us in the very tones and measures to which
our ears were once accustomed?"

This was a "private communications" forecast for the phonograph; in Bell's Canada, the future of everything new, including
communication and actual visits to other planets, was being tak-
en for granted, and there were many "public communications"
forecasts for it. "There will follow a revolution in public singing
and speaking."[25] Private or public, the "revolution" loomed
large, and a reviewer of Oliver Lodge's *Modern Views of Electric-*

*ity* could write confidently in the centenary year of the French
Revolution: "Progress is a thing of months and weeks, almost of
days. The long line of isolated ripples of past discovery seems
blending into a mighty wave, on the crest of which one begins
to discern some oncoming magnificent generalization."[26]

It is through such reactions, as well as through the history of
technology itself, that change is registered. Such comments were
not just reactions or "impact effects": they carried an apprehension of promise or threat. *Before* the invention of new devices to
produce "signals in time and signals in space,"[27] there were already significant changes in attitudes toward time and space
themselves. Regular hours of work preceded regular hours of
entertainment; a railway system reducing distance preceded a
system of telegraphy promising to annihilate distance; the global system of telegraphy preceded the global system of radio
without wires; transportation metaphors preceded metaphors of
nervous systems or of circuitry. It could finally be claimed in
1884, although there was, of course, no finality about it, that
"the world now lives like a lumbering whale whose nerve centers and brain are slow to tell him that he has been harpooned,
but all the members respond to influences exerted on any one of
them with the sensitiveness and promptitude of the most highly
wrought organisms."[28]

The continuities between late nineteenth century and twentieth century communications history were emphasized by John
Logie Baird, the pioneer of television, from the vantage point of
the 1920s. When he first advertised his "televisor" in 1926, he
began his long and somewhat flamboyant advertisement with a
reference not to Marconi (with whom he has often been compared) but to his fellow Scotsman Bell. A few "wags," he noted,
had claimed in 1876 that "seeing by telephone" would follow
naturally from "hearing by telephone." Now, Baird said, the
claim had been justified. Ignoring all the rejected alternatives of
communications history—and the "pleasure telephone" by then
was one of them—and ignoring all the blocked paths in the way
of particular technical advances, Baird concentrated on the es-
sentials. He was proud to advance a kind of Whig interpretation
of communications history (naturally with a dose of the history
of entertainment thrown in), an ingredient not to be found in
Whig political history, which directly linked what he himself

was doing with what his predecessors had done.[29] He did not
point to the fact that 1922, the year of Bell's death, was also the
year of the founding of the BBC, for he felt that he had skipped
the sound radio phase: he was already thinking of systematiz-
ing vision at a time when the BBC was seeking to systematize
sound.

Less prophetic figures fitted sound radio into the story as a
natural development from telephony. Thus his friend Frank Gill,
who was Engineer-in-Chief of the National Telephone Company
from 1902 to 1913 (and later of the International Western Electric
Company), treated "radio telephony" as one branch of telephone history in his preface to Baldwin's *History of the Telephone in the United Kingdom*. "Telephony," he stated, "has
something of the properties both of the letter and of the newspaper: it can be clothed with privacy given to one individual
only, or it can be broadcast to millions simultaneously."[30] Gill
was one of many European telephone engineers who bemoaned
the slow development of the private telephone system in Europe
before 1922; he was also, however, one of the participants in the
crucial talks arranged by the British Post Office in 1922, which
led to the foundation of the BBC, and when he surveyed the future of communications in Britain he put his trust (rightly as it
turned out) in a very rapid increase in the number of wireless
set owners.[31]

Baird and Gill, both with practical ends in view, were seeking
to summarize half a century of unfinished communications history, and neither of them drew a sharp contrast at that moment
between "messages" and "programs" or between television exchanges with switchboards and broadcasting studios with mi-
crophones at the nub of the communications system. Nor did
the Danes draw such a distinction; until 1925 they did not reject
the idea, widely canvassed in Denmark, of hitching broadcasting to the existing telephone network.[32] For such reasons, historians of communications must accept, even if critically, Baird
and Gill's sense of a continuum.

The first refinement involves looking for beginnings in an
earlier period than they had considered: the transportation
phase of communications history, the early railway age before
railways themselves became "pleasure vehicles" for passengers
or at least vehicles for passengers en route for "pleasure" as well

as carriers of freight. Telegraphy, which in its early stages required substantial business promotion before it became acceptable, was closely related to railway development, and it took time to foresee its social consequences. The recognition that it could transform the content of newspapers, for example, was quick but not immediate, even though the transformation of content was a condition of the transformation of the widespread distribution of newspapers by railway. Daily newspapers could provide daily news, but they could also—and this took more time to realize—provide daily entertainment as well.

The time lags between prediction and performance varied. There were suggestions during the very early period, predictions not fully realized until the age of the telephone, that telegraphy was "too good a thing to be confined to public use" and that it should be introduced not only into the office but into "the domestic circle."[33] The idea of the "singing telegram" had a long history. "It appears that songs and pieces of music are now sent from Boston to New York by electric telegraph," *Punch* reported in 1848, adding that "it must be delightful for a party at Boston to be able to call upon a gentleman in New York for a song."[34]

The implications could be and were pushed much further. If "popular vocalists" could sing in four or five places at once (note that ubiquity, the idea of being heard "everywhere," was not yet fully developed[35]), might not their incomes be trebled or quintupled? "Our own Jenny Lind, for example, who seems to be wanted everywhere at the same time, will have the opportunity of gratifying the subscribers to Her Majesty's Theatre and a couple of audiences many hundred miles off at the same moment."[36] Although the idea of ubiquity was not yet fully developed, the conception of instantaneity was already there.

Concerning the long run, there were warning voices as there had always been about the railways. "Listeners" might not do as well as performers, however. "With a house telegraph it would be a perpetual tête-à-tête. We should all be always in company. . . . The bliss of ignorance would be at an end."[37] Nor would "ignorance" alone be shattered; truth would be in serious jeopardy also. Long before the adjective "phony" was coined, *Punch* complained of telegraphic "fibs":

What horrid fibs by that electric wire
Are flashed about! What falsehoods are its shocks!

Oh! rather let us have the fact that creeps
Comparatively by the Post so slow
Than the quick fudge which like the lightning leaps
And makes us credit that which is not so.[38]

Instantaneity could be described as "quick fudge," of course,
not because of the inherent characteristics of the telegraph but
because of the system of reporting. In other words, one of the
questions concerning the control of "content," which was to be
forced to the forefront in later phases of communications history, was already being posed at the beginning of the story. So,
too, was the question of triviality. When Emerson was told that
the installation of the Trans-Atlantic Cable would in the future
permit someone in London to speak to someone in New York,
he is said to have remarked, "But will he have anything to
say?"

The term "medium of communication" was slow to emerge,[39]
yet even before the development of the telephone there was
clearly a sense of there being "media," both printed and electronic. The development of the "pleasure telephone," the phonograph, and other inventions of the late nineteenth century
extended, enriched, and complicated this sense.

In this context, the history of the Budapest Messenger or Gazette (Telephon Hirmondo) is exceptional but revealing. It was
more than experiment: it worked regularly over several years,
increasing the number of subscribers (including hotels) and offering them a genuine program service. Its "inventor," Theodore Puskas, who is said to have been a friend of Edison, was
present at the International Exhibition of Electricity in Paris in
1881; in the same year he obtained exclusive rights to develop
the telephone in Hungary. Thereafter he publicized his own
way of developing it at international conferences.[40] His company started regular "broadcasting operations" in 1893; within
five years it was said to be using 220 miles of wire and to have
attracted 6,000 subscribers.

It was in relation to Puskas' operations that Arthur Mee, a
young journalist and later editor of the Children's Encyclopedia,

coined the phrase "the Pleasure Telephone." In future, "no element in our social life would be unprovided for: The Pleasure
Telephone opens out a vista of infinite charm which few prophets of radio have dreamed of . . . and who dare to say that in
twenty years the electric miracle will not bring all the corners of
the earth to our own fireside?" Baird would certainly have approved of this particular prophet. "If, as it is said to be not unlikely in the near future, the principle of sight is applied to the
telephone as well as that of sound, earth will be in truth a paradise, and distance will lose its enchantment by being abolished
altogether."[41]

Mee foresaw broadcasting of football and cricket, sport, politics, and religion. Indeed, many of these activities were already
developed in the city of Budapest, which then had a population
of 500,000 and was divided by Puskas into thirty circuits, each
connecting 200 to 300 subscribers. Within each house "long
flexible wires" made it possible to carry the receiver to the bed
or to any part of the room.[42] The receiving apparatus occupied a
space of five inches square and included two tubes so that two
members of the family might listen simultaneously. Subscribers
could not talk to each other "on their own account" and remain
connected to the system—thus ruling out "private communication"—and the company installing the apparatus had the right
to introduce it into any house in the city without the landlord's
permission.

Each day a schedule or program was announced to the subscribers.[43] The day began with a news bulletin and with summaries of the newspapers. In the midmorning, summaries of
stock exchange prices were repeated at regular intervals while
the Exchange was open. There were hourly news summaries for
those who had missed earlier bulletins, and at noon there was a
report on proceedings in Parliament. During the afternoon,
"short, entertaining stories" were read, "sporting intelligence"
was transmitted, and there were "filler items" of various kinds.
In the evening, there were theatrical offerings, visits to the opera,[44] poetry readings, concerts and lectures—including repeats
of Academy lectures by well-known literary figures. There were
also "linguistic lessons" in English, Italian, and French which
were hailed "as a great benefit to the young generation." ("Each
telephone subscriber who cares to listen holds a copy of the

book before him, and the teacher speaks into the double telephone transmitter at the central office.") At that office—and
note that it was still not called a studio—there were over forty
"reporters and literary men" in addition to "the persons who
actually speak to or transmit the news to the subscribers, and
who are chosen on account of their good voices and distinct
articulation."[45]

The very language, stilted though it sounds, anticipates the
language of sound broadcasting by radio. Yet the term which
was felt to describe the wide-ranging operation most closely
during the 1890s was "telephonic messenger," "gazette," or
"newspaper," admittedly "for want of a better name."[46] The no-
menclature must be related, of course, to the changing format
and role of newspapers themselves during the 1890s.[47] They
were becoming more "entertaining," priding themselves on
their "scoops," extending their advertising, and appealing
frankly to "mass audiences." The year 1896 saw the founding
of Harmsworth's new *Daily Mail* in London as well as the
demonstration of the Marconi wireless patents.

To experienced British observers, the Budapest service "completely fulfilled the functions of the daily paper," from the leading article to stock exchange news, "from the agony column to
the advertisement of the latest panacea." It constituted "a spok-
en instead of a printed record of the world's doings." Yet the
parallels were not complete. The "listener," if he wished, could
"spend the whole day at the telephone," and he could end by
lulling himself to sleep by "the latest music of Strauss" if he
and the program makers so chose.[48] No newspapers, however
new, could offer such distractions. Moreover, the Budapest
"telephone messenger" was a newspaper not for the masses,
but for an elite.

Most commentaries on the Budapest service concentrated on
the news rather than on the music, and certainly the handling
of the news was very sophisticated.[49] It was collected during the
day and night, edited before transmission, updated in as many
as twenty-eight editions a day, presented in "a telegraphic
style, clear, condensed and precise," and summarized in short
bulletins. Special "flashes" brought news to subscribers long
before they could read it in the newspapers. There was a wide
range of reference; foreign news was presented in the late even-

ing, but there were also detailed city reports. Since subscribers often found themselves "lying in wait for the news," arrangements were suggested to record it for them in their homes. A phonograph was to be fixed to the telephone receiver "in such a way that the first sound over the wire would start the phonograph, which would then record the news, and make it available for the subscriber at his convenience."[50]

When British observers tried to explain why London could not outdo Budapest, they gave economic reasons: higher costs, legal reasons (the law did not confer the same rights of access), and social differences. Budapest was a city of pleasure; in London "time was everything," and "a man could not sit the whole day with the apparatus to his ear waiting for some particular news or exchange prices."[51] Yet the Budapest service may have been promoted with relative success because the telephone system in that city could not be quickly developed along "American lines." The very first mention of Puskas in the *Scientific American* refers also to the visit to Hungary of D. H. Washburn, who was trying to introduce the telephone into that country. Far from appreciating the flexibility of Hungarian law, Washburn stressed its restrictiveness. Before one could subscribe to the telephone exchange, one's name and business had to be sent to four different government offices. Budapest might be a city of pleasure but it was also a city of bureaucracy. The telephone company, Washburn added, had to report to the authorities the cost of everything and the wages of every employee. In fact, he concluded comprehensively, "everyone that lives here is but a slave of the government."[52] Although there is no evidence that *Telephon Hirmondo* was "a slave of the government," it certainly did not encourage free or private communication but instead bound together an entrenched social group.

By 1910 Herbert Webb, reporting on telephone's development in Europe, did not mention Puskas or *Telephon Hirmondo*. His summary of the state of the telephone in Hungary was admirably concise:

Outside of Budapest there is little development of the telephone, and the development of the whole country is somewhat lower than that of Austria, and only slightly exceeds that of Italy, where the telephone is practically non-existent. In Budapest the telephone service was originally started by a company, but the State purchased the system after a few years and assumed

the monopoly of telephone work. The system was reconstructed
in 1903. . . . The low development and the slow progress of the
telephone in Hungary are principally due to lack of capital, absence of commercial policy, and to the retention of the old flat
rate tariff.[53]

There were still subscribers to *Telephon Hirmondo* in 1918 and
after the First World War. They paid an installation fee (said to
be about twenty-five shillings or five dollars) and an annual
rental of thirty shillings. The company derived additional revenues from advertising. This was already a deterrent in the Britain of the 1890s as it was to be in the Britain of the 1920s and
the early 1950s. "From time to time during the intervals between the different items of news, sometimes, alas, mixed up,
in American fashion, with the news itself, the patient listener is
assured that A's soap is unrivalled for the complexion or is en-
treated in dulcet tones to buy B's pills and avoid all others."[54] It
was easy to be disdainful about advertising in a decade when
there were innumerable advertising schemes, including "signs
upon the clouds," "celestial advertising," and advertising by
gramophone; innovation was to be diverted to the "wrong purposes" when the United States Gramophone Company could offer "as a novel form of advertising" to record any "musical
selection" along with a sponsor's advertisement. "Nobody
would refuse," the company claimed, "to listen to a fine song or
concert piece or an oration—even if it is interrupted by a modest remark, 'Tartar's Baking Powder is Best.'"[55]

Advertising's role in the development of the media was a
matter of debate long before the question of how best to finance
radio programs arose during the 1920s. It was certainly a matter
of debate within the Harmsworth empire, and it also affected
the development of the cinematograph, which according to one
British magazine was being "vulgarized" like the phonograph.
"Nestlés, of condensed milk renown, and Levers, of soap renown, have put their hands and purses together . . . and have
put up free machines behind hoardings in various parts of the
town. . . . If the experiment is a success, i.e. if a marked increase in the business transactions of Mr. Lever and Nestlé reasonably attributable to the cinematograph takes place, it will be
continued on an increasing scale."[56]

It is interesting to note that both Lever (in the form of Unilever) and Nestlé became staunch supporters of commercial tele-

vision in Britain fifty years later. In advertising, therefore, there
was as much of a continuum as in technology, and it, too, goes
back well before the 1890s to the development of consumer
goods industries and new systems of retailing earlier in the century. There are also links with entertainment. The British provincial firm of David Lewis deliberately set out to imitate P. T.
Barnum's methods: Lord Woolton, its most active manager between the two World Wars, was to play a major political part in
the launching of British commercial television.[57] But that was in
the distant future. The British "electrophone" venture of the
1890s never had powerful commercial or political backing, nor
did it attract the interest of the state.[58] The Electrophone Company was founded in 1894, later than *Telephon Hirmondo*, although there had been far earlier metropolitan provincial ven-
tures and the transmission "with entire success" two years
earlier of theater and concert performances in Birmingham,
Liverpool, Manchester, and other cities to audiences gathered at
an Electrical Exhibition in the Crystal Palace (a completely
appropriate link with 1851).

The new company's initial capital was £20,000 in £5 shares:
this may be compared with the initial capital of £100,000 in £1
shares of Marconi's Wireless, Telegraph and Signal Company in
1897, and the £100,000 of the British Broadcasting Company in
1922. Its object was defined as "the hiring out of an instrument
designed to enable subscribers to hear at their own homes or at
central offices [an interesting concept but a strictly limited term]
the performances at theatres, concert halls, etc."[59] It was explicitly stated that there was no novelty in this objective,[60] and it
created relatively little interest.

Although the Electrophone Company would continue to provide a service into the Edwardian period,[61] it never developed
the idea of a regular program as *Telephon Hirmondo* had done.
By the end of its first year of operations it had only 47 subscribers, and by the end of ten years, only 600. At best this was minute communication within minority communication. What was
said about it is more interesting to the historian, therefore, than
what it did. Thus, a comment in *Answers* is interesting because
it went much further in defining the electrophone's purpose
than the company's own prospectus had done. The electrophone, according to *Answers*, was "an invention by which per-

sons may have laid on in their houses, with the water, the gas and the other usual accessories, a regular supply of the most up-to-date music, the most recent plays, the latest *cause célèbre*, or the best passages from a sermon by one of the most eminent divines."[62] The concept of the wired city was already here, and there was great stress on topicality, not only the up-to-date but "the most up-to-date."

There was another remarkable half-ancient, half-modern journalistic note of 1892 on the "dangers" of such a "laid-on service." "It seems to us," wrote the *Electrician*, "that we are getting perilously near the ideal of the modern Utopian [back to Bellamy] when life is to consist of sitting in armchairs and pressing a button. It is not a desirable prospect; we shall have no wants, no money, no ambition, no youth, no desires, no individuality, no names and nothing wise about us."[63] *Electrical World* had been even more uneasy in commenting on E. J. Hall's remarks in 1890: "With the success of the first telephonic musical organization there will spring into being rival organizations, the very names of which would make incipient distress bliss. Imagine the awful devastation that would be wrought by 'The Organ Grinders' Telephonic Manual'. Fancy the horrors of having one's disposition wrecked by a 'popular program' headed by a memorial to the late McGuinty."[64]

The omission of news from the London service was the biggest contrast with Budapest, and it was significant. Newspaper interests during the 1890s were strong in Britain, like the telegraphic interest, and were to remain strong enough during the 1920s and 1930s to prevent the BBC from developing a news service of its own.[65] In Budapest, a solution was found that was not found in Britain until the 1940s and 1950s. "The newspapers of Budapest persistently boycotted the invention on its introduction, but they recognized now that, instead of being taken as a substitute for a newspaper, its effect is to whet the appetite of the public for events announced briefly through the telephone."[66]

The presentation of music and drama posed fewer such problems, for in Budapest the theaters realized that "to give the public a snatch or two from a favorite opera" would not adversely affect their receipts. They knew, too, that *Telephon Hirmondo* could always "organize concerts and entertainment in the editorial office" There might be some difficulty concerning performing rights,[67] but the most serious difficulty concerning music was the technique of transmission. One early injunction speaks for itself: "The transmitters should not be fitted in close proximity to the bass drums or the trombones of the orchestra."[68]

Religion always figured prominently in the bill of fare, as it would in the first bill of fare of the BBC after 1922.[69] Here again, the social and cultural comment was wide-ranging; as early as 1878, a short-story writer in an Australian magazine had contemplated a telephone installed back in Britain between Abney Hall and the village church of Mortham "so that the Hall people could have the benefit of Mr. Earle's pulpit oratory without going outside their own doors."[70] Mr. Earle was not very pleased, but he would have been less pleased had the Hall people decided to link up with a fashionable church in London. This was the theme of a later dialogue:

"I hope this cold weather agrees with you, Mr. Meteor," said the vicar to me the other day.

"No," I replied, "I have the service laid on by electrophone from St. Margaret's, Westminster, to my study."

The vicar looked very grave and his pretty daugher turned her head to hide a smile. Then I began to realize that my £5 for installing the instruments and £10 per annum for using them was quite thrown away. Neither the vicar nor Mrs. Grundy would be satisfied with this substitute for personal attendance.[71]

Mr. Meteor, the *Lightning* editor's nom de plume, remained convinced, however, of the telephone's advantages for this purpose. "You could listen to just as much of the service as you felt inclined for, and put the tube down when you'd had enough." He knew that it was precisely this "convenience" to which the vicar was objecting. "It is only too easy to salve one's conscience by the reflection that you can't really afford to suffer from 'telephone ear'. And besides there's the collection to consider."

Of course, such religious "broadcasts" would be of special value, it was felt, to the infirm—an argument to be made many times in the future, when the "consoling" power of the media was stressed. Budapest was praised also because its programs reached the hospitals, and the Electrophone Company supplied its apparatus "gratuitously" to London hospitals on Hospital Sunday so that patients could listen to sermons.[72] The writer who coined the phrase "the pleasure telephone" once again found his own balance. "It may be objected, perhaps," he wrote "that religious worship by telephone is not calculated to inspire reverence or include virtue, but at any rate the system is an inestimable boon to the aged and infirm, the patients in hospitals, and the women who are unable to leave their houses."[73]

Some doubt was expressed as to whether people would want "only to listen" to concerts, operas, and sermons. Would they not want to see what was going on as well? It was convenient that chronologically the phonograph (offering sound alone) pointed in the right initial direction long before the advent of sound broadcasting. Music need not be seen, and it might not be desirable to see Parliament debating even if it was right to hear it.[74] Yet seeing was very quickly brought into the reckoning too. "What if Edison and his followers advance so far as to bring out newspapers whose moving illustrations furnish their own descriptions . . . when readers might see . . . every detail and action?"[75] The properties of selenium had been known since the 1880s—they were very well known to Bell[76]—and it did not require special gifts of imagination to forecast that "before the next century shall expire the grandsons of the present generation will see one another across the Atlantic and the great ceremonial events of the world as they pass before the eyes of the camera will be enacted at the same instant before all mankind;"[77] or that "the theoretical possibilities are far beyond the reach of the wildest dreams of the imaginations of the most visionary of such quasi-scientists as Flaumarion and Jules Verne. What are we coming to? At this rate is there any necessity for limiting ourselves to such a mere bagatelle as space? Interplanetary communication will be carried on at fixed rates."[78]

This last forecast was conceived as a fun prophecy with a gentle touch of satire;[79] yet it is easy to understand why within this social and cultural milieu Marconi's wireless inventions of 1896 did not create quite the surprise they might have. One year before Marconi arrived in London, a writer summed up the situation very concisely: "Telephones are daily becoming more and more used, and hence more and more useful. The future for them should mean still further usefulness by reason of more perfect service. In the future, too, we may hope to have communications by telephone without wires. Shall we also see by electricity without wires?"[80]

Paradoxically, there was less talk between 1900 and 1920 about wireless providing the opportunities for a new "medium of communication" than there had been in relation to the telephone before 1900. It was seen at first mainly as a substitute for line telegraphy; its "broadcasting" element was underplayed, not least by Marconi himself. Only after the invention of the thermionic valve did wireless telephony become possible, and even thereafter, despite the work of Fessenden and de Forest,[81] the idea of a "program service" was very slow to take shape.

Business was less interested in broadcasting, by "wireless telephony" or other means,[82] than a handful of inventors and a growing group of radio amateurs. Although by 1914 the world of "amateur wireless" was a world of lively activity, it was certainly not controlled by entrepreneurs, the "enterprising individuals" willing "to work the vein," whose existence had been taken for granted by *Nature* in 1876. There were, after all, other veins for such entrepreneurs to work, very profitable ones, like the cinema and the press. Meanwhile, the telephone industry developed on the basis of a service to customers carrying out their own exchanges, and much of the early talk about theater-phones and electrophones was forgotten, buried in a rejected past. Fortunately, the idea of "the pleasure telephone" never completely disappeared. As the business uses of the telephone became obvious, more and more was made by its own advertisers of its possible private uses for "pleasure"; social chitchat, family drama, more convenient travel, dialing a joke or a prayer. Telephone culture, more lush in the United States and Canada than in Europe, caught all the moods from "Hello Central, Give Me Heaven" to "All Alone by the Telephone," "All Alone Feeling Blue," and "Find My Baby's Number." Meanwhile, Austin C. Lescarboum, Managing Editor of the *Scientific American*, made the best of every "pleasure world" in a book written in 1922; it surely has one of the largest titles in any library catalog:

Radio for Everybody, Being a popular guide to practical radio-phone reception and transmission and to the dot-and-dash transmission of the radio telegraph, for the layman who wants to apply radio for his pleasure and profit without going into the special theories and intricacies of the art.

## NOTES

1. *The Phonogram*, Jan. 1891; G. Seldes, *The Movies Come from America* (1931), p. 18.

2. The art critic Roger Fry, then aged thirteen, was rebuked by his father for trying to take away a piece of wire left lying on the floor in the palace. See D. Sutton (ed.), *The Letters of Roger Fry* (1972), Vol. I., p. iii. Bell was presented to Queen Victoria in January 1878 and demonstrated his invention. "Songs and recitations were transmitted," including "Comin' Through the Rye," and there was a link-up between Osborne and London. See G. E. Buckle (ed.), *The Letters of Queen Victoria*, second series, Vol. II (1925), p. 594; *The Times*, 15 Jan. 1878; F. G. C. Baldwin, *The History of the Telephone in the United Kingdom* (1938), p. 15.

3. Lord David Cecil, *The Cecils of Hatfield House* (1973), p. 239. Lord Kelvin used the same nursery rhyme in demonstrations to the British Association in 1876.

4. S. McKechnie, *Popular Entertainment through the Ages* (1931), p. 177. The early phonograph would not only "speak" in Dutch, German, French, Spanish, and Hebrew but would "imitate the barking of dogs and the crowing of cocks." See R. Gelatt, *The Fabulous Phonograph* (1956), p. 9.

5. "Those magnificent men in their flying machines," as in the film, often hated the idea that "stunts" would have to go if an "aviation system" were to operate. The electric light, however, may have been one invention which reduced fun. Or at least, it was so claimed:

Oh, Mr. Edison, whatever have you done?
Oh, Mr. Edison, you've gone and spoiled our fun!
No more can we ramble with the girls all night!
The people they will see us all by the Electric Light.

(Quoted in *Blue Bell*, 1960)

There were some critics, however, who claimed that the telephone itself reduced "romance." "Who would emulate Lord Lochinvar or attempt a ride like Paul Revere's?"

He (the foe) would have no chance to pillage, in truth,
For someone'd rush to a telephone booth
And call "Long Distance". We'll never hear
Of another ride of Paul Revere.

*The Daily Witness*, Montreal, 20 Jan. 1906.

6. See his remarkable address at the dinner of the Institution of Electrical Engineers in 1889 as reported in *The Electrician*, 8 Nov. 1889.

7. *The Electrician*, 8 June 1894.

8. Ibid., 11 June 1897. Preece brought the first practical pair of telephones to Britain and introduced Bell to his first British Association audience in 1877. (Baldwin, *History of the Telephone*, p. 14).

9. C. Mackenzie, *Alexander Graham Bell* (1928), p. 158; H. N. Casson, *The History of the Telephone* (1911), p. 45. For the early history see also

R. V. Bruce, *Bell: Alexander Graham Bell and the Conquest of Solitude* (1973), Chapters 1–4.

10. Casson, *History of the Telephone*, p. 44.

11. *Montreal Gazette*, 3 Jan. 1878. Cf. *Ottawa Citizen*, 16 Dec. 1881. It was shrewdly pointed out that listeners to music by telephone were more interested in "the marvellous feat" the telephone was performing than in the "musical treat" itself.

12. *Blue Bell*, April 1973.

13. It is interesting that earlier in the nineteenth century the transmission of music had been attempted before that of speech. See, among others, Baldwin, *History of the Telephone*, p. 267; for Elisha Gray and others, see E. N. Simowitch, "Musical Broadcasting in the Nineteenth Century," *Audio*, June 1967, pp. 19–23.

14. *Nature*, 24 August 1876.

15. *Scientific American* 10 Dec. 1891; *The Telephone*, 2 Sept. 1869. *Electrical World*, 3 Feb. 1894, describes similar installations in Bordeaux.

16. *Electrical Review*, 16 Dec. 1896. See also A. M. Schlesinger, *The Rise of Modern America*, 1805–1951 (1951), p. 66. McKinley talked from his Ohio home with campaign managers in thirty-eight states. The first White House telephone was installed in 1878.

17. "Extension and Improvement of the Telephone Service," *Electrical World*, 20 Sept. 1890.

18. Talk of the dangers of the "telephone ear" had been preceded by talk of the dangers of the "electric light eye."

19. George Bernard Shaw objected to the first London telephones being of such "stentorian efficiency" that they broadcast private messages.

20. Baldwin, *History of the Telephone*, p. 267, describes a broadcast of a play from the Comedy Theatre to the Hotel Bristol.

21. For the testing phase, see the *Brockville Reporter*, 14 Feb. 1878. "So perfect was it (the telephone) in acoustic qualities that the song from a gentleman separated by two or three rooms from the tin tube . . . was distinctly heard by those at the other end of the line, and he was not much of a singer either."

22. G. Seldes in E. Carpenter and M. McLuhan (eds.), *Explorations in Communications* (1966), p. 197. Schlesinger, *Rise of Modern America*, p. 142, quotes a contemporary: "vaudeville belongs to the era of the department store and the short story." See also Asa Briggs' Fisher Memorial Lecture of 1960, "Mass Entertainment: The Origins of a Modern Industry."

23. *Electrical World*, 12 April 1884. By 1890 between thirty and forty periodicals in Britain alone dealt with "electricity." (*Electricity*, 15 Nov. 1890).

24. *The Times*, 20 March 1872.

25. *The Daily Expositor*, 23 March 1878.

26. *Engineer*, 8 Nov. 1889.

27. Colin Cherry, *On Human Communication* (1962), p. 125.

28. *Electrical World*, 12 April 1884.

29. Ironically Baird's own contribution to small-screen television, relying as it did on mechanical scanning, proved one of the rejected alternatives of history. The electronic system developed by EMI, without any publicity, prevailed during the early years of "the age of television." Much has been made of the contrast between Baird's improvisations and EMI's highly organized laboratory teamwork, but far less of the fact that EMI was itself a merger (1931) of gramophone interests whose history stretched back to a prelaboratory period. One year after Baird issued his advertisements, the Bell Telephone Company staged the first American demonstration of public television over a considerable distance—from Whippany, New Jersey, to Washington and New York.

30. Baldwin, *History of the Telephone*, p.v. Section 5 of Baldwin's Chapter XXII, "Miscellany," deals with "radio telephony." "Radio-telephony," Baldwin writes, "was really born of the father science, wireless telegraphy, in a manner somewhat analogous to that in which line telephony was derived from telegraphy." (pp. 641–642). For an interesting study of some of the links, see E. N. Sirowitch, "A Technological Survey of Broadcasting's Pre-History, 1876–1920," in the *Journal of Broadcasting*, Vol. XV, Winter 1970–1971, pp. 1–20.

31. For the slow development, see ibid, pp. 616–617, with statistics of telephone usage, and H. L. Webb, *The Development of the Telephone in Europe* (1910). For Gill's part in the 1922 talks, see A. Briggs, *The Birth of Broadcasting* (1961), pp. 107ff. Gill was a President of the Royal Institution of Electrical Engineers.

32. See R. Skovmand (ed.), *D-R.50* (Denmarks Radio, 1975), Chapter 1.

33. *Punch*, Vol. XI, 1846, p. 253. Cf. an interesting article in ibid, Vol. XXXV, 1858, p. 254, where the idea of the telegraph being brought "within a hundred yards of every man's door" is explored.

34. Ibid., Vol. XV, 1848, p. 275.

35. By 1858 it was. "Electricity is now fairly taking the circuit of the entire globe . . . .We do not see what there is to prevent a pianist, who holds this electric accomplishment at his finger's ends, from performing in every capital of Europe at the same time." (Ibid, Vol. XXXV, 1858, p. 165).

36. It is interesting not only to log communications history in strict chronological sequence, but to note parallels in the history of different media and periods. In June 1920, another "popular vocalist," Dame Nellie Melba, gave a broadcast from London by courtesy of the *Daily Mail* which did much to publicize the possibilities of a new medium. Cf. publicity in relation to crime detection and arrest in the history of the telegraph (a murderer's arrest at Paddington station), the telephone (see, for example, the *Montreal Witness*, 23 Feb. 1881, for the arrest of a young "ruffian" after a telephone call), and radio (the arrest of Crippen).

37. *Punch*, Vol. XXXV, 1858, p. 254.

38. M. McLuhan in *Understanding Media* (1964) quotes a 1906 example of use of the word "phony" in the *New York Telegram*: it referred to the fact that "a thing so qualified has no more substance than a telephone talk with a suppositious friend." For "telegraphic fibs," see *Punch*, Vol. XXVII, 1854, p. 143. For *Punch's* "prophecies" about communications, which continued throughout the century, see A. Briggs and S. Briggs, *Cap and Bell* (1973). It is interesting to compare them with the remarkable prophecies of Albert Robida in Paris and the later philosophies of Tesla. See A. Robida, *La Vie Electrique* (1887), and *Le Vingtième Siècle* (1883) and J. J. O'Neill, *Prodigal Genius, The Life of Nikola Tesla* (1944).

39. Francis Bacon had referred in 1605 to "the Medium of Words," and the word "medium" was applied, usually somewhat indirectly, to the nineteenth-century newspaper. See R. Williams, *Keywords* (1976), pp. 169–170.

40. *Scientific American*, 2 July 1881. See also G. Richards, *Hungary Yesterday and Today* (1936), p. 234, and F. Erdei (ed.), *Information Hungary* (1968), p. 645. An International Telegraph Conference was held in Budapest in 1896. (*The Electrician*, 14 August 1896).

41. Arthur Mee, "The Pleasure Telephone" in the *Strand Magazine* (1898), pp. 339–369.

42. *Scientific American*, 26 Oct. 1895. There are good accounts of the system in the *Electrical World*, 4 Nov. 1893, and *Electrical Engineer*, 6 Sept. 1895.

43. The term "program" was used in *Invention*, 26 March 1898, referring to "a time-table of the various items which will be telephoned during the day." (*Strand Magazine*, p. 361).

44. The first opera was transmitted from "editorial buildings, in rooms specially adapted for the purpose," but later arrangements were made to connect the opera house and some music halls with the office. (*Answers*, 16 Nov. 1895).

45. *Newspaper Owner and Manager*, 4 May 1898. The *Scientific American* (26 Oct. 1895) referred to the announcers as "ten men with strong voices and clear enunciation."

46. *Newspaper Owner and Manager*, 4 May 1898.

47. See among others J. H. Holmes, "The New Journalism and the Old," in *Munsey's*, April 1897, and R. Pound and G. Harmsworth, *Northcliffe* (1959), Book II.

48. There is a brief account of the system in J. Erodess, "Le Journal Téléphonique de Budapest" in *Radiodiffusion*, Oct. 1936. See also P. Adorian, "Wire Broadcasting" in the *Journal of the Royal Society of Arts* (1945). This article by a prominent supporter of twentieth-century wire broadcasting links two periods.

49. Listeners to the service could learn of the German Kaiser's toast to the Hungarian nation within a half hour after the toast was proposed. "Without this apparatus they would have had to wait until next day." (*Invention*, 26 March 1898).

50. *Siftings*, 15 July 1893.
51. *Invention*, 26 March 1898.
52. *Scientific American*, 2 July 1881.
53. H. L. Webb, *Development of the Telephone*, pp. 69–70.
54. *Newspaper Owner and Manager*, 4 May 1898.
55. Quoted by W. Abbot and B. L. Rider, *Handbook of Advertising* 4th ed. (1957), p. 387.
56. See R. Pound and G. Harmsworth, *Northcliff*, and P. Ferris, *The House of Northcliffe* (1971). *Answers*, founded in 1891, was at the heart of the Harmsworth enterprise; in 1894 (26 March) it had an interesting article on "People who suggest advertising ideas." A manager of a firm well known for its advertisements was said to have told a reporter that his firm was "besieged with suggestions for novel advertisements, every article under the sun being offered (note the term) as a 'medium of publicity.'" See also *Invention*, 22 Jan. 1898.
57. See the evidence given by a group of firms, including Unilever, to the Beveridge Committee on Broadcasting (Cmd. 8117, 1951). For Lewis, see A. Briggs, *Friends of the People* (1956), and for Woolton's twentieth-century role, see H. Wilson, *Pressure Group* (1961).
58. As early as 1882, in a remarkably percipient article, the *Electrician* (4 Feb.) suggested that while there was doubt about what the telephone system might eventually develop into, it was "problematical" whether any government department could ever enter into "tapping the stage and the pulpit telephonically, or of laying on operatic music like gas for the use of every householder." Already by the agreements of 1880, the Post Office in Britain (using the Telegraph Act of 1869) had imposed licensing regulations and controls on telephone companies. See A. Hazlewood, "The Origins of the State Telephone Service in Britain" in *Oxford Economic Papers*, Vol. 5, March 1953.
59. *Lightning*, 10 Jan. 1895.
60. Ibid. Certainly by 1895 several European opera houses were wired through to homes and other centers.
61. Queen Victoria heard the electrophone for the first time just before the end of the century when she listened at Windsor Castle, first to boys from naval and military schools singing "God Save the Queen" and later to a concert from St. James's Hall (*Electrician*, 9 June 1899).
62. *Answers*, 4 Dec. 1897.
63. *Electrician*, 11 Feb. 1892.
64. "Music on Tap" in the *Electrical World*, 20 Sept. 1890.
65. See A. Briggs, *The Birth of Broadcasting* (1961), pp. 162–166, and *The Golden Age of Wireless* (1966), pp. 153–160; A. Clayre, *The Impact of Broadcasting* (1973), Chap. 4.
66. Mee, *Strand Magazine* (1898), p. 366.
67. The Performing Rights Society had not been founded, but there were already discussions on "mechanical music." As early as 1893 there had been a legal case in Paris about the use of a singer's voice on a theaterphone. (*Lightning*, 19 Jan. 1893).

68. Quoted in Baldwin, *History of the Telephone*, p. 268.

69. For the early role of religious broadcasting along with information and entertainment, see J. C. W. Reith, *Broadcast over Britain* (1924).

70. "The Days of the Telephone: A Tale of the Future" in *The Australian*, Oct. 1878. This short story may have appeared first in a British magazine.

71. *Lightning*, 10 Jan. 1895.

72. *Newspaper Owner and Manager*, 4 May 1898; *Electrician*, 9 June 1899.

73. Mee, *Strand Magazine* (1898), p. 362.

74. *Invention*, 10 Dec. 1898; Baldwin, *History of the Telephone*, p. 269. McLuhan, *Understanding Media*, refers to very early New York comment on "the terrors of the telephone" as an instrument of political communication for demagogues.

75. *Referee*, 12 Jan. 1896. This article, "Sporting Notions," was inspired by a cinematograph show in Paris.

76. *Lightning*, 7 Sept. 1893. According to Bell, at least a dozen scientists were then working on "vision." His own "photophone" had been ready as early as 1880.

77. Ibid., 20 April 1893.

78. *Siftings*, 1 Jan. 1893.

79. "Dream after dream will be realized, until we wake up to the fact that the whole scheme may possibly be some well-conceived and choice morsel for the delectation of plethoric-pursed investors." The speculative theme was strong during the 1890s.

80. S. F. Walker, "The Future of Electricity" in *Electrical Engineer*, 4 Jan. 1895.

81. See H. Fessenden, *Fessenden: Builder of Tomorrows* (1960), and L. de Forest, *Father of Radio* (1950). An early Fessenden "broadcast" included, much as a Bell telephone demonstration might have done, Handel's *Largo* and the inventor himself singing "Holy Night." De Forest used the contralto Madame von Boos Farrar to sing "I Love You Truly."

82. For the other means, including Telharmonium, see Sirowitch, "A Technological Survey of Broadcasting's Pre-History."

# Editor's Comment

The two preceding chapters have left us with a question. At the turn of the century there were two conceptions as to the most fruitful use of electronic voice transmission either for broadcast or conversation. As it turned out, broadcasting was deferred for half a century while networks for conversation grew with remarkable speed, especially in the United States. Why did the conversational alternative win out?

That is a question of a classic genre; in another form it appears in the crucial point of difference between the theories of Karl Marx and Max Weber. Was the outcome determined, as Marx would explain, by configurations of technology and economics, or, as Weber might suggest, with the intervention of human will and values? Did the telephone become a point-to-point service because the physical devices that were first available lent themselves to that use? Did radio emerge only when the vacuum tube provided a cost-effective technology for broadcasting? Or did the developers of telephony have a particular conception of a particular grand system, which motivated them to solve those specific technical problems obstructing its realization? Did point-to-point networks precede broadcasting because of value choices about society's needs?

To tease out the historical interaction in telephony between what was technically possible, economically profitable, and socially desirable is a puzzle worthy of scholarship. The choice among those rival determinisms, which have permeated social history since the mid nineteenth century, is not a simple either/or.

In Pool's chapter, we partly return to the question. It is also

present in Colin Cherry's chapter, which begins with the technology of the switchboard and sees that as shaping the social
character of telephonic conversation. Cherry's thesis is that the
technology of the phone network favored democratic social relations—in contrast to broadcasting, which lends itself to top-
down exercise of authority.

The relationships noted in those articles make a strong case
for some technological determinism. Aronson has already argued that Bell's development of a two-way instrument was a vital step in moving from a broadcast to a point-to-point strategy
in marketing the telephone. He also notes that costs were such
as to attract business subscribers but few pleasure seekers. The
initial quality of reproduction permitted a spoken message to be
understood with only occasional repetition, but the power required to produce truly enjoyable music posed problems. The
switchboard was a device inherited from the telegraph era in
primitive form; it quickly suggested the possibility of linking
subscribers to each other, as well as to the central telephone office. Community telephone exchanges thus developed easily and
naturally.

The use of copper wire later, instead of iron, and the development of the loading coil made long-distance telephony a practical albeit expensive proposition. The automatic switch made
privacy possible, as Brooks notes later on, and thereby helped
promote attachment to it as a value. All these technical developments may have helped determine that the promising market
for telephone entrepreneurs was in conversational exchanges between individuals engaged in private commerce.

There might also be a strong case made that traditions and
ideas played a crucial role in determining telephone use. First,
it did not emerge with equal speed in all countries. Attali and
Stourdze for France and Perry for England show in the next two
chapters how governmental policies retarded telephone growth.
In France, from the days of telegraphy, telecommunication was
seen primarily as an instrument for governmental use rather
than for commerce; in Britain, the telephone was seen as a
threat to the publicly owned telegraph system. A telephone system, Pierce argues below, is the most complicated machine built
by man and is easily degraded by mistaken policy. Technology
is not impervious to human decisions.

Perhaps, as the authors of Chapter 6 suggest, there is much self-fulfilling prophecy in the way that Bell, Hubbard, and Vail foresaw the phone system's growth. Perhaps the grand design of universal service, with access, privacy, and freedom of discourse by all, stemmed equally from the technology and its creators' idealism. They saw, in a public communication system, a goal worth striving for.

# 3

## The British Experience 1876–1912: The Impact of the Telephone During the Years of Delay

### Charles R. Perry

In Great Britain the telephone never became a symbol for a particular era. While the early Victorian period is often called "The Age of the Railway," it would be a misnomer to label either the late Victorian or the Edwardian years "The Age of the Telephone." One gets no impression that the telephone's impact on British society remotely approached the railways' influence; no Edwardian novelist could have accurately written of the telephone as Dickens did of the railways in *Dombey and Son*.[1] Subsequent social and economic historians have also minimized or ignored the history of the telephone in Great Britain. For example, Ashworth, in a study covering the years from 1870 to 1939, referred to the telephone only four times, while he devoted many pages to the railways.[2] Court, in his commentary on the economy between 1870 and 1914, failed to mention the telephone at all.[3] Even when the telephone is mentioned, it is often lumped with the telegraph and wireless in vague statements about the growth of communications.[4]

On an a priori basis, one might assume that the telephone should have had a greater impact. After all, its promise in some ways closely resembled the potential of the railway. Both could lessen the vicissitudes of time and space, bring loved ones together, and facilitate trade and commerce—in short, bind a still remarkably divided and provincial society. My purpose is to explore the failure of telephone development in Great Britain from

Without holding them responsible for the final product, I would like to thank H. J. Hanham and John Clive for their advice and criticism of an earlier version of this article. Part of the research was funded by fellowships from Harvard University and the Whiting Foundation.

its introduction in 1876 until the nationalization of the industry in 1912.

To consider this problem, one must first assess how the telephone was perceived in Great Britain—what it meant to the public. Of course, this question cannot be answered. Opinions regarding the telephone's usefulness and contributions to society varied enormously from one individual to another and from one year to another, but it can still be argued that the telephone never suffered from a lack of exposure in Great Britain. This invention caught the public eye early, and continued to be of interest throughout the period leading to nationalization. In September 1876, only a few months after Bell's invention was shown at the Centennial Exhibition in Philadelphia, Sir William Thomson (the future Lord Kelvin) introduced the telephone to the United Kingdom at a meeting of the British Association in Glasgow. Thomson was one of the telephone's most enthusiastic advocates and called it "the greatest by far of all the marvels of the electric telegraph."[5]

The next year brought increased publicity in the scientific community and the general public. *The Times* closely covered events connected with the telephone. In July it reported experiments in communication between the Queen's Theatre and Canterbury Hall, south of the Thames.[6] Bell's lecture tour also aroused interest. After he spoke before the Society of Arts in November, *The Times* noted that "if any proof were wanting of the universal interest this remarkable instrument is now exciting, it was shown by an assembly of members which not only filled the hall and staircases of the building, but overflowed into the street outside."[7] On a more popular level, *The Illustrated London News* gave prominent attention to Bell's invention "for the eléctrical transmission of distinctly articulate sounds to great distances."[8] Explanations such as this probably went a long way in helping the lay public understand exactly what the telephone could do, even if most of them could not comprehend the more technical aspects of operation. The audience that Victoria granted to Bell and his agent, Colonel W. H. Reynolds, in January 1878 increased public awareness of the telephone. *The Times* reported that the festivities at Osborne House continued almost until midnight with Kate Field singing "Comin' Thro' the Rye"

and reciting the epilogue to *As You Like It* over a telephone.[9] All
in all, the telephone received adequate attention by the press.

Later, the popular press contained descriptions of what an exchange actually looked like. In December 1883, the *Pall Mall Gazette* reported on an East India Avenue exchange in the City of
London:

From the lofty roof of one of the houses of that sombre court
rises a derrick, a square structure of wrought-iron bars 30 or 40
feet high by 8 or 10 feet, and looking like the upper portion of a
skeleton lighthouse. . . . This edifice is surrounded by a lightning conductor; you ascend it by a perpendicular ladder, and,
pausing on its upper story . . . you look round from your airy
perch to find that what appear innumerable wires radiate from
your transparent cage in every possible direction over the dirty
housetops of the City. Most of these wires are bare and unen-
closed; others are in cables containing each twenty wires. . . .
Below, in the attic, is a room occupied by eleven young ladies.
The 271 wires, which represent the subscribers of the East India
Exchange with 46 trunk and other direct wires, are guided
down from the derrick above into neat mahogany cabinets or
cases, in front of which the young ladies are seated. The alert
dexterity with which at the signal given by the drop of a small
lid . . . the lady hitches on the applicant to the number with
which he desires to talk is pleasant to watch.[10]

*The Times* had a similar report when the National Telephone
Company opened its Gerrard Street Exchange in 1908.[11] Of
course, not all the attention the telephone gained was positive.
There were numerous complaints that wires were disfiguring
the countryside and destroying the atmosphere of places like
Runnymede.[12] In any case, the telephone did not remain hidden
in darkness.

One must therefore look elsewhere to explain the failure of
the telephone's development from 1876 to 1912. The answer cannot be found in any vague claims that Britain failed to understand or appreciate the potential of the telephone, especially its
practicality. Too much has been made by previous writers of the
initial negative reception some periodicals gave the telephone.[13]
To be sure, early problems with the invention were reported as
well as its successes. For instance, when induction from nearby
wires ruined an experiment in communication between the
Adelphi Theatre and Fleet Street in November 1877, the difficulties were duly reported.[14] One can quote examples like the following letter to *The Times* to substantiate British indifference to
the telephone:

In America, with long lengths of single wire, and a fine dry climate, the telephone may perhaps come into use practically. But in England, with most of the telegraph wires already overweighed, it is hardly likely to become more than an electrical toy, or a drawing-room telegraph, or at most a kind of electrical speaking tube.[15]

One can also cite a lack of appreciation for the telephone among those who should have known better. For example, John Tilley, who was the chief permanent official in the Post Office in the years immediately after the telephone's arrival in England and thus directly involved in formulating government policy on the telephone, was unconvinced of the practicality of the telephone. After an encounter with an Edison model, he reported his feelings. "The circumstances were ordinarily favourable," he wrote, "and the result, as far as I am concerned, was that I could not understand a single word. The noise, for I can only call it a noise, strikingly resembled an exceedingly bad Punch [and Judy show], and the working is most inconveniently cumbrous."[16]

Such attitudes are almost invariably expressed about any new invention, and these examples of skepticism were not unique to England. This hostility, moreover, was by no means the only or even the dominant outlook on the telephone. As early as July 1877, two and one-half years before Tilley's encounter with the telephone, *The Times* pointed out:

Few of the recent applications of science have attracted so much popular curiosity [as the telephone], and few, perhaps, have been the subject of such extravagant and erroneous statements as the telephone. . . . *Yet the invention is a most startling one—too remarkable, indeed, to be discredited by any amount of exaggeration* [italics mine].[17]

For example, the telephone was taken very seriously by the Post Office. This department had become involved in communication by electricity in 1870 when the telegraph companies were nationalized. An even more direct link between the telephone and telegraph forced the Post Office to study the telephone. This was the ABC telegraph, by which one could spell out messages with no knowledge of Morse or any code. The Universal Private Telegraph Company had been established in 1864 to provide local intercommunication between customers, and its operation had been purchased by the government along with the other telegraph companies.

The telephone appeared to be an alternative to the ABC instrument. By March 1877, the Post Office had already instructed its chief engineer to report on the capabilities and practical utility of the telephone. Studies were made of telephone technology in America.[18] It was thought that the telephone posed no immediate threat to the telegraph, particularly for long-distance communications. By February 1878, the Postmaster General, Lord John Manners, had announced that the Post Office had no plans to introduce the telephone as part of the long-distance telegraph branch.[19] Given the primitive state of telephone development, especially the problems of amplification over distances, it would be unfair to fault the Post Office for any lack of prescience.

The telephone did seem suitable for use, however, in certain local private-wire situations where the ABC instrument had been used exclusively. William Preece, the Post Office Engineer-in-Chief and one of the most learned electrical engineers in the country, reported: "Although in its present form, the telephone is not generally applicable, there are a great many instances. . . where the instrument is perfectly practicable, and it will be generally demanded by our renters. More than that, I believe it will lead to a large accession of private wire business."[20] The department thus viewed the telephone as both a practical invention that could serve the public and a source of additional revenue. Moreover, in contrast to its lethargic reputation, the Post Office was acting in advance of concerted public opinion.

Negotiations with Colonel Reynolds were begun, and by December 1877 an agreement was successfully concluded.[21] The Post Office would act as an agent for the Bell telephone and would lease it to the public at annual rates of £5 for short circuits and £10 for long circuits. As compensation the Post Office would receive 40 percent of the gross rental income. Although the Post Office was not to be the developer of the telephone during its early years in Great Britain,[22] by no means did the government consider the invention an impractical toy. Manners' comments in a letter to the Treasury indicate the respect and fear that the telephone invoked. He wrote:

I was compelled to adopt the invention for private wire purposes; and I think you will see that the matter was not one in which the Post Office could remain passive, for if renters of private wires, on the expiration of their agreements, had applied to have their wires fitted with telephones, and their application had been refused, they would, I need scarcely say, have employed contractors to erect fresh wires, and the Post Office would have had old wires thrown on its hands.[23]

The country became aware of the telephone's practical uses faster than it recognized the telegraph's utility. The period between the first experiments with each and the formation of companies to manage and develop the industry was much shorter with the telephone. William Fothergill Cooke first demonstrated the potential of the telegraph for signaling and other safety measures along railway tracks in 1836.[24] Yet not until 1845 was the Electric Telegraph Company provisionally registered. Only two years elapsed between the exhibition in Philadelphia and the registration of the Telephone Company in June 1878. Moreover, the practical utility of the telephone was heightened by the formation of the United Telephone Company in 1879. An amalgamation of the Telephone Company and its rival the Edison, the United used the best features of both the Bell and Edison telephones.[25] That England led the world in the manufacture of cables should also have contributed to the development of the telephone industry.[26] By the early 1880s the telephone appeared to have a future. As *The Spectator* noted:

There are few inventions which have more impressed the public mind than the telephone. To scientific men it is no doubt merely one—though by far the most original—of many improvements in telegraphy which are continually being made. But to the outside public, the possibility of actually speaking to a person miles away seems to be an altogether new gift of science. . . . [The telegraph ticks in code.] But this, after all, is a very different thinking, from hearing the voice of your friend speaking not only in his native tongue, but in tones and manner which are peculiar to him individually.[27]

Yet by 1912, the potential had not been fulfilled, although the public knew of the telephone and its practicality was proved. Because of the social attitudes, economic conditions, and political complications that the telephone encountered in Great Britain, it was by no means considered a necessity but indeed, something of a luxury. People and institutions to whom the telephone would later prove a necessity simply saw no need to have one. The chief constable of Exeter, for instance, did not bother to get one until 1901,[28] and Harrods department store did not install pay telephones for the public until 1908.[29] Arnold

Morley, Postmaster General from 1892 to 1895, expounded the view of the telephone as something of a frill. "Gas and water were necessities for every inhabitant of the Country," he argued in 1895. "Telephones were not and never would be. It was no use trying to persuade themselves that the use of the telephone could be enjoyed by the large masses of the people in their daily life. . . . "[30] This outlook was not confined to short-sighted bureaucrats who regarded penny-pinching more highly than the needs of the public. *The Times* adopted a very similar attitude toward the telephone. In 1902 the London County Council was exorcised over the poor state of telephone service in London and especially over what it felt were unnecessarily high rates. *The Times* was not sympathetic to their complaints:

When all is said and done the telephone is not an affair of the million. It is a convenience for the well-to-do and a trade appliance for persons who can very well afford to pay for it. For people who use it constantly it is an immense economy, even at the highest rates ever charged by the telephone company. For those who use it merely to save themselves trouble or add to the diversions of life it is a luxury. An overwhelming majority of the population do not use it and are not likely to use it at all, except perhaps to the extent of an occasional message from a public station.[31]

Two years later an article in the *Edinburgh Review* presented almost exactly the same reasoning.[32]

The telephone was considered a luxury by some because Great Britain already possessed the finest communications system in the world. Rowland Hill's successful campaign for a Penny Post had marked a turning point in the 1830s, and the nationalization of the telegraph industry had further added to the strength of the Post Office. By 1872, for example, there were 5,000 telegraph offices with 22,000 miles of line and 83,000 miles of operation.[33] Further, the telegraph system was an especially strong competitor because the government subsidized its use. In 1883, Dr. Charles Cameron forced a reduction in telegraph rates upon an unwilling Gladstone ministry; the new tariff was six pence for twelve words, including the address, plus one-half pence for each additional word.[34] The number of telegrams sent rose from thirty-three million in 1884–1885 to fifty million in 1886–1887.[35] For long-distance communication, the telephone was more expensive than the telegraph; when the Post Office took over operation of the trunk lines in 1896, the following charges were made for a three-minute call:[36]

| | | |
|---|---|---|
| For 25 miles and under | 0s. | 3d. |
| Above 25 and under 50 miles | 0 | 6 |
| Above 50 and under 75 miles | 0 | 9 |
| Above 75 and under 100 miles | 1 | 0 |
| Every additional 40 miles or fraction thereof | 0 | 6 |

As a result, the telegraph continued to be heavily relied on, and long-distance calls made up only a small percentage of the total calls placed.[37] The efficiency of the postal and telegraph systems did much to reinforce the attitude of civil servants who formed telephone policy that the old and the new were essentially different. In 1911, the Post Office approached the Treasury on the possibility of opening unprofitable but needed telephone exchanges. Roland Wilkins replied that "telephones were not in the same position as telegraph and postal facilities; they were not exactly to be regarded as a luxury; but while postal and telegraph facilities might be provided at a loss to the taxpayer, telephone extensions should . . . as a general rule pay their own way."[38]

The conviction that not everyone needed the telephone was manifested in the way the industry developed; the telephone was to be extensively used only by certain areas, groups, and classes in the country. Its greatest impact, for instance, was in urban areas. The early promoters in the industry naturally regarded the cities as the place to begin because arrangements were easier to make than in rural areas, and a return on investment seemed more likely. Table 1 indicates the cities where the telephone was initially promoted.[39]

More remarkable is that this attitude of relegating the telephone to cities and towns hung on. As late as 1901, Michael Hicks Beach, the Chancellor of the Exchequer, persisted in believing that "telephonic communication is not desired by the rural mind."[40] As late as 1913, London accounted for over one third of the telephones in the entire country.[41] Some, however, saw the need to bring the telephone to all parts of the country regardless of cost. John Lamb, a Post Office expert on the telephone, argued before a Select Committee in 1898 that it could be of much greater benefit:

In the postal and telegraph systems the whole of the country is treated as one, and the richer districts lend support to those which are not so prosperous; but if, in the telephone exchange business, the prosperous cities and towns be empowered to establish exchanges for their own benefit, who is to find the money for development in the rural districts?[42]

Lamb's question remained unanswered in the years before nationalization, and only gradually were rural areas affected by the telephone. Here again, fiscal philosophy was more important than technology. The Post Office believed that as long as the total system ran at a profit, losses in underpopulated areas were acceptable.[43] The Treasury was much stricter and insisted that exchanges should be opened only when the projected deficit was 3 percent or less.[44]

Emphasizing urban telephone use is not enough, because it affected various social classes within the cities and towns in profoundly different ways; to have a telephone in one's home was simply beyond the economic reach of the laboring masses. One of Mrs. C. S. Peel's correspondents reported that it was "usual" for people to have a telephone by the early 1890s,[45] but the correspondent's husband had an income of £1500 per year. Even to assume that everyone of middle- or upper-class background had a telephone would be foolish. It is very difficult to

**Table 1**
Cities Where the Telephone Was Initially Promoted

| Town | Proprietors | Probable Date of Opening |
|------|-------------|--------------------------|
| Glasgow | D. and G. Graham | March 1879 |
| London | The Telephone Co., Ltd. | Aug. 1879 |
| London | Edison Telephone Co. | Sept. 1879 |
| Manchester | Lancashire Telephonic Exchange, Ltd. | Oct. 1879 |
| Liverpool | Lancashire Telephonic Exchange, Ltd. | Oct. 1879 |
| Sheffield | Tasker, Sons and Co. | 1879 |
| Halifax | Blakey and Emmott | 1879 |
| Birmingham | Midland Telephone Co., Ltd. | Dec. 1879 |
| Edinburgh | Scottish Telephonic Exchange, Ltd. | Oct. 1879 |
| Belfast | Scottish Telephonic Exchange, Ltd. | 1880 |
| Sunderland | United Telephone Co. | 1880 |
| Bristol | — | Dec. 1879 |

establish an exact figure dividing working class life and middle-class gentility, but Laski estimated that in the Edwardian period £200 per year would assure a fairly comfortable life for a single woman.[46] Certainly at this level the telephone would seem a luxury to many people. Perhaps an examination of another income at a slightly higher level might also be helpful. Take, for example, the budget of a young Edwardian couple living in a London apartment on £700 per year.[47] Their budget might allocate £50 for savings, £80 for clothing, £25 for life, fire, and accident insurance, and £20 10s for the entire year's coal, wood, and electricity. In 1901, the Post Office and the National Telephone Company charged London subscribers £17 per year for unlimited service; for measured rate service, they charged £6 10s per year within the County of London and £5 10s per year in outer London. (These last two figures included £1 10s worth of calls charged at 1d or 2d depending on the distance.)[48] The telephone was thus expensive compared to other items. When one could employ a maid for £20 per year, having unlimited telephone use for £17 did not seem to be a bargain. Incidentally, none of the budgets presented by Laski mentioned the telephone, which again indicates its somewhat peripheral status.

How did the telephone affect those who could afford one? Again, there is no one answer. Mrs. C. S. Peel, looking back on the years before World War I, wrote that the "telephone has helped us whirl along faster and to make life less formal."[49] Certainly one should not discount this assessment. The world probably did seem to move a little more rapidly but perhaps more striking is the telephone's limited use for social purposes.[50] One simply did not print a telephone number on one's calling card; no one of good breeding would be so crass as to extend or accept an invitation by telephone. According to Laski, "the telephone, usually inconveniently located in the flower-room, in a corner of the hall, or in a lobby between the smoking-room and the gentleman's lavatory, was seldom used for chats."[51] There seems to have been an initial reluctance on the part of even those who had telephones to work them into the elaborate etiquette of the late Victorian and Edwardian years. The telephone may have been too direct and abrupt a means of communication to be fully accepted. "Calling" continued to mean visiting friends between three and six P.M. rather than using the telephone.

Part of this reluctance and resentment may also have been due to the telephone's impingement upon one's privacy. Although the long-range effect of the telephone may have been to increase privacy,[52] the average Englishman did not recognize this result before World War I. A. H. Hastie was the leader of the Association for Protection of Telephone Subscribers and an advocate of increased telephone use, but he was still very much aware of the problem of privacy. Hastie analyzed the situation in this way. People "complain that when they are busy they are continually being rung up about trivial matters. A man might as well complain that he has to open his door to see unwelcome visitors, and his back door to admit the sweep."[53] To Hastie, the answer was obvious: "The telephone should be primarily answered by a servant and there should be further internal connection . . . with other rooms."[54]

Another cause of resentment may have been its anonymous, often annoying nature. In ordinary commercial life, if a clerk were not properly respectful or attentive, one could complain immediately to the management. If one felt cheated or overcharged, the management similarly was on the spot to make amends, but such courses of action and redress were not so available to telephone customers. It was hard to get satisfaction from a machine, and telephone operators were often singled out for criticism because they simply did not know their place. The Times reported that "too many of them [operators] seem to regard the telephone user as their natural enemy and treat him with utter nonchalance, if not with an insolence and impertinence which are all the more irritating, because there appears at present to be no remedy for them."[55] Others complained about the difficulty of recovering one's money when calls placed from pay phones did not go through[56] and about how provoking it was for the other party in a phone conversation to break off in the middle to speak with someone else.[57] In many ways, the telephone modified traditional business and social conventions, and to some the modification was synonymous with a deterioration in standards and form.

If the use of the telephone in private life could be avoided, in business affairs it could not. The telephone had its greatest impact as a business machine and an aid to commerce before 1912.

In 1880, a correspondent to *The Times* presented the classic statement of the telephone's usefulness in the professional world. He predicted that the telephone would "allow business transactions to be carried on *viva voca* between different parts of the country. It will thus produce a saving which is incalculable of time, of trouble, of anxiety, and therefore of wear and tear, as well as actually in money, to every business man who adopts it."[58]

Thus, the business community, more than any other group in society, continually demanded an efficient telephone system at reasonable rates. In February 1899, a meeting was held at the Guildhall to discuss the needs of business in relation to the telephone.[59] In 1888, 1908, 1910, and 1911, representatives from the Associated Chambers of Commerce were involved in protests over the condition of telephone service in Britain.[60] Individual firms did not hesitate to bring their complaints before the public. Two examples are the letters from Thomas Goode and Sons on the National Telephone Company's inability to restore service rapidly after a snowstorm in 1900 and from David C. Pinkney and Company on a two-hour delay in placing a call from London to Bristol.[61] Also, businessmen in Liverpool demanded a voice in making policy.[62]

Professionals were even more vexed because they recognized that the telephone was not the only weak spot in the British economy. The early years of telephone development in Great Britain were the same period in which the country began to lose ground to the United States and Germany in industrial supremacy. One constantly recurring note in contemporary comments on the telephone is almost an echo of this larger theme of relative decay and decline. As early as 1882, *The Times* began to publish statistics indicating the regrettable state of telephonic development at home;[63] there was one telephone for every 200 people in Chicago, but in London the ratio was one to 3,000. In 1890, *The Economist* pointedly asked, "What hope was there of our getting something like continental standards?"[64] A correspondent to *The Times* displayed this point of view in 1906:
Surely it is high time these medieval methods of conducting business should be replaced by modern methods. In such places as Germany and the United States they would certainly not be tolerated, and we do not see why the British public should be allowed to suffer as it does. Business is difficult enough to carry through in these days of hot foreign competition, without hav-

ing it blocked by the unpardonable stupidity of the call regulations of the National Telephone Company and the Post Office.[65]
The grievances of the business community were not satisfied
even after nationalization. In 1914, C. S. Goldman charged that
the telephone system "has not expanded to half the extent that
it would if the Post Office recognized their responsibilities toward *the great commercial community*"[66] [italics added].

How bad was the state of British telephone development?
Measured in terms of the relationship between the number of
telephones and the size of the population, it was serious indeed.[67] (See Table 2.) Even though the demand for telephones
was limited to businesses and the more prosperous, the system
at times could not fulfill requests for telephones. At the National
Telephone Company's annual meeting in 1904, its general manager reported that 3,300 people were waiting for service in London and another 7,700 in the provinces.[68] In 1914, two years
after nationalization, the average time to connect a new customer varied from 18½ days in London to 51 days in Birmingham;[69]
how many potential customers became discouraged by such delays and did not request telephones is impossible to determine.
An often expressed opinion was that the nation had failed to
develop a valuable invention; as one *Quarterly Review* writer
put it, "by common consent we have the worst telephone
service in the civilised world."[70]

Much of the explanation for this rests not in the intrinsic usefulness of the telephone, but in the political-economic situation
in which the telephone was developed. Wider implementation
of the telephone was handicapped by a lack of consensus on
what the structure of the telephone industry should be. In 1881,
*The Electrician* pronounced: "It seems axiomatic to us that telephony must necessarily be a monopoly."[71] The question was by
no means axiomatic to the parties concerned with the telephone—individual private companies, the Post Office, the Treasury, and municipal governments. Competition was regarded
by many as the best means to secure a cheap, efficient service.
Even those who believed in monopoly disagreed over who
should manage it. Should it be private or public? If public,
should it be nationally or locally controlled? The inability to
achieve any fundamental agreement did much to delay the impact of the telephone and may help to account for the lack of at-

**Table 2**
Telephones as a Percentage of the Population

| Year | Number of Telephones | Related as Percentage of the Population |
|------|---------------------|----------------------------------------|
| 1890 | 45,000 | 0.12 |
| 1895 | 99,000 | 0.25 |
| 1900 | 210,000 | 0.51 |
| 1905 | 438,000 | 1.02 |
| 1910 | 663,000 | 1.48 |
| 1915 | 818,000 | 1.85 |

tention paid to the social effects of the telephone.[72] What may be even more paradoxical is that if the telephone had been less desirable and useful, the struggle over its development might have been less prolonged and heated.

Of the parties involved with the telephone, the Post Office has probably received the largest share of criticism for the slow rate of British telephone development. It is charged that the department saw the telephone as a threat to its telegraph business and, therefore, deliberately restricted the growth of a competitor. It is true that the Post Office enforced the monopoly on communication by electricity, which it had received in 1869, to control the private telephone companies,[73] and it is true that the Post Office collected a 10 percent royalty on the earnings of those companies.[74] It is also true that the size of exchange areas was originally limited to preserve long-distance communication for the telegraph system. The Post Office's reputation suffered greatly from such decisions; the 1884 comments of *The Spectator* were not atypical:

The New Yorker of means is understood to be no more able to do without his telephone than the Englishman without his Penny Post. As we are the most letter-writing country in the world . . . it is most probable that had it not been for the hateful effects of state monopoly we should have been the most wire-speaking country in the world.[75]

But the situation is more complicated than merely a story of government ineptitude and interference that destroyed private initiative. While the Post Office was always concerned with the telephone's impact on telegraph revenue, it is less than accurate to maintain that this concern totally shaped the department's telephone policy. There were always men inside the Post Office who believed the telephone could be best developed through a government-operated system. They did not, however, envision a telephone in every house, nor did they think telephone rates should ever be subsidized from other taxes. As one Postmaster General said, "It is a false policy to pay low fees, if the result of the low fees should be a service which is slow, untrustworthy, and troublesome. . . . [Both] trunk and local service must be maintained at the highest level of speed and accuracy."[76] What the Post Office believed it could offer to the public was an *efficient* system to meet the needs of customers, especially business customers.

S. A. Blackwood, secretary from 1880 to 1893, was one Post Office man who believed the state should operate telephone systems in competition with private industry. He wrote:

There is good ground for believing that it is the object of the Companies, not so much to meet a public want in an efficient and lasting manner, as to establish a system which may compel the Government to purchase. They are straining every nerve to give their business an appearance of success by carrying their wires to the premises of people who have never asked for them and by offering to these people a year's use of the system free.[77]

He proposed that the Post Office buy 5,000 telephones from Frederick A. Gower, who held his own patent; thus, the opposition would be outflanked. The Post Office, however, could not move without Treasury approval; the Treasury, fearful of losses to the exchequer, hesitated to support a state telephone system. Although the Treasury did agree to allow the Post Office to establish a telephone-exchange network, it did so only to force the existing telephone companies to cooperate more amicably with the government in working out licensing procedures.[78]

Throughout the years before nationalization the Post Office found itself in an extremely awkward position vis-à-vis the telephone. Not only did it possess a monopoly on communication by electricity, but it also licensed new companies to enter the field. For example, problems arose in March of 1881 when the National Provincial Telephone Company applied for a license to work in the Midlands, Yorkshire, Scotland, and Ireland. The Post Office already had private ABC wire systems in Newcastle, Leeds, Bradford, and several towns in Scotland, which it was converting to telephone service. How should the department respond to a request to invade its own territory? The answer proposed by Blackwood indicates that the Post Office was more willing than its critics have allowed to accommodate the demands of private industry as well as the needs of the public. The formula was a simple one.[79] Where the Post Office had an exchange, any request from a private company to establish a competing system would be denied. Where a company already had an exchange, the department would refrain from competition and further would deny others the right to compete. In towns without existing systems, "steps should be taken to ascertain whether the Public desired that the Post Office or a private Company should establish such communication and to act accordingly."[80] Again, the Treasury, preferring private enterprise, rejected Blackwood's plan.[81]

The eventual structure of the telephone industry was far from settled however. In 1882, the London-Globe Telephone Company applied for a license to operate in the capital, where the United Telephone Company was already in business. Naturally the latter opposed the entry of any competition. The decision on the request was made by the Postmaster General, Henry Fawcett, who was one of the leading economists of the late Victorian period and a firm believer in the merits of competition and free trade.[82] While against Blackwood's vision of one system per area, Fawcett opted for competition in the telephone industry—not only between private firms, but also between private firms and the government.[83]

The Treasury had no objection to competition between individual firms, but it was less sanguine that the Post Office could compete effectively with private enterprise.[84] As Leonard Courtney of the Treasury wrote, "There will be obvious difficulty in the future in examining the prudence of any proposed expenditure by your Department when its operations in any particular town or district would be exposed to the chances of speculative rivalry."[85] In essence, the Treasury was objecting to the uncertainty of competition between state and private industry. This uncertainty would never have been a problem if Blackwood's original plan of limiting local service to one system—state or private—had been approved. The Treasury felt it a bit unseemly for the Post Office to engage in hard competition. For instance,

Fawcett considered canvassing and advertising essential to establishing a strong Post Office system. Yet twice in 1882, the Treasury implied that canvassing was beneath the dignity of a government clerk and refused to allocate the necessary funds.[86] They were so bent on limiting the expansion of the Post Office telephone system that they even suggested raising the fees charged to discourage applications.[87]

The Post Office submitted more requests for canvassing staff and continued to push for expansion of its telephone system for another entire year, but the Treasury refused to back down. It cited the problem of paying a constantly increasing staff which resembled a "large Army."[88] It is striking how much ideological attitudes actually influenced telephone development in Great Britain. The philosophy of the Treasury on expansion of state services was well expressed by a clerk who reminded the Post Office that:

The sound principle in the opinion of My Lords is that the State, as regards all functions which are not, by their nature, exclusively its own, should, at most, be ready to supplement, not endeavor to supersede, private enterprise, and that a rough but not inaccurate test is, not to act in anticipation of possible demand.[89]

The Post Office, on the other hand, believed that demand for telephones existed and that canvassing would tap this demand.[90]

It was all to no avail. The department could not rely on public support; in fact, by the spring of 1884 the Post Office's telephone policy was coming under increasing attack. The state of telephone development at this time can be summarized by a few statistics. The department had a total of 748 subscribers in 17 towns, whereas the major opposition, the United Telephone Company and its subsidiaries, had over 10,600 subscribers in 66 towns.[91] Allowing competition had not proved successful, since larger companies tended to swallow newcomers; the United Telephone Company, for example, bought out the London-Globe in June 1884.

In the early 1880s, the public incorrectly blamed the Post Office for restricting telephone development; the refusal to allow companies to erect their own trunk lines and the decision to limit exchange systems to small areas were seen as a part of a Post Office plot. Further, the department's insistence on the right to buy any potential instrument from any licensed company had discouraged companies from applying for new licenses. The *Manchester Guardian*, which had earlier supported the nationalization of the telegraphs, now vehemently opposed any national state-run telephone system.[92] *The Times* blamed the situation on "the fear of the officials from the first that any great increase in the use of the telephone would be at the expense of the telegraph, and that it would serve, by diminishing the receipts of the Department, to direct attention once more to the improvidence of the bargain by which the telegraphs were taken over by the nation."[93] (The telegraph companies had been purchased for over £5,800,000 versus an original estimate of £2,400,000.)

The force of this criticism had two results. Within the Post Office itself a split developed over telephone policy. Blackwood, disagreeing more and more with Fawcett, continued to argue for a more aggressive stance toward the private companies and the Treasury and against concessions in giving away Post Office privileges (such as buying telephones from licensed companies). He pointed out that private initiative was not meeting the public's needs and that "Exeter, Falmouth, Limerick, Waterford, and West Hartlepool would be without Telephone Exchanges if it had been left to the companies to supply them."[94] Blackwood specifically criticized the United Telephone Company for seeking only short-run profits, especially in its handling of subsidiary companies.[95] Blackwood's reasoning was supported by W. H. Preece and John Lamb, who were both convinced of the usefulness of the telephone; Preece and Lamb suggested that the best solution would be to nationalize the telephone companies.[96]

Fawcett disagreed.[97] He considered nationalization inappropriate then; the government would be forced to buy costly patents which had only seven years to run. Moreover, much of the private companies' wire was aboveground, and after purchase the department would have to place it underground. The second result may be seen in Fawcett's decision to abolish some of the Post Office restrictions that were previously criticized. He moved in effect even closer to a laissez-faire position. Radius restrictions on the size of exchange areas were abolished, thus allowing companies to construct their own long-distance systems.

In addition, the government option to buy telephones from companies holding patents was dropped. In return for these concessions, the Post Office was to collect a 10 percent royalty on the gross receipts of each company. New licenses, which would lapse in 1911, were to be issued, with the government having the option to nationalize in 1890, 1897, or 1904.

The policy was a disaster, since the public was not satisfied with the service offered by the United Telephone Company and its subsidiaries. Given their control over the patents, they should have been able to provide an efficient system. They did not. *The Electrician* attacked the companies in the following way:

In defiance of all warning . . . the telephone companies pursue their reckless course, with seemingly no guiding principle but the exigency of the moment or the forecoming shareholders meeting. Their overhead wires have assumed proportions which constitute them a public danger, as well as a national disgrace.

The apparatus is miscellaneous and in some instances a long way from being the best, so that customers very often find themselves in front of an instrument they do not know how to use. The boon conferred upon the commercial classes by the rapid extension of telephonic communication is undeniably very great; but it has been hurried forward in such a fashion that disappointment is very close at hand. . . . [These] private companies have run up a system as bad as could well be devised, and one that is certainly doomed to fail.[98]

After 1884, public disenchantment with the service offered by the private companies grew. Between 1885 and 1887, a mounting tide of public opinion called for nationalization of the telephone under the aegis of the Post Office. In 1886 and again in 1887, the Convention of Royal and Parliamentary Burghs of Scotland petitioned the Postmaster General to take such action.[99] In 1888, the Associated Chambers of Commerce similarly voted for a national telephone system.[100] The Duke of Marlborough, who was extremely interested in telephone development and was himself later a proponent of the New Telephone Company, made a similar suggestion in the House of Lords in 1889. Marlborough argued that the Post Office had both the past record of successful expansion and the present ability necessary for proper employment of the telephone. He stated: "the Government alone is able to undertake this system and develop it and work it properly. It was on such a principle that the Post Office acted with regard to the Parcels Post, and most successful and admirable that arrangement had been."[101] Even *The Econo-mist* favored the creation of a state monopoly.[102]

The reputation of the private firms was not helped by the for-mation of a monopoly—the National Telephone Company—from the United, National, and Lancashire and Cheshire telephone companies in 1889. The planning of the National Telephone Company always depended on the need to reap maximum prof-its before the eventual lapse of its license. A company official once admitted before a Parliamentary Select Committee that it planned to accept no new subscribers after 1904 in order to achieve the largest net income possible.[103] As Table 3 indicates, the National Telephone Company did do quite well.[104]

Public discontent once again gave the Post Office the oppor-tunity to propose nationalization,[105] but the Treasury refused on the grounds that "My Lords are not prepared to embark upon another enterprise gigantic in itself, while the developments it might lead to are beyond their powers of prediction."[106] An ex-asperated clerk at the Post Office replied, "My Lords' 'powers of prediction' are very limited. They cannot see beyond their nose at present."[107]

In the 1890s, two piecemeal expedients appeared to offer some improvement for the telephone system. The first was the deci-sion to nationalize the trunk-line system. It was believed that if the government operated long-distance communications, the National Telephone Company might be encouraged to open lo-cal systems in areas previously considered unprofitable.[108] The trunk-line purchase was completed in March of 1896 with the government paying £459,114 for approximately 29,000 miles of trunk lines.[109] Part of the reasoning behind the trunk-line na-tionalization, however, involved a return to the pre-1884 policy of one exchange system per local area—the establishment of small private monopolies. As G. H. Murray of the Treasury pointed out, the arrangement would establish "a partnership between the Post Office and the National Company. The Com-pany confines its operations to several exchange areas, while the Post Office connects these areas with each other."[110]

There was much opposition to this partnership. The decades of the 1880s and 1890s were the high point of "municipal trad-ing," as A. L. Lowell called it.[111] Birmingham had led the as-sault on private industries such as gas, electricity, and tramway

**Table 3**
Financial Record of the National Telephone Company

| Year | Income | Post Office Royalties | Net Income | Working Expenses | Net Result |
|------|--------|----------------------|------------|------------------|------------|
| 1890 | £ 380,075 | £ 30,493 | £ 349,581 | £ 157,590 | £ 191,991 |
| 1895 | 819,034 | 74,674 | 744,360 | 424,164 | 320,195 |
| 1900 | 1,432,696 | 140,074 | 1,292,621 | 808,180 | 484,441 |
| 1905 | 2,212,358 | 206,455 | 2,005,903 | 1,275,161 | 730,742 |
| 1910 | 3,422,423 | 329,494 | 3,092,928 | 1,987,356 | 1,105,572 |

companies, and it appeared logical that the telephone would attract the proponents of the Civic Gospel. Led by Glasgow, municipalities began to consider establishing their own telephone systems in competition with the National Telephone Company. The National Telephone Company greatly feared this prospect since municipalities could obstruct private construction through denial of wayleave rights. In the national government there was again no consensus over the desirability of such competition. Some, like Lamb of the Post Office, contended that the telephone was a natural monopoly and the competition would only lead to needless complications. Others, like Murray of the Treasury, felt it offered a means to coerce the National Telephone Company into giving better service. Murray especially favored municipal competition as it offered no risk to the Exchequer. He frankly pointed out, "If they like to waste their money, I do not see why we should get in their way."[112]

Although the government introduced a bill in June 1899 enabling municipalities to raise money for establishing exchanges, the hope that municipal competition would improve the poor state of telephone development was not fulfilled. Of 1,334 local bodies, only 56 sought information from the government on the matter.[113] Of these only 13 ever applied for a license, only 6 ever opened exchanges,[114] and only one (Hull) can be said to have been successful. The municipalities' lack of interest again indicates that the telephone was not seen as essential by society as a whole. In retrospect, the movement for municipal exchange systems was a smokescreen that only delayed the inevitable creation of a single system under Post Office management. Again, this takeover came in stages. In 1901, the Post Office and the National Telephone Company agreed that the state would purchase the National Telephone Company's London plant at the end of 1911 at valuation and with no compensation for goodwill. In 1905, the two parties agreed to extend the purchase to the entire country,[115] and the policy suggested by Preece and Lamb in 1884 was finally accepted. The cycle was completed in 1912 when the Post Office bought out the National Telephone Company for £12,500,000.

This brief summary of the telephone's history in the years before nationalization serves as a reminder that inventions, like ideas, seldom encounter a neutral environment. Preexisting conditions, outlooks, and prejudices had more to do with the impact of the telephone than its intrinsic features. The confusion over who should develop the telephone not only deflected the press' attention from the telephone's social effects, but also actually shaped those same social effects. One contemporary observer saw this point quite clearly when he wrote: "The financial and administrative difficulties are much greater than the technical; telephone engineering has become fairly standardized, but rapid development is mainly a question of large capital development, and high efficiency of service is mainly a question of sound and efficient organization."[116]

All of these necessities were absent before 1912. Neither the National Telephone Company nor the Post Office had adequate funds for capital expenditure, and the rapid shift from one policy to another—competition, private monopoly, state control of trunk lines, municipal competition, nationalization—prevented the formation of an efficient organization and confused the public. In London after 1901, for example, the National Telephone Company had a monopoly in the eastern areas, the Post Office a monopoly in the west, and the two competed in the central districts. In 1905, *The Times* pronounced: "The thing is to make the best of it [the telephone], to make ourselves, as of any other of the appliances of modern civilization, its masters and not its slaves."[117] The irony is that, because of the particular situation it encountered in Britain, the telephone was neither a master nor a slave during the years of delay.

## APPENDIX 1
## THE GROWTH OF THE TELEPHONE INDUSTRY IN GREAT BRITAIN, 1890–1920

Number of Telephones in Great Britain, 1890–1920 (Exchange and Private Wire Systems)

| | P.O. | N.T.C. | Guernsey | Hull | Other Municipalities | Total |
|---|---|---|---|---|---|---|
| 1890 | 5,000 | 40,000 | — | — | — | 45,000 |
| 1895 | 7,000 | 92,000 | — | — | — | 99,000 |
| 1900 | 8,800 | 200,200 | 1,000 | — | — | 210,000 |
| 1905 | 54,100 | 362,500 | 1,400 | 1,900 | 18,100[a] | 438,000 |
| 1910 | 121,000 | 534,000 | 1,900 | 3,100 | 2,500[b] | 663,000 |
| 1915 | 804,500 | — | 2,100 | 11,400 | — | 818,000 |
| 1920 | 970,000 | — | 2,600 | 13,200 | — | 986,000 |

[a] Glasgow    12,300
Portsmouth   2,500
Brighton     1,900
Swansea     1,400

           18,100

[b] Portsmouth

Source: Parliamentary Papers, Select Committee on the Telephone Service, 1921, v. 7, p. 302.

## APPENDIX 2
## CHARGING POLICY

Paradoxically, the very strength of the business lobby on the telephone may have hindered its diffusion and wider development in Great Britain. The business community was committed to the flat rate—one charge regardless of how many calls were made. The flat rate naturally favored large users, like professional firms. As early as 1892, W. E. L. Gaine, General Manager, proposed the establishment of a measured rate, but the business community presented such great opposition that the proposal was dropped.

As F. G. C. Baldwin pointed out, large users

neglected to take into account the communal advantages which undoubtedly would have accrued from the universal application of the measured rate and the additional channels of communication opened up by a large increase in the number of subscribers. The argument that the cost of operating their service was actually greater than the price paid for the facilities rendered, the difference being made up by the flat rate small user who, in consequence, paid a higher rental than was equitable, had no weight with them.[118]

The rate structure was not completely overhauled until 1915,[119] when the Post Office made a concerted effort to discourage the flat rate system by raising its charges and by charging business firms more than private citizens for its use.

## NOTES

1. See, for example, the graphic description of the impact of the railways in Chapter Six, Charles Dickens, *Dombey and Son*.
2. William Ashworth, *An Economic History of England* (London, 1960).
3. W. H. B. Court, *British Economic History 1870–1914* (Cambridge, England, 1965).
4. See Pauline Gregg, *Modern Britain: A Social and Economic History Since 1760* (New York, 1965), p. 383.
5. Quoted by F. G. C. Baldwin, *The History of the Telephone in the United Kingdom* (London, 1925), p. 14.
6. *The Times*, 2 July 1877, p. 4.
7. Ibid., 29 November 1877, p. 6.
8. *The Illustrated London News*, v. 71 (15 December 1877), 581. See also, ibid., v.75 (15 November 1879), 462, 465 for details of Edison's version of the telephone.
9. *The Times*, 16 January 1878, p. 9.
10. Quoted by Baldwin, pp. 56–57. *The Pall Mall Gazette* also reported that "a higher class of young women can be obtained for the secluded career of a telephonist as compared with that which follows the more barmaid-like occupation of a telegraph clerk."
11. *The Times*, 8 April 1908, p. 8.
12. Ibid., 29 August 1912, p. 6, and 19 September 1912, p. 4. For complaints about the proliferation of overhead wires in London, see *The Times*, 5 December 1911, p. 14.
13. See, for example, the old popular account in Herbert N. Casson, *The History of the Telephone* (Chicago, 1910), pp. 245ff.
14. *The Times*, 29 November 1877, p. 6.
15. Ibid., 21 August 1877, p. 8.
16. Post Office Records, Post 30/398, E3497/1881, Tilley to Manners, 13 January 1880.
17. *The Times*, 14 July 1877, p. 7.

18. Post Office Records, Post 30/542, E13627/1889, Minute 3013/1878.
19. *Hansard's Parliamentary Debates, Third Series*, v. 238, 58, 21 February 1878.
20. Post Office Records, Telephone History to 1889, E33372/1877, Preece's Memorandum, 19 September 1877.
21. Post Office Records, Post 30/542, E1327/1889.
22. See above, pp. 81ff.
23. Post Office Records, Telephone History to 1889, Post Office to Treasury, 17 May 1878.
24. See J. L. Kieve, *The Electric Telegraph* (Newton Abbot, England, 1973), Chapter One, for the following.
25. A. Hazelwood, "The State Telephone Service in Britain," *Oxford Economic Papers*, New Series, v. 5 (1953), p. 14.
26. See Ashworth, *An Economic History of England*, p. 150, for an account of the cable industry.
27. *The Spectator*, v. 57, 16 August 1884, p. 1068.
28. Robert Newton, *Victorian Exeter 1837–1914* (Leicester, England, 1968), pp. 208–209.
29. *The Times*, 6 April 1908, p. 14.
30. *Hansard's Parliamentary Debates, Fourth Series*, v. 31, 1 March 1895, pp. 219–220.
31. *The Times*, 14 January 1902, p. 7.
32. "Telephones in Great Britain," *Edinburgh Review*, v. 199 (January 1904), p. 73.
33. *Historical Summaries of Post Office Services to 30th September 1906* (London, n.d.), p. 53.
34. Ibid.
35. Ibid., p. 54.
36. Ibid., p. 66.
37. In 1912–1913, 36,000,000 trunk calls were placed. Annual Report of the Postmaster-General, 1912–1913, p. 69.
38. Post Office Records, Post 30/3281, File 16, Conference of Post Office-Treasury Officials, 20 October 1911.
39. Baldwin, *History of the Telephone*, p. 119.
40. Post Office Records, C.7, Hicks Beach to A. Chamberlain, 11 June 1901.
41. Annual Report of the Postmaster-General, 1912–1913, p. 69.
42. British Parliamentary Papers, Select Committee on Telephones, 1898, v. 12, p. 410.
43. Post Office Records, Post 30/3281, File 16, Conference of Post Office-Treasury Officials, 20 October 1911.
44. Ibid., File 8, Treasury to Post Office, 4 March 1912.
45. Mrs. C. S. Peel, *A Hundred Wonderful Years: Social and Domestic Life 1820–1920* (New York, 1927), pp. 155–156.
46. Marghanita Laski, "Domestic Life" in *Edwardian England 1901–1914*, edited by Simon Nowell-Smith (London, 1964), pp. 170–171.
47. Ibid., pp. 169–170.

48. Post Office Records, Post 84, Telephone Rates Committee, 1920, p. 5.

49. Peel, *A Hundred Wonderful Years*, p. 8.

50. See Laski, p. 190, for the following.

51. Ibid.

52. See Orientation Memo #7, MIT Telephone Technology assessment, (unpubl.)

53. A. H. Hastie, "The Telephone Tangle and the Way to Untie It," *The Fortnightly Review*, v.70 (1898), p. 894.

54. Ibid.

55. *The Times*, 27 December 1905, p. 7.

56. Ibid., 27 July 1908, p. 17.

57. Ibid., 27 December 1905, p. 7.

58. Ibid., 24 January 1880, p. 10.

59. Ibid., 2 February 1899, p. 10.

60. Post Office Records, Post 30/603, E4522/1892; Post Office Records, Post 30/1616, E29062/1908; Post Office Records, Post 30/22231, E6813/1912.

61. *The Times*, 15 February 1900, p. 8; ibid., 6 June 1906, p. 3.

62. Ibid., 14 July 1911, p. 7.

63. Ibid., 1 December 1882, p. 3.

64. *The Economist*, v.48, 12 July 1890, p. 889.

65. *The Times*, 6 June 1906, p. 3.

66. *Hansard's Parliamentary Debates, Fifth Series*, v.64, 3 July 1914, p. 770.

67. British Parliamentary Papers, Select Committee on Telephones, 1921, v.7, p. 302.

68. *The Economist*, v.62, 26 March 1904, p. 524.

69. Post Office Records, Post 30/3112, File 7.

70. Leonard Darwin, "Municipal Socialism" in *Quarterly Review*, v.205 (1906), p. 434.

71. *The Electrician*, 8 January 1881, a clipping in Post Office Records, Post 30/603, E13267/1889.

72. See Orientation Memo #14. MIT Telephone Technology assessment.

73. Kieve, *Electric Telegraph*, pp. 158–159.

74. See below for details.

75. *The Spectator*, v.57, 6 September 1884, p. 1167.

76. Post Office Records, Post 30/2223, E6813/1912. The Postmaster General was Herbert Samuel.

77. Post Office Records, Post 30/542, E13267/1889, Blackwood to Fawcett, 10 December 1880.

78. Ibid., Treasury to Post Office, 16 December 1880.

79. Ibid., Blackwood to Fawcett, 4 March 1881.

80. Ibid.

81. Ibid., Treasury to Post Office, 3 May 1881.

82. See, for example, Fawcett's "Modern Socialism" in *Essays and Lectures* (London 1872).

83. Post Office Records, Post 30/542, E13267/1889, Fawcett to Patey, 4 July 1882.

84. Ibid., Treasury to Post Office, 17 July 1882.

85. Ibid.

86. Post Office Records, Post 30/603, E4522/1892, Treasury to Postmaster General, 24 July 1882.

87. Post Office Records, Post 30/542, E13267/1884, Treasury to Post Office, 11 November 1882.

88. Post Office Records, Post 30/603, E4522/1892, Treasury to Postmaster General, 25 June 1883.

89. Ibid.

90. Ibid.

91. The *Manchester Guardian*, 30 June 1884, a clipping in Post Office Records, Post 30/542, E13267/1889.

92. Ibid.

93. *The Times*, 13 June 1884, p. 9.

94. Post Office Records, Post 30/603, E4522/1892, Blackwood to Fawcett, 7 February 1884.

95. Ibid.

96. Post Office Records, Post 30/542, E13267/1889, W. H. Preece to Fawcett, 18 February 1884, Telephone History, 1879–1913, Lamb's Memorandum of 22 May 1884.

97. Ibid., Fawcett to Treasury, 28 July 1884.

98. *The Electrician*, 20 October 1883, quoted in Baldwin, *History of the Telephone*, p. 568.

99. Post Office Records, Post 30/542, E13267/1889, Petitions of 17 May 1886 and 27 April 1887.

100. J. H. Clapham, *An Economic History of Modern Britain* (New York, 1938), v.3, p. 392.

101. *Hansard's Parliamentary Debates, Third Series*, v.337, 4 July 1889, p. 1432.

102. *The Economist*, v.46, 12 May 1888, p. 593.

103. Ibid., v.62, 26 March 1904, p. 524.

104. Baldwin, *History of the Telephone*, p. 657.

105. Post Office Records, Box File 1, H. C. Raikes to W. H. Smith, 18 February 1890.

106. Post Office Records, Bundle R, File 70, Treasury to Post Office, 27 June 1890.

107. Post Office Records, Telephone Policy 1879–1898, S. A. Blackwood's Memorandum, 4 December 1890.

108. Post Office Records, Bundle F, File 17, J. Fergusson's Memorandum, 21 January 1892.

109. Post Office Records, Bundle N, File 33A, J. C. Lamb's Memorandum, 15 February 1898.

110. Public Record Office, T1/8982A,17448, G. H. Murray's Memorandum, 4 December 1895.

111. A. L. Lowell, *The Government of England* (New York, 1908), v.2, Chapter 44.

112. Public Record Office, T1/9133B/4495C, G. H. Murray to J. C. Lamb, 27 January 1896.

113. Post Office Records, C.7, Unsigned Memorandum, June 1911.

114. Ibid.

115. Post Office Records, Post 30/1216, E13364/1905.

116. *The Times*, 21 August 1907, p. 8.

117. Ibid., 27 December 1905, p. 7.

118. Baldwin, *History of the Telephone*, p. 607.

119. Post Office Records, Post 84, Report of the Telephone Rates Committee, 1920, pp. 5–6.

# 4 The Birth of the Telephone and Economic Crisis: The Slow Death of Monologue in French Society

Jacques Attali
and
Yves Stourdze

On July 13, 1889, upon the hundredth anniversary of the French Revolution, the French State bought up the privately owned General Telephone Company.

The telephone first made its appearance in France in 1877–1878. Private organizations were quick to ask the government for the right to develop this new means of communication, the administration agreed, and the General Telephone Company was created in 1881. It was the beginning of a unique chapter in the history of French communication because it introduced two-way dialogue into a domain that up to that point had been entirely conceived in terms of one-way monologue. A society built on order and hierarchy—such was the French society in which the telephone appeared. This accounts for the long, hard development of the phone network. The basic ideas underlying that network—reciprocity, equality, easy access, dialogue—were precisely those denied by the previous systems of communication that had developed in France, the visual telegraph and the electric one.

Communication, as understood by the French centralized state, was primarily a lecture which the State, with professorial wisdom, delivered to society. C. Chappe, the inventor of the visual telegraph, well understood this when he wrote in an open letter, dated Fructidor 22, the sixth year of the Republic (i.e., early September 1798): "The day will come when the Government will be able to achieve the grandest idea we can possibly have of power, by using the telegraph system in order to spread *directly, every day, every hour,* and *simultaneously,* its influence *over the whole republic.*"[1] Here was clearly expressed the basic idea of the great monologue the State was to have "with" its citizens.

In his history of the visual telegraph that his brother had invented, I. Chappe wrote in 1840:

Lack of interest in new inventions has always existed, even in France. Their usefulness can be proven only through experience, and no one is in a hurry to try it if it involves time and money. Trying new means runs counter to habits, and it hurts the interests of those who profit from the old methods; with the exceptions of the inventors, few people are prone to exploit projects the success of which looms always uncertain.[2]

Yet Chappe succeeded in financing his invention. The revolutionary assembly, the "Convention," recognized the advantage of being able to communicate quickly with the Northern Army; they granted Chappe a subsidy of 60,000 Francs which allowed him to build the first line of visual telegraph towers between Lille and Paris. Thus the *military use* of Chappe's invention had carried weight. Paris had to send messages quickly to the troops in the field. The visual telegraph gave the Central Government the device that linked it rapidly with the farthest boundaries.

Indeed, the system was not only swift but safe, for the Chappe telegraph transmitted messages in code. The employees who worked in each "station"—a tower with movable arms— were illiterate and reproduced signals without understanding them.[3] Thus only specialists at the end of the line could decode the messages. In addition, the administration limited the use of the Chappe telegraph to official ones. Thus, from the start, telecommunications were placed under the strict watch of the central authorities. From then on throughout the nineteenth century, the government kept for itself the absolute monopoly on fast transmission of messages. The fundamental principle, asserted until 1850, was that the tools of communication were essentially *political* instruments not to be open to the public. No other use could be envisioned although Chappe himself, in the face of financial difficulties in the late 1790s, had suggested other uses which were only made possible one hundred years later.

First Chappe urged that the telegraph service be made available to industrialists and merchants. To accomplish this, Chappe imagined the construction of a European network, putting Paris in contact with Amsterdam, London, Cadiz.[4] Around 1800, he also proposed using the telegraph to make up a newspaper, or rather, he imagined a telegraph bulletin summing up the news of the day, to be approved by the First Consul. This bulletin could then be added, in every city equipped with the telegraph, to the official newspaper printed in Paris and mailed to the provinces. Finally, Chappe suggested using the telegraph system to send out the national lottery results, the only proposal that took. Thus, for many years, the Chappe telegraph was subsidized not only by the War, Navy, and Interior departments, but also by the lottery.

Thus, since 1793, the telegraph system in France has been a solid instrument of power. Before 1850, when occasionally and for financial reasons only it was put at the disposal of the public, it was strictly for betting and amusement. For over a half century, the visual telegraph functioned within these limits. A bill proposed under the regime of Louis-Philippe (1830–1848) clearly stated the basic viewpoint:

Governments have always kept to themselves the exclusive use of things which, if fallen into bad hands, could threaten public and private safety: poisons, explosives are given out only under State authority, and certainly the telegraph, in bad hands, could become a most dangerous weapon. Just imagine what could have happened if the passing success of the Lyons silk workers' insurrection had been known in all corners of the nation at once.

In that candid summary of the official doctrine on communication, the instantaneous transmission of news is compared to poison and explosives, the swift diffusion of news to insurrection and disorder. Its use, therefore, must be reserved exclusively to the State. Such was indeed the intent of a May 3, 1837, law which imposed "jail sentences of from one month to one year, and fines of from 1,000 to 10,000 Francs on anyone transmitting unauthorized signals from one place to another by means of the telegraph machine or any other means."

Thus, even before the birth of the electric telegraph or the telephone, the State asserted its monopoly on telecommunications and provided itself with a legal code designed to dissuade in advance any attempt to put them to private use. With the Chappe telegraph, an official legal framework was established, which later served as reference for the electric telegraph and the telephone,[5] not to mention future media.

The governments between the Convention and the end of the reign of Louis-Philippe (1793–1848) progressively defined the main patterns of a communication policy in France, based on the visual telegraph precedent—a truly significant precedent in itself. Around 1844, a total of 534 stations, fanning out from Paris, linked twenty-nine cities; a message took twenty minutes to go from Paris to Toulon, some 840 kms (525 miles) away. Between 1840 and 1852, the annual rate of telegrams sent between Paris and Montpellier was 120.

The decline of the Chappe telegraph is not without interest in itself. The Chappe system had gradually imposed its logic—administratively, technically, and politically. It had instilled habits, fashioned mental attitudes, imposed its own specific organization. It did not give way easily to the coming of the electric telegraph. Far from it. It fought fiercely for its place for many years. In 1846, a Chamber of Deputies committee, having to decide between the two systems, opted for the Chappe one.

At the same time, a bill that would have opened up the telegraph to the public was rejected. How are we to explain such rejections? The fear of change or innovation? Perhaps, but above all, nervousness about the rise of an authentic communication *network*. Dr. Barbay, the inventor of a night lighting adaptation of the Chappe telegraph, expressed this fear:

No, the electric telegraph is not a sound invention. It will always be at the mercy of the slightest disruption, wild youths, drunkards, bums, etc. . . . The electric telegraph meets those destructive elements with only a few meters of wire over which supervision is impossible. A single man could without being seen cut the telegraph wires leading to Paris and in twenty four hours cut in ten different places the wires of the same line without being arrested. The visual telegraph on the contrary has its towers, its high walls, its gates well guarded from inside by strong armed men. Yes, I declare, substitution of the electric telegraph for the visual one is a dreadful measure, a truly idiotic act.

For a half-century, the Chappe telegraph was able to combine the Central power's need for communication with its concern for "fortification."

A world of sequestered communication was fashioned where the fundamental orientation was to seek isolation, to escape as much as possible the dangerous pulsations of the social body. In the ten-year confrontation between the visual and electric telegraphs, conflicts emerged which would reappear at the birth of the telephone. Technical innovation in communication runs counter to already recognized and accepted technologies. To break through with its own new forms and constraints, it must enlist the support of social forces which it, in turn, confirms and strengthens.

Thus is explained the amazing resistance of the Chappe telegraph, not only on the administrative and financial planes, but even on the technological one. French engineers desperately tried to build an electric telegraph with certain characteristics of the visual one. The result was a still-born hybrid dial telegraph. The game was up. The last visual station was dismantled in 1856, not without regret, as witnessed the nostalgic poem of the popular G. Nadaud:

What's with you, my old telegraph
At the top of your old steeple
As stern as an epitaph
As still as a stout boulder
. . . . . . . . . . . . . . . . . . . . .
Life has passed you by
Scientists had warned you
And when the pomp is gone
Neither flatterers, nor friends abound
In the old days you were the marvel
And we stood amazed
To see Marseilles in a single day
Dispatch a few words to Paris
You were the wonder of our age
Children in awe, we wished
We could decode that silent tongue
When your bewitching arms
Carried over the paling horizon
Diplomatic lies
Lost in the fog
Now in only one second
North converses with South
Lightning crosses the world
On a twig or rounded wire[6]

In 1850, the telegraph was at last opened up to the public, but the liberalization was hardly full scale. First, the administration had the right to stop messages contrary to good morals and public order; next, the users had to prove their identity. A reader of the newspaper, *L'Univers*, complained about the new system:

Finally I land, but it's too late to get the train. I'm impatient to get to Paris where my worried family awaits me. I want at least to let them know I have arrived and hurry to the telegraph office. A stern character who believes himself to be an administrator because he is picayune tells me:
"Do you have a passport?"
"It is in the hands of the immigration police."
"Do you have an authorization signed by His Honor the Mayor to use the telegraph?"
"The Mayor must be in bed."
"Do you have a certificate, signed by two upright citizens, vouching for your impeccable morals? Do you have two witnesses to establish your identity?"
"Two hours ago, I was still on the high seas and I know no one here."
"In that case, sir, come back tomorrow."
(Belloc, *La Télégraphie Historique*, 1888)

Whether the incident is real or imagined matters little; it still reveals the relation of the telegraph to the public. Access was difficult and under the control of local authorities. Once again, communication was reserved to a selected few and was censored. The electric telegraph did not spontaneously remove the constraints put on communication by bourgeois society. With the advent of the electric telegraph, the freedom of communication was still limited and closely watched.

With the siege of Paris (1870) and the Commune insurrection (1871), the problems of communication with Paris became evident, as reflected in the popular image showing Gambetta leaving Paris in a balloon. Thus, one cannot underestimate reactions such as those of the little town of St. Savin (in the department of Vienne, some 200 miles southwest of Paris), which, after denouncing the Parisian insurrection and reaffirming its attachment to national unity, one month later voted the necessary funds to build a telegraph line.

Thus is understood the new impulse given by the Third Republic to the telegraph network. With the school system and the railroads, the telegraph became its essential weapon. It was truly a republic of one-way media; over the whole national territory, it sowed railroad tracks, schoolmasters, and the telegraph.

In 1881, the General Telephone Society was formed. The State made the telephone an exception to its rule of monopoly on communications for two reasons. First, the State's immediate concern at that time was to merge the Postal system with the telegraph. Begun in December 1873 with the merger of those two services in municipal offices, it was fully achieved in 1879. Also, the Post Office and Telegraph Administration admitted with hindsight eleven years later that they did not "at that time [1879] fully appreciate this new and marvelous use of electricity, suspect the importance it would have in everyday life, or foresee the cost of implementing the building of telephone networks. Thus the State could not think of taking over responsibility for the cost of such an endeavour."[7]

Under the management of the General Telephone Society, the development of this new method of communication remained very slow, because its originality was not apparent to most. It was usually presented as a new channel for "monologue," meant to "transmit speeches, sermons, lectures,"[8] as a forerunner of radio, a new means of diffusion, another one-way system. "The day will come," wrote a judge, "when thanks to this wonderful apparatus everyone will be able to enjoy concerts, theatrical performances, to attend meetings, lectures, sermons, without having to leave his own hearth."[9]

These were not idle dreams. The General Telephone Society doubled its installations in Paris following the Electrical Exhibition, at which opera was relayed by telephone to a room in the Industrial Palace where music lovers could hear the whole show. The repercussions were considerable. On October 1, 1881, Paris had 2,442 subscribers, twice as many as the preceding year. Thus, as Aronson and Briggs have noted in other contexts, during the first years of its development, the telephone was seen only as an extension of the existing means of transmission. The "push-button telephone" (i.e., one-way intercom) is typical in this respect. In the same way as Briggs and Cherry describe for England, the Telephone Society in France proposed to replace summoning bells with telephone networks, and they did. "While you ring someone, you can also talk to the person who's been called."[10]

To careful observers it should be clear that installing bell-pulls in houses had been a sign of deep changes in the urban socia-

bility of the eighteenth century. Bells had been an answer to the
new demands resulting from the functional specialization of
rooms within a house. Sebastien Mercier pointed out the habit
of ladies ringing for their servants. Bells were installed so that
they could be rung from any distant place, whereas they formerly were heard only in the room where they were rung. Nothing
could be more telling of the new need to keep servants apart
and also to keep strangers away.

Toward the end of the eighteenth century, one no longer
called on a friend or business associate at any time of the day or
without warning. Special days were set aside to entertain callers, and visiting cards were exchanged through servants acting
as messengers. "Nothing is easier, no one is to be seen, everybody has the good taste of keeping his door closed. . . . People
used to live in the public eye, as actors on a stage, and everything was accomplished through oral conversation. From now
on, walls separate social life, professional life, and private life;
each one of these activities has its own specialized place: the
bedroom, the office, the drawing room."[11]

Thus, the "button-telephone" reproduced the pattern of com-
munication/separation which extended deeply into the social
and cultural changes of the eighteenth century and thereafter.
What's left of a dialogue, when the telephone is considered an
improved summoning bell, or a forerunner of radio? "Perhaps
the day will come when the sound of musicals will be sent
through wires into branch stations, when people will be able to
hear it as soon as they turn on a switch, just as we get water
when we turn on a faucet."[12] Thus, in a paradoxical way, the
telephone appears as a technological "parenthesis," an intruder
as it were, sandwiched between the telegraph and radio. French
society of the late nineteenth century did not seem to need it. It
might even have seemed an encumbrance, a nuisance, posing
problems to lawmakers and judges, such as the problem of
the identification of the speaker; in matters of counterfeiting, a judge remarked sadly, we can deal only with "written
counterfeiting."[13]

Contemporaries vied in describing the telephone in its most
innocuous, least scandalous social and cultural aspects: on the
preacher's pulpit, at the disposal of the police, on prison walls
to spy on the prisoners, in the hospital. Never does it appear as

a tool of reciprocal communication. People in authority, in State offices, in business or industry can use it to give orders rapidly. The public and private institutionalized networks that developed quickly used the instrument to their own ends. Business firms in Paris and the big provincial cities had telephones. In Marseilles, Lyons, and Bordeaux, municipal networks linked up firehouses, excise tax offices, police stations, and various administrative offices. The same was true of Le Havre and Rouen. But those were closed circuits; the public was not admitted. In 1883, the railroads adopted the telephone, and that same year the Decauville Corporation (makers of narrow-gauge tracks) set an example. It had twelve stations in the department of Seine-et-Oise and the stationmasters could, thanks to the network, announce to Mr. Decauville in person the arrival of merchandise.

In 1883 the State, acknowledging the importance of the telephone, decided to develop its own network rivaling that of the General Telephone Society. Still, both networks remained modest ones. "Whether through the indifference of the public who looked upon the telephone as a mere toy without practicality or because of the many obstacles raised by the State which feared the absorption of the telegraph by the telephone, this new means of communication, up until 1882, had made little progress in France."[14]

The State's entrance into the telephone operation is telling indeed. As early as 1883, practices were followed in telephone operation which are found again in 1976 with respect to cable TV. Experimentation is an example; the State chose a "pilot" city in which to install a network. In 1883, the city of Rheims was chosen because a group of 150 would-be subscribers wishing to have the telephone had been organized. The mail rooms, phone booths in railroad stations, post offices, and city offices were all connected and further linked with the telegraph network as well as with the Paris Stock Exchange. On January 1, 1885, eighty-two public phone booths were opened in Paris, and seventy-seven in provincial cities.[15] Finally, in 1885 the first long-distance line was set up between Rouen and Le Havre, a distance of 192 kms (120 miles).

In its agreement with the city of Limoges, the State defined the relations it wished to establish with local communities in matters of the telephone. The State would set up new networks only if they had first been financed by the collectivities (towns, Chambers of Commerce, etc.) wishing to use the new system. These advance payments made to the State would be reimbursed without interest from the proceeds. The system of advance payments put the development of the telephone in the hands of the local leaders. Yet they generally preferred investing in the local papers which they already controlled rather than expanding the telephone.[16]

One-way communication won over reciprocity. On July 13, 1889, the State bought up the General Telephone Society, but that gesture of authority did not change anything. Because of the advance payment system, the development of the telephone remained under the control of the local power structure; the telephone declined because it was an instrument poorly adapted to that orderly social hierarchy. There was little room in France for an instrument of reciprocity. In 1900, there were only 30,000 phones in the whole of France; nine years later, there were 27,000 for the 100 biggest hotels in New York alone.

The conclusion is obvious. The French power structure did not need the telephone. An undersecretary in charge of the Post Office, Mr. Clementel, in a report to the prime minister, after the First World War, observed: "The telephone network did not completely fulfill its function in our country, as often noted: international statistics put France behind most of Europe, due to the slowness and precariousness of long distance communications, the poorly equipped local lines and the lack of subscribers."[17] He went on to analyze the deep reasons for such a "telephone crisis": "Experience has shown that [the advance payment system] does not allow the planning or execution of a vast program. The incredible number of steps and agreements to take and pursue, the inevitable limitation of the funds provided by the system of advance payments, all prevent a rational and fast development of the telephone network. Thus far, the State's initiative has been paralyzed by the system."

A lucid diagnosis, but what did he propose? "Just as in the case of the international long distance network, it would be desirable for the State to finance, if not all, at least an important part of the future development of the local networks. This however does not seem possible because of the heavy burden the State will have to assume, due to the need for prompt action in the development of the international lines."

The rural populations the Minister had just recognized "were getting more and more familiar with the telephone," and the town networks answered a real need, but these populations were once more sacrificed.

The development of the telephone in France during this early period therefore reveals that the dynamics of business which could have helped diffuse the medium were hampered by the State.[18] The private interlude of the General Telephone Society lasted but ten years. The takeover by the national State, however, was a formality, since it could not directly intervene in telephone system development, which was still in the hands of the local power structure. The latter did not feel attracted to technologies that introduced reciprocity and dialogue in their living space. The interplay of forces that surrounded the telephone in 1889 looks very much like the one that plagues cable TV or the teletypewriter seventy-five years later.

The relinquishing by the national State of its responsibility to the local power structure has usually escaped those who have tried to analyze the slow development of the telephone in France. Fayol's treatment is a good example. In a small booklet, he tried to answer the following question: "Given the fact that, today, a powerful telephone system is indispensable to industrial nations, why is France more than ten years behind the United States . . . and even a defeated Germany whose network is more powerful than our most ambitious projects?"[19]

His point, like that of most observers who followed him, was to denounce "administrative neglect by the State." Here are the faults he listed:
1. An unstable and incompetent undersecretary at the head of the Post Office department.
2. No plan of action.
3. No balance sheet.
4. Too much meddling by deputies and senators.
5. Lack of motivation and no clear-cut responsibility.
His conclusion is simple: the State, by its very nature, is unable to manage an industrial enterprise. Hence he proposed "that the telephone be set up as an autonomous corporation." However, the telephone's slow development was primarily due not to a ponderous bureaucracy, as Fayol thought, but to the refusal of a

whole segment of the social body. The French local power structure did not want the telephone. Fayol did not grasp this fact and go beyond a critique of the bureaucratic system. As early as 1920, the Post Office organization had over 150,000 employees. That bureaucratic organization of the State acted primarily to prevent an extension of telephone technology that could endanger the balance to local powers.

The telephone provided employment for men and women who, having just finished their schooling, could pass the Post Office, Telegraph and Telephone competitive admission examination. It allowed the State, in 1918, to hire women, veterans, and especially the handicapped, a policy which Fayol, as a spokesman for the industrialists, strongly condemned. That policy reflected the reticence of the men in power to enter fully into the industrial era. On the other hand, as Fayol noted, the number of engineers heading the institution was still very low, and the meddling of elected representatives continued. The telephone was thus literally absorbed by the Post Office bureaucracy, a fascinating revenge by one of the oldest administrations of the old monarchy. One of the rare "technical" departments from before 1789 had succeeded in swallowing two technological innovations, the telegraph and the telephone.

The encounter of the Post Office and the telephone was the original encounter between the centralized State and one of the first technologies of reciprocity and free access. The State negotiated its use with the local men in power since they were the only ones who could serve as mediators and relays in the hierarchy of social structure. The result was a frail balance. The industrial interests were badly injured. In 1886 the telephone was still looked upon as only "an indispensable auxiliary in the big public administrations, the police, the military, the navy, and (secondarily) in the big industries."[20]

Should we then marvel that in 1971, fifty years after Fayol, the General Director of Telecommunications should write: "Since the creation in 1947 of the Committee to Modernize Telecommunications within the General Commissariat of the Five-Year Plan, our service again and again urgently called the nation's attention to the basic necessity, in a modern country's economic infrastructure, to have a good telecommunications network—all in vain."[21]

Can it be that the French State is only now discovering the sorry state of the telephone situation that it created by sacrificing telephone development for many decades on the altar of communication networks run by and for the local powerholders? It probably means that the French State was willing to begin promoting the telephone only when industrial interests became prominent in economic and social relationships. In the meantime, a slow and painful death of monologue in French society has taken place since 1947, or rather 1889!

We may propose a hypothesis to analyze what happened. The 1880s were a period of major economic crisis, followed by deep changes in the social fabric and especially in the field of communication. Before 1880, the economic expansion of the most developed capitalistic countries of Europe (France and Great Britain) rested on the export of goods and existing communication technologies were sufficient. In the 1880s, the competition of new capitalistic countries on the international market (the U.S.A. and Germany) provoked a grave economic crisis in Europe.

During that period of critical changes in capitalistic production, the U.S.A. developed its own power structure, banking system, and telecommunications network at home and abroad. A new strategy of growth appeared: in times of expansion, foreign markets were increasingly exploited not only as consumers but also in part as sources of production (the beginning of the multinational aspects of societies). Europe had shrunk; to keep production profitable, she had to develop large and highly modern units of production with very fast channels of distribution and sale, resulting in much more competitive management.

The growth of enterprises needing world markets made them more difficult to manage: the chain of production became unwieldy, requiring more time and space. To remain competitive, time lags in distribution had to be kept brief. The long technical processes of production and distribution had to be mastered, something that the local power structures unfamiliar with the workings of big industrial firms could not grasp. From then on, the telephone acquired vital importance. It helped bring the 1880 crisis to an end by winning new markets, and it played its part in the rampant competition in national and international markets until the 1929 depression. The technology of the telephone however could develop only if, at a given time, three conditions were met:

1. The invention was technically perfected.
2. A powerful segment of society demanded its development.
3. An adequate system of financing channeled the necessary funds toward that segment.

As shown, the adequate financing system (namely, the use of public funds by the State) developed very late in France. Industrialists needed the telephone earlier to create new means of communication that would have increased productivity, conquered foreign markets, speeded distribution, and restored profitability.

Unfortunately for France, the men who controlled public funds then were not interested in industrial expansion. On the other hand, the industrialists still had no say over the channeling of public expenditures. The year 1925 was a pivotal year when France at last achieved the means of solving the 1880 crisis and partially replaced those formerly in power with industrialists. By that time, however, the first symptoms of a new economic crisis had appeared—a crisis which would impose more complex structures on the State and more complex means of communication on social relationships. After being organized only briefly around industry, the telephone network after 1925 came to be organized around the national State which had begun to regulate the economy.

The financial requisites will become evident only later in the 1970s when the critical need for new technological advances will necessitate the financing of decentralized intercommunication networks. We may anticipate the end of an era in which the social apparatus could be telecommanded from a single node of power.

## NOTES

1. Claude Chappe, *Lettre sur un nouveau télégraphe* (An VI).
2. Ignace Urbain Chappe, *Histoire de la télégraphie* (1840).
3. Alexis Belloc, *La télégraphie historique* (1888), p. 98.
4. Edouard Gerpach, *Histoire administrative de la télégraphie aérienne en France* (1861).

5. During a debate in the Chamber of Deputies some 40 years later, a deputy affirmed that the telegraph ought to be a "political institution, not a commercial instrument," leading a civil servant to exclaim: "Just look at all the services rendered by the visual telegraph in the fields of politics, law and defense." Cited in E. Renard, *La télégraphie et la téléphonie dans le départment du Gard* (1896).

6. Quoted by L. Barray, *Histoire de la télégraphie à Argentan* (1912).

7. Julien Brault, *La téléphonie historique* (1890).

8. Ibid.

9. G. Vidal, *Le Téléphone au point de vue juridique* (1886).

10. Brault, *La téléphonie historique* (1890).

11. Philippe Ariès, *Histoires des populations Françaises* (1971).

12. Vidal, *Le téléphone* (1886).

13. Ibid.

14. Brault, *La téléphonie historique* (1890).

15. Instructions posted for their use read as follows: Only one person at a time. Talking with people outside the booth is forbidden. Leaving and reentering is forbidden. Letting anyone take the user's place is forbidden.

16. Between 1876 and 1914, all newspapers, but especially local papers, enjoyed a prodigious expansion. In 1868, the local press had a circulation of 250,000 copies; in 1914 that figure was 4 million.

17. A report presented by Mr. Clementel, Undersecretary for the Post Office, Telegraph and Telephone, to the Prime Minister (1920).

18. What economic interests the State defended through the international long distance network, as early as 1918, were those that had multinational overtones.

19. H. Fayol, *L'incapacité industrielle de l'Etat, les PTT* (1921).

20. Brault, *La téléphonie historique* (1890).

21. *Le Monde*, Dossiers et Documents, no. 26.

# 5

## The Telephone System: Creator of Mobility and Social Change

### Colin Cherry

### INVENTION AND "REVOLUTION"

There are certain rare moments in history when, through some remarkable human insight, discovery, creative work, or invention, human life and social institutions take a great leap. The invention of the wheel or the discovery of fire are sometimes quoted as such moments, but of their introduction and of life before them we know little. But the invention we are now celebrating—the telephone—is fully documented, as is the social history of the industrialized world before and after its introduction.

Inventions themselves are not revolutions; neither are they the cause of revolutions. Their powers for change lie in the hands of those who have the imagination and insight to see that the new invention has offered them new liberties of action, that old constraints have been removed, that their political will, or their sheer greed, are no longer frustrated, and that they can act in new ways. New social behavior patterns and new social institutions are created which in turn become the commonplace experience of future generations.

Such realization does not come easily, quickly, or even "naturally," for the new invention can first be seen by society only in terms of the liberties of action it currently possesses. We say society is "not ready," meaning that it is bound by its present customs and habits to think only in terms of its existing institutions. Realization of new liberties, and creation of new institutions means social change, new thought, and new feelings. The invention alters the society, and eventually is used in ways that were at first quite unthinkable.

The computer is a present-day example of this process; it was seen at first as a "robot," an "electronic brain," which could play chess and challenge human expertise (or could it?). The workaday uses for accountancy, industrial process control, airways bookings, and innumerable functions within business, industry and commerce, so familiar in today's world, came later. The computer, like the telephone, and other *radical* inventions are seen, at the very first, as "adult toys." (See Chapter 2.)

It would be hazardous today for anyone reared in an industrialized country to imagine what personal life was like before the telephone, or what feelings people had then. I do not refer merely to the domestic telephones (for many people in Britain and other industrial countries have no home telephones); rather, I mean life before the creation of hosts of social organizations in the economic and public spheres, which today utterly depend upon the telephone: business, industry, government, news services, transport systems, police. The list is endless, and today it embraces increasing numbers of international organizations.

These form our modern world; the telephone was invented within a very different world. We can try to imagine ourselves living in that world, but we shall be deceived. We may find amusing the uses to which the people of 1876 first applied their new toy, "the Speaking Telephone," but we cannot *be* one of those people.

## ORGANIZATION: THE HALLMARK OF INDUSTRIAL SOCIETY

The telephone (or, rather, telephone *system* or network) is no ordinary invention, not just another desirable consumer product. It handles "message traffic" and indeed, creates this traffic (the very stuff of which social intercourse is made) by its very existence. For this reason, its importance to industrial life is not just that it is another machine of production, like a spinning jenny or a steel rolling mill. Its importance is as a contribution to the organized bureaucracy that is the hallmark of modern industrial society. It was Max Weber, years ago, who argued that organized bureaucracy forms the essential characteristic of an industrial society; rather than capital or machinery, it is *organization—*

above all, systematic recording and good accountancy—that makes industrialization possible. (After all, the early Chinese had both science and technology, centuries before Europe, but they did not develop industry.)

In the economic sphere, the telephone service is essentially organizational in function; it creates *productive* traffic.

It could be objected that other systems of rapid communication preceded the telephone. Were they not equally important? The telegraph?[1] The heliograph? Or even the Talking Drum (to which the telephone is blood brother)?[2] Of course, these were of great importance; yet the telephone system stands above all, and has been far more profound in its effects, for two essential reasons. First, the telephone *system* allows us to move about the country (or, today, over much of the world) and yet appear to stay in one place (thereby adding security to mobility). Second, it offers all the psychological values of the human voice.

## THE CRITICAL IMPORTANCE OF THE TELEPHONE EXCHANGE

Having so far engaged in what may seem to be hyperbole, I shall now do the opposite and indulge in apparent denigration, by arguing that the telephone itself, the talking instrument that stands upon your desk, would not have been very important by itself. It might have lived in posterity's mind alongside electric doorbells, ships' telegraphs, and Gramophones. Perhaps we ought not to celebrate the telephone at all, in 1976, but should wait a couple of years more and celebrate instead the telephone *exchange*. It was the exchange principle that led to the growth of endless new social organizations, because it offered *choice* of social contacts, on demand, even between strangers, without ceremony, introduction, or credentials, in ways totally new in history.

The exchange principle led rapidly to the creation of *networks*, covering whole countries and, since World War II, interconnecting the continents. Anybody, without special training, can move about the geographic areas covered by the network and yet appear to another person on the network to be stationary. (In Britain, we often say, when opening telephone conversation: "Hello, are you *there*?" Where is "there," pray? It really doesn't matter.) Whole new forms of social institution and organization eventually became possible, new forms that no longer required people to be located at a fixed point. Today we accept as natural that business people may travel without appearing to leave their offices; that industrial branches can be located in scattered places, yet remain as operative units; that diplomats may fly around the world (perhaps they might do better staying at home); that government departments no longer must be concentrated in the metropolis; that police whizz about in cars; that shoppers may stay at home or go elsewhere at choice. The field telephone did as much for army tactics as radio later did for naval warfare. Such freedom of movement was not at first understood, though the early telegraphs and railways had paved the way.

The telegraph, however, had not previously offered such liberty, for two basic reasons. First, it required expertise to operate. Second, since its "codes" required trained operators, public demand did not grow sufficiently to justify installation of public telegraph exchanges or the setting up of a public network. On the contrary, telephone exchanges rapidly came into public use.

The early telegraphs operating point-to-point enabled railways to run on planned schedules; they also connected stock exchanges (e.g., the London and Paris exchanges were connected in 1851). Telegraphs found institutional use rapidly, but even today they are not found in homes; for a very long time they were used without exchanges to serve pre-assigned functions (such as connecting railway signal boxes).

Telegraph exchanges eventually came into use, but for the reasons mentioned they served economic (institutional) usages, with trained operators, rather than private (domestic) needs.

Nevertheless, these telegraph exchanges provided a conceptual model when telephones sprang to public attention. They were rapidly introduced for use within the domestic and economic spheres; the telephone network, from its early days, served both.

The first public telephone exchange appeared at New Haven, Connecticut, in 1878, followed by another of eight lines, in London in 1879. From that time, the telephone network grew rapidly and spread throughout Europe and across America, taking migrants and pioneers with it.

## TELEPHONES: THE ECONOMIC AND THE DOMESTIC SPHERES OF USAGE

In the domestic sphere (homes) the telephone is a "consumer product"; the home has only a finite disposable income, and if money is spent on a telephone that same money cannot be spent on anything else. Certain home telephones serve economically productive functions; e.g., doctors or businessmen working from home. On the other hand, telephones in the economic sphere (e.g., business, industry) eventually come to serve economically *productive* functions, as organization is increasingly based upon telephone usage. This distinction between the domestic and economic spheres is frequently recognized by the application of different tariff systems by the telephone authorities.

When the telephone first appeared 100 years ago, its productive function was not understood. As Chapters 1 and 2 have explained, the telephone was first seen as a one-way "broadcast" service, anticipating radio broadcasting by some forty years. Once the two-way *conversational* function became clear, exchanges were introduced. This set the future of the telephone system as an *organizing* instrument, and it began to be adopted within the economic sphere. Then, as wealth steadily increased, it became increasingly used in homes.

This principle may be demonstrated today by statistical comparison of countries having different economic conditions. Figure 1 shows the high correlation between (a) the ratio of telephones per 100 population in the domestic and economic spheres and (b) the Gross National Products of various countries in 1968.

It was the introduction of the telephone exchange principle and the growth of the *network* that finally converted Graham Bell's invention from a toy into a social instrument of immense organizational and economic power. The network served both the domestic and economic spheres of usage, in changing ratio, as time passed. (The same thing has been said of roads—that they serve the needs of both sport and transport.)

The telephone network called for something else that was new—some form of *subscriber organization* involving rights of usage and identification (i.e., a numbering plan). Telephone di-

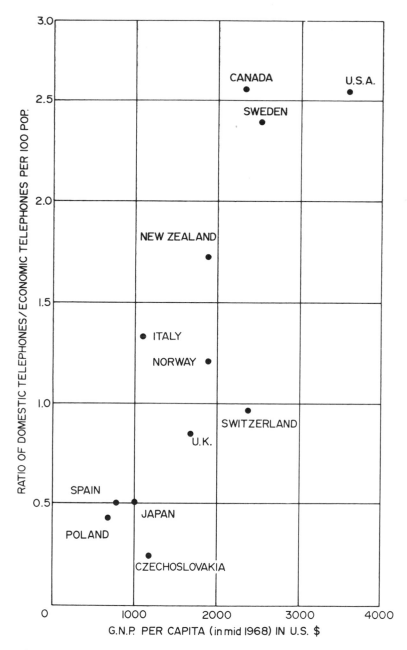

**Figure 1**
As a country becomes richer, the more the telephone becomes a
domestic ("consumer") product.

rectories appeared, and in principle any of those people listed could reach any of the others as a *right* of "membership." "Privacy" was apparently threatened and new law was called for. We have become completely accustomed to the telephone today and tolerate the way it enables others to barge into our offices and homes, in voice and spirit, preceded only by the ritual ringing of a bell, whereas we would not permit such forced entry otherwise (say by climbing through a window).

## THE TELEPHONE IN THE SOCIETY OF 1876

What did people think about the telephone when it was first exhibited and became known through popular articles and advertisements? We can gain some idea of their attitudes to the new wonder and of their imaginings with regard to its uses by reading some early accounts. An article[3] reprinted from the 1877–1878 edition of the *Journal of the Telegraph Electrical Society*, of Melbourne, describes some Australian telephones and experiments of the day. The same issue contained two advertisements. Figure 2 shows the "articulating telephone" alongside an electric bell and a hotel room indicator; the advertisement reads: "Estimates for fitting Telephones or a series of bells in hotels, mansions, etc. furnished on application." It then explains that "the ringing of house bells, by means of Cranks, movable Wires and Oscillating Pendulums, which are uncertain of action, confusing to persons attending them [i.e., servants] and, in short, clumsy contrivances, are altogether things of the past . . . all pulling, tugging, and the grating noises occasioned by the cranks, and consequent breaking of the wires, are done away with." The electronic era had arrived!

It may be surprising to learn that the house where I was born in 1914 was fitted with such mechanical bell-pulls. Pulling a handle in the drawing room would move iron wires through the rooms and stairway and operate an indicator and swinging bell in the kitchen. Not surprisingly, the telephone was first seen as a better bell-pull, whereby master could summon a servant, or as an improved speaking tube, through which orders could be sent.[4] The *conversational* power of the telephone was revealed only later,[5] when it became clear that the servant now had enhanced power for answering back or even for summoning the

## ARTICULATING TELEPHONES,

# ELECTRIC BELLS *and* INDICATORS,

### FOR PUBLIC AND PRIVATE BUILDINGS.

ELECTRIC INDICATOR.

# J. E. EDWARDS, Patentee,

## MANUFACTURER OF

### ARTICULATING TELEPHONES,

Telegraph Instruments, Electric Bells, Indicators

Galvanic Batteries. Galvanometers, &c.

## 35 & 37 ERSKINE STREET,

### HOTHAM HILL, MELBOURNE.

J. E. EDWARDS' NEW PATENT INDICATOR is very simple and reliable, which enables him to supply a first-class Indicator at a lower price than any other of even inferior workmanship. Estimates for fitting Telephones or a series of bells in hotels, mansions, etc., will be furnished on application.

All orders, accompanied with cash remittance, will receive immediate and careful attention.

It is now conceded that the ringing of house bells by means of **Cranks**, moveable **Wires**, and **Oscillating Pendulums**, which are uncertain of action, confusing to the persons attending them, and, in short, clumsy contrivances, are altogether things of the past. On the **Electric Bell System**, all pulling, tugging, and the grating noises occasioned by the cranks, and consequent breaking of the wires, are done away with; for, however distant **The Electric Bell** may be, or however tortuous its course, it can be rung by the slightest pressure of the finger on a little ivory button; and, as the electric wires do not move, the wear and tear, as on the old system, is entirely avoided. In order to secure an entirely satisfactory result three things are necessary, viz., the exclusive use of first-class and highly finished materials—the employment of skilled workmen—and the superintendence of a person of long practical experience.

With these at command J. E. EDWARDS (who was connected with the Melbourne Post and Telegraph Department for upwards of Ten years) is able to execute in a first-class manner any orders entrusted to him. He has had the honor to execute work on a large scale, and attained results highly satisfactory at the following public and private buildings:—

| | | |
|---|---|---|
| The Treasury | General Post Office | The London Chartered Bank |
| Parliament Houses | Crown Law Offices | Registrar-General's Buildings |
| Beechworth Asylum | Melbourne Town Hall | New Government Offices |
| Ararat Asylum | Crown Law Offices | W. J. Clarke's, Esq. |
| Royal Mint | Messrs. Sands & McDougall | And many other buildings. |

The Bells and Indicators may be seen in operation in Melbourne.

## Figure 2

An early advertisement, dated 1878, for "Articulating Telephones"—as means for summoning the servants. (From the *Telegraph Electrical Society Journal*, Melbourne, 1877–78. Figures 2 and 3 are reproduced here with the permission of the General Secretary, the Telecommunication Society of Australia.)

master; it became a major force for social leveling!

A second advertisement (Figure 3) reveals this attitude and shows the telephone in use. It used neither a battery nor any amplification, so the resulting signals were said to be very feeble. There was no form of exchange, and it is shown here being used "person-to-person." The small print says,

These instruments are of great practical value. They can be used for any purpose, and in any position, without technical training, wherever communication or conversation is required from a distance, as between the principals and employees in commercial houses—between central and branch banks—in mining operations, between the manager's office and the employees in the mine—in large hotels or mansions—in factories of every description between the manufacturer and his factory, and between the superintendent and his leading men; and, in fact, it may be considered as an ordinary speaking tube, with all the advantages of telegraphic communication.

Note the social implications of these words. Every use suggested serves for master/servant relations. The telephone was seen as an improvement over the speaking tube—for sending orders or for extending powers of authority. As the chapter by Attali also notes for France, the bilateral use for conversation came later, as did the exchange principle that offered *choice* of one's partner to converse with—even strangers.

The article describes the telephone as being "like a pocket telescope with one joint drawn out."

Everyone around should be perfectly quiet . . . the listening operator should not even allow his mind to be distracted from his work, but must concentrate it on listening alone. The sound comes in a singularly weird-like manner . . . when a ghostly "co-o-o-e" is heard, coming from fifty or a hundred miles away, the receiver is almost awe-struck.

The comment is then made that

The Australian *cooey*, the sound that the simple savages of Australia [the Aborigines!] pitched on to communicate with each other from long distances, is found to be the one which, of all others, travels best over long distances between telephone and telephone. What a jump to make in a few years in this country of ours! *Then*, some perfectly uncivilized natives co-oeying to each other in the forest; now, over the same ground, men of high civilization uttering the same sound, from city to city, through one of the most wonderful inventions of modern or of any times.

I have given this quotation at length, because of the implied social commentary of such words—written only 100 years ago. It

# THE
## ARTICULATING OR SPEAKING
# TELEPHONE,

DURING the past few months MR. J. E. EDWARDS has made a great variety of experiments to test the powers of the Telephone. It is found possible to speak with facility through a hundred miles of wire, and persons using them on lines of a few miles long can converse with ease in the same manner as though they were in the same room ; even the voice of the person speaking can be recognised at the other end of the line.

In the early part of February a pair of these Instruments was fixed on Messrs. M‘LEAN BROS. AND RIGG's line, which communicates between their Warehouse, 69 Elizabeth street, and their Store, 190 Bourke street west, near Spencer street, about one mile by wire, by which all their business is transmitted, the Telephone having superseded the Wheatstone instruments previously used. The Telephone can be seen in use, and also on sale, there at any time during business hours.

These Instruments are of great practical value. They can be used for any purpose, and in any position, without technical training, wherever communication or conversation is required from a distance, as between the principals and employes in commercial houses— between central and branch banks—in mining operations, between the manager's office and the employes in the mine—in large hotels or mansions—in factories of every description between the manufacturer and his factory, and between the superintendent and his leading men ; and, in fact, it may be considered as an ordinary speaking tube, with all the advantages of Telegraphic communication.

The above arrangement needs but a wire between the points of communication, a pair or two pair of Telephones, and two alarm-bells to call attention to either end of the line, though a hundred yards or miles apart.

Further particulars and estimates may be obtained on application to M‘LEAN BROS. & RIGG, 69 ELIZABETH ST,; or

## J. E. EDWARDS, the Manufacturer, 37 Erskine st., Hotham Hill, Melbourne,

Figure 3
Another early advertisement for the ''Articulating or Speaking Telephone''—seen as a better speaking tube.

is also interesting to note that other experiments described in this article consist of singing hymns, and "God Save the Queen," and playing musical instruments. "What is the future of this wonderful discovery," it asks "now in its infancy, and not much more than a scientific toy . . . it is impossible to for-see," and then suggests that it could be used for relaying hymns and sermons to a hundred people simultaneously. Even Bell himself (as Aronson has already told us) conceived of the early telephone first as contributing to entertainment, for its real so-cial value as an *organizing* instrument did not become clear until the telephone exchange was imagined.

## THE TELEPHONE NETWORK: THE LARGEST AND MOST COMPLEX MACHINE IN THE WORLD

It is easy to be amused by such early visions, but we should not feel superior in our hindsight. As argued already, the future us-ages of any invention are almost impossible to predict, because the nature of the future society is totally unknown. In fact, Bell and his financial backers (who rapidly came forward) did have remarkable vision, which the invention of the manual telephone exchange clarified. The enormous potential for creation of new social relationships and organizations was envisaged very early.

As the number of lines to a manual exchange increased, so the number of *possible* interconnections between pairs of sub-scribers increased combinatorially; that is, as the number of subscribers increased linearly, so the size and complexity of the exchange needed to increase binomially. (See Chapter 15.) Even-tually there came a limit to the practical size of an exchange, beyond which manual operators could no longer cope. Trunk exchanges and local exchanges developed to spread the load, but the steady growth of subscribers gradually revealed an in-creasing problem. In 1889 Almon Strowger (an American under-taker), annoyed by his apparently inefficient operator, designed his automatic selector switch. The first automatic exchange was opened in LaPorte, Indiana, in 1892; the Strowger selector was adopted subsequently throughout the world, until 1926 when the crossbar switch was first introduced in Sweden.

Few members of the public have the slightest idea of the com-plexity and scale of modern national telephone networks when they are critical of lost calls or wrong numbers. The network is a technological miracle; its global reach, scale, and capital value are staggering. The telephone network now interconnecting the continents is by far the largest integrated machine in the world.

## THE PSYCHOLOGICAL IMPORTANCE OF THE TELEPHONE

I should like to turn next to my second reason for claiming that the telephone service has had a much more profound social effect than other technical modes of communication; namely, because of its psychological values.

The telephone continues the verbal tradition because it operates with the human voice and requires no special codes, training, or skills as did the telegraph. Phoning is as easy and natural as talking. For such reasons, it has great value for developing countries where literacy may be low and the verbal tradition dominant. Radio has similar value for the same reason; indeed President Nasser of Egypt once remarked, "Radio now counts for more than literacy."

The psychological importance of human speech is not always appreciated, at least by people of normal hearing. Every time we hold a conversation we relate to another person, not to a thing; their replies to our remarks reinforce the sense of our existence. We are challenged and valued—our partner is a mirror in which we see our own images. We exist as persons, through society.

The telephone operates as an extension of nature in this way. We have today become so accustomed to the telephone that it rarely strikes us as odd that when we want advice, consolation, or information we walk toward this little black thing on the table, talk to it, and listen to what it says. Our trust is really extraordinary. We *consult* the telephone, much in the same way that an Ancient Greek would have consulted the Oracle. If we want to know the time we dial TIME and *believe* what the golden voice tells us—we trust it more than we trust our own watches!

I should like to end by saying a little more about the social values of the telephone. With it, I shall link radio (for that, too, uses Graham Bell's inventions, the microphone and earpiece as they were in his day, or derivatives of these).

## ON THE POLITICAL SIGNIFICANCE OF THE
## TELEPHONE SERVICE

Emile Durkheim, the great sociologist, once asked the question:
How can it be that we feel more free, as the powers of the State
have grown? He answered that it is only when the numerous
state institutions are efficient and clearly defined that we know
what is expected of us and what constraints bear upon us. Law
may constrain us; yet only if it is clearly defined do we know
where we stand and how we may act. Anarchy is not liberty,
but slavery. It is very much through the extensive network of
communication in industrial countries, reaching from the metropolis into every town, village, home, and back again, that we
can operate in ways that enable us to feel "free" as individuals,
though knowing that we are socially constrained. Liberty rests
not only on a foundation of defined authority but also upon the
operation of a two-way communication service.

If a country or a scattered empire does not possess an adequate communication service, how can it be held together? How
can its various scattered institutions of local government operate
in concert? How can a centralized authority know what is happening in outlying regions? It can only be done by the creation
of a strong ideological system, some rigid system of education
through which local governors are trained into intense loyalty
to principles, so that when they are sent out to govern, their
behavior will be predictable.

The public schools of nineteenth-century Britain ideologically
trained those going out to govern scattered regions. The telegraph eventually made a difference! The Roman Empire covered
a vast area in its day, yet it had only the slenderest lines of
communication. Caesar could send orders or receive advice from
the Governor of Londinium only through a postal service, using
horses and caravans. There were relay stations every forty miles
where horses could be changed, but the trip took about a month
and a lot could happen in that time to render those messages
obsolete.[6] Caesar needed a telephone service! Long delay of
messages is likely to make social organization more unstable because counteraction may become outdated by the turn of events.

A further social consequence follows from the introduction of
a telephone service. Because a *network* offers choice of contact

between subscribers over larger geographic areas, these contacts are likely increasingly to be between strangers. In tribal society the social institutions are local and formed of kinsmen or close personal associates, but those of industrial societies force us to associate with abstracts whom we call "managers" "representatives," or "officers," people not necessarily known to us as *persons*. This means that new forms of trust have to be developed; not trust in kinsmen but trust in abstract "representatives" whom we have never met. It is a major psychological change.[7]

A highly developed, two-way communication service is an essential pre-requisite to any form of "democratic" state (if that means one rendering social change possible in acceptable and stable ways). It is discourse, the conversational mode, which is needed; the telephone has made an immense contribution, not merely in the home but also in the functioning of our great institutions.

The introduction of the telephone and the growth of the nationwide network and service that evolved over the last century have contributed greatly to our changing concept of both central and local government. Subsequent developments like radio and television, which also operate with speech, are available to the whole public, and require no special skills, have given momentum to the process (although they are "one-way"). Authorities, ministers, trade union leaders, experts, the great and the famous are no longer remote, awful, or charismatic figures. They have been cut down to size as human beings, feeble or sinful at times, like any of us. The popular watchword now is *participation*, a word which suggests that the public wants more say in affairs. Whether such a watchword has any substance may perhaps be questioned; but as an *idea* there is no doubt that it owes a very great deal to the technology of speech communication through the invention of the telephone.

## NOTES

1. The telegraph has, of course, evolved into the Telex service today, whose rapidly expanding traffic is essential to the economic sphere, especially to the airways.
2. The Talking Drum actually mimics the human voice and is listened to as though it speaks. It does not merely send coded signals.

3. "The Telephone - Australian Experiments in 1878," *Australian Telecommunication Research, 8*, No. 2, 1974; reprinted from the *Telegraph Electrical Society Journal*, Nov.–Jan. 1877–1878, Melbourne, with the permission of the Telecommunication Society of Australia.
4. It is noteworthy that the very first words ever spoken over a telephone, by Bell himself, were an *order*: "Mr. Watson, come here. I want to see you."
5. Bell demonstrated the two-way telephone and realized its conversational function in October 1876.
6. H. G. Pflaum, "Essai sur le *Cursus Publicus* sous le Haut-Empire Romain," Paris:*Imprimerie Nationale*, 1950; A. M. Ramsey, "The Speed of the Roman Imperial Post", *Journal of Roman Studies, 15*, 1925, p. 60.
7. Colin Cherry, "World Communication—Threat or Promise?" London: Wiley & Sons, 1971.

# 6

### Foresight and Hindsight: The Case of the Telephone

Ithiel de Sola Pool
Craig Decker
Stephen Dizard
Kay Israel
Pamela Rubin
Barry Weinstein

Let us try to go back to the period from 1876 until World War II, to ascertain how people perceived and foresaw the social effects of the telephone. By taking advantage of hindsight in 1976, we can ask which forecasts were good, which went askew, and why.

## FORECASTERS: GOOD AND BAD

Some sensationally good forecasts were made. In 1878, a letter from Alexander Graham Bell in London to the organizers of the new Electronic Telephone Company outlined his thoughts on the orientation of the company; it is such a remarkable letter that we quote it in full in an appendix to this chapter. The letter describes a universal point-to-point service connecting everyone through a central office in each community, to in turn be connected by long-distance lines. Aronson noted in Chapter 1 that when the telephone was first invented, it was not obvious that it would be used in that way. Bell briefly considered a path that others were to pursue after he had given it up—using the device for broadcasting in a mode like that of modern radio. The reader can refer to Aronson's interesting analysis of why Bell originally pondered that alternative and why he instead came to a clear perception of the telephone as a conversational rather than broadcasting device. The result was a prevision of the phone system as it exists today, a century later. The technology

This is the first report of a "retrospective technology assessment of the telephone" being done at MIT, supported by a grant from the National Science Foundation.

that existed then did not permit universal, switched, long-distance service; Bell's description of such a system was a prescient forecast.

The small group of men who created the telephone system did share Bell's vision. Theodore N. Vail and Gardiner Greene Hubbard, in particular, worked to implement "the grand system" they had in their minds—a system in which the monopoly telephone company would provide service in virtually every home and office, linking local systems throughout the nation and the civilized world. They visualized the device as one that everyone could afford and saw it organized as a common carrier eventually surpassing telegraph usage.

Their optimism was not shared by all. In 1879, Sir William Preece, the chief engineer of the British Post Office, testified to a special committee of the House of Commons that the telephone had little future in Britain.

I fancy the descriptions we get of its use in America are a little exaggerated, though there are conditions in America which necessitate the use of such instruments more than here. Here we have a superabundance of messengers, errand boys and things of that kind. . . . The absence of servants has compelled Americans to adopt communication systems for domestic purposes. Few have worked at the telephone much more than I have. I have one in my office, but more for show. If I want to send a message—I use a sounder or employ a boy to take it.[1]

In 1878, Theodore Vail, then a young railroad mail superintendent, quit the U.S. Post Office to join the newly organized Bell Telephone Company. "Uncle" Joe Cannon, a young Congressman, expressed surprise and regret that the company had "got hold of a nice fellow like Vail."

The Assistant Postmaster General could scarcely believe that a man of Vail's sound judgment, one who holds an honorable and far more responsible position than any man under the Postmaster General, should throw it up for a d--d old Yankee notion (a piece of wire with two Texas steer horns attached to the ends with an arrangement to make the concern bleat like a calf) called a telephone.[2]

Not all the forecasting errors were on the side of stodgy conservatism. There were also wild dreams of the wide blue yonder. General Carty, Chief Engineer of AT&T since 1907, predicted there would be international telephony one day, and also forecast that it would bring peace on earth.

Some day we will build up a world telephone system making necessary to all peoples the use of a common language, or common understanding of languages, which will join all the people of the earth into one brotherhood. . . . When by the aid of science and philosophy and religion, man has prepared himself to receive the message, we can all believe there will be heard, throughout the earth, a great voice coming out of the ether, which will proclaim, "Peace on earth, good will towards men."[3]

There is a striking contrast in the quality of forecasting and analysis between the small group of initial developers of the telephone and other commentators. Is it because the founders were particularly intelligent men? Perhaps the same mental powers that made them succeed as inventors and entrepreneurs were at work in their forecasts. Or, second, is it because they were living with the subject, eighteen hours a day, year after year? Should we expect the same insight in a congressman's or journalist's quick comments? Third, the telephone pioneers brought together a combination of scientific knowledge and business motivation; Bell was not just a scientist nor Vail only a businessman. They belonged to that remarkable American species of practical technologists, including Morse, Edison, Ford, and Land, who were both inventors and capitalists. They were interested not only in what might be theoretically possible but also in what would sell; the optimism of their speculations was controlled by a profound concern for the balance sheet.[4] Perhaps such concerns are among the crucial ingredients for good technological forecasting.

The activism of the early developers suggests a fourth possible reason for their success as forecasters: they fullfilled their own prophesies. They had the inventions, a vision of how the inventions could be used, and they controlled the businesses that implemented those visions. This theory does not exclude other propositions. The self-fulfilling prophesy of planners can work if, and only if, it accounts for technical and economic realities. Yet the weight we attach to each of the possible reasons for successful prediction and the conditions under which they apply make a great deal of difference to how we would make a technology assessment.

## SOME FORECASTS

We shall offer further speculations about the factors contributing to accurate forecasts, but first we shall examine closely a few predictions, who made them, and how they look in retrospect.

### Universal Service

The telephone was an expensive device in its early years; a subscriber paid a flat amount for unlimited service. Furthermore, the combinatorial nature of a network meant that linkage complexity increased faster than the number of nodes. Until fully automated switching, the company's cost of serving each subscriber was greater the larger the number of other subscribers.[5]

With switchboards, the problem was partially solved. Still, as the number of subscribers grew, the operator's job in making the connection grew more than proportionally; a manual switchboard could only be of a certain size. Bell understood this; in *Financial Notes* in 1905, he is quoted as arguing that as the number of people in an exchange increased, the operator's work increased exponentially; hence, the exchanges would eventually all have to be automated:

In the telephone of the future I look for all this business to be done automatically. . . . If this can be accomplished, it will do away with the vast army of telephone operators, and so reduce the expense that the poorest man cannot afford to be without this telephone.

One early telephone manager commented that "so far as he could see, all he had to do was get enough subscribers and the company would go broke."[6] Telephone service in large communities was therefore very expensive.[7] In 1896, the fee for service in New York was $20 a month. The average income of a workman in that year was $38.50 a month, a six-room tenement in New York rented for about $10 a month, and a quart of milk sold for 5¢. The first subscribers to the telephone were business offices, not ordinary homes. Residential phones in the AT&T National Telephone Directory of 1896 were almost 30 percent in Chicago, one in six or seven in Boston and Washington, and only about one in twenty in New York and Philadelphia.

Yet Bell's letter of 1878 mentioned connecting "private dwellings, counting houses, shops, manufactures, etc. etc.," in that order; he was not thinking of the phone as just a business device. He talks of "establishing direct connection between any two places in the city," not just a system limited to industrial or affluent neighborhoods. Perhaps it was too early for him to assert categorically, as he did slightly later, that the telephone could even become an instrument for the poor. But the goal of universality, which became one of the watchwords of the Bell System, was there from the beginning.

Though in its first two decades the telephone's growth had largely been in business or among the rich, by 1896 several factors led to a rapid expansion of service. There was the continued acceleration of an ongoing exponential process of growth, and (with the expiration of the initial Bell patents in 1893) competitors sought to discover and occupy parts of the market not yet served—for example, the fast growing Midwest and rural areas. Perhaps the most critical change, however, was the introduction of message charges. The problem was to reduce the price of phone service for the small user, while still collecting adequate amounts from businesses and large users.[8] In 1896, the New York phone company abandoned flat rate charges and introduced charges by the message.

Between 1896 and 1899, the number of subscribers doubled; in six years it quadrupled; in ten years it increased eight times. By 1914, there were 10 million phones in the United States, 70 percent of the phones in the world. As Burton Hendrick's October 1914 article in *McClure's Magazine* stated:

Until that time [1900] the telephone was a luxury—the privilege of a social and commercial aristocracy. About 1900, however, the Bell Company started a campaign, unparalleled in its energy, persistence and success, to democratize this instrument—to make it part of the daily life of every man, woman, and child.

In AT&T's annual report of 1901, President Frederick P. Fish expressed its ethos of growth.

That the system be complete and of the greatest utility, it is necessary that as many persons as possible should be connected to it as to be able to talk or be talked to by telephone. . . . [The user's] advantage as a telephone subscriber is largely measured by the number of persons with whom he may be put in communication.

At the turn of the century the telephone was clearly still a luxury, but the leaders of the industry in the United States foresaw its becoming a mass product, and their forecast was at least partially self-fulfilling: they had planned for growth. In 1912, AT&T published a manual of the urban planning process of telephone systems. It prescribed and explained how to do a "development study" for a city; the result would be a document known as the "Fundamental Plan." Since the average life of materials entering a telephone system was estimated to be fifteen years, the plan was to provide adequate capacity for that period,[9] with a planning goal for an ordinary city of at least one telephone for every eight inhabitants.[10] Typically, ducts were only half filled to allow adequate room for growth.

Among the four hypotheses about why the telephone pioneers' forecasts succeeded, the self-fulfilling prophecy seems most apposite to explaining the success of the forecast of growth. Vail and his colleagues had a dream and they made it happen. It was, as it had to be, a realistic dream or it would have failed. But abroad, where the same technology and equally talented men existed, governments and phone companies were structured differently, they followed a different perspective, and growth was slower.[11] The movement of the telephone system to rapidly become a universal low-cost service was more an entrepreneural decision than a foregone outcome of social processes.

## Long-Distance Service

The original telephones of the 1870s could only operate over a range of about twenty miles. Yet Bell in his 1878 letter had already declared "I believe in the future wires will unite the head offices of the Telephone Company in different cities, and a man in one part of the country may communicate by word of mouth with another in a different place."

Even earlier, long-range communication was assumed by Sir William Thomson (later Lord Kelvin), when judging the technical exhibits at the Philadelphia Exposition in 1876, though most of his report covers which words he had been able to understand and which not, on the primitive device at hand.[12]

Vail also anticipated a far-flung global telephone network and saw that strategic control would lie with the company running the long-line interconnections. "Tell our agents," he wrote to one of his staff in 1879, "that we have a proposition on foot to connect the different cities for the purpose of personal communication, and in other ways to organize a grand telephonic system."[13] Yet he daringly forecast: "We may confidently expect that Mr. Bell will give us the means of making voice and spoken words audible through the electric wire to an ear hundreds of miles distant."

Burlingame describes his perception:

As general manager of the American Bell Telephone Company formed in 1880 (for extension of the telephone outside of New England), Theodore Newton Vail saw with an extraordinary prophetic clarity the development of a nationwide telephone system. This prophecy was expressed in the certificate of incorporation of the American Telephone and Telegraph Company formed in 1885 which certified that "the general route of lines of this association . . . will connect one or more points in each and every city, town or place in the State of New York with one or more points in each and every other city, town or place in said state, and in each and every other of the United States, and in Canada and Mexico; and each and every other of said cities, towns and places is to be connected with each and every other city, town or place in said states and countries, and also by cable and other appropriate means with the rest of the known world.[14]

The first long-distance line was built between Boston and Lowell in 1880 with Vail's encouragement.

This success cheered Vail on to a master effort. He resolved to build a line from Boston to Providence, and was so stubbornly bent upon doing this that, when the Bell Company refused to act, he organized a company and built the line (1881). It was a failure at first and went by the name of "Vail's Folly." But one of the experts, by a happy thought, doubled the wire. . . . At once the Bell Company came over to Vail's point of view, bought his new line, and launched out upon what seemed to be the foolhardy enterprise of stringing a double wire from Boston to New York. This was to be a line de luxe, built of glistening red copper, not iron. Its cost was to be $70,000, which was an enormous sum in those hard-scrabble days. There was much opposition to such extravagance and much ridicule. But when the last coil of wire was stretched into place, and the first "Hello" leaped from Boston to New York, the new line was a success.[15]

By 1892, there were lines from New York to Chicago; by 1911, from New York to Denver; and by 1915, from New York to San Francisco. Experiments with overseas radio telephony took place in 1915, but the first trans-Atlantic commercial service began only in 1927.[16] While long-distance telephony grew rapidly, Bell's and Vail's predictions preceded its reality. There were many technical difficulties, and not everyone anticipated (as did Bell and Vail) that they would be overcome.

Much of the effort to make long-distance telephony work focused on repeaters, devices that rebuilt the deteriorating and fading signals as they passed through long lengths of wire. Berliner developed one. When Vail was out of the company (before 1907), Hayes—the Director of the Mechanical Department—decided that the company could most economically abandon its own fundamental research and instead rely on "the collaboration with the students of the Institute of Technology [MIT] and probably of Harvard College." On research concerned with long lines, however, Hayes made an exception. He employed George A. Campbell, who had been educated at MIT and Harvard, to study the essentially mathematical problem of maintaining transmission constants over long lines of cable. By 1899, Campbell had outlined the nature of discretely loaded electrical lines and had developed the basic theory of the wave filter.

Around 1900, Pupin at Columbia University developed the loading coil which greatly improved the capabilities of long-distance cable. Before 1900, long-distance lines demanded wire about an eighth of an inch thick; the New York–Chicago line consumed 870,000 pounds of copper wire. Underground wires in particular had to be very thick. One fourth of all the capital invested in the telephone system before 1900 had been spent on copper. With the Pupin coil, the diameter of the wire could be cut in half. Then, still later, the vacuum tube and other developments in repeaters made long-distance communication increasingly economical.[17]

Vail wrote in the 1908 annual report (p. 22):

It took courage to build the first toll line—short as it was—and it took more to build the first long-distance line to Chicago. If in the early days the immediate and individual profit of the long-distance toll lines had been considered, it is doubtful if any would have been built.

One obvious speculation as to why the forecast of long-distance communication was so successfully made by Bell and Vail is that the telegraph shaped their thinking; the telephone's invention, after all, had been a by-product of telegraphy. Bell had been employed to create a harmonic telegraph which would carry messages at different pitches simultaneously, and the telegraph's great achievement had been the contraction of distance. It was not surprising, therefore, that when a way was found to make voice travel over wires, a realizable goal seemed to be its transmission over distances. Quite rightly, telephone enthusiasts saw the technical problems as temporary difficulties.[18]

## Video

Experiments on transmission of pictures over wires go back long before Bell's telephone.[19] After that invention, many people felt that since a voice could be captured and sent over wires, transmission of pictures was an obvious next step. The difficulties in going from conception to realization were frequently underestimated by nontechnical people. The more naive and less scientifically sophisticated the writer, the more immediate the extrapolation from telephone to television seemed.

Kate Field, a British reporter associated with Bell, projected in 1878 that eventually, "while two persons, hundreds of miles apart, are talking together, they will actually *see* each other!"[20] That belongs in a class of journalistic whimsy along with the *Chicago Journal* suggestion:

Now that the telephone makes it possible for sounds to be canned the same as beef or milk, missionary sermons can be bottled and sent to the South Sea Islands, ready for the table instead of the missionary himself.[21]

In 1910, Casson made a passing comment in "The Future of the Telephone" to the effect that "there may come in the future an interpreter who will put it before your eyes in the form of a moving-picture."[22] A more serious discussion, *The Future Home Theater*, by S. C. Gilfillan, appeared in 1912.[23] In some respects it is a remarkable forecast, in others a dismaying one.

There are two mechanical contrivances . . . each of which bears in itself the power to revolutionize entertainment, doing for it what the printing press did for books. They are the talking motion picture and the electric vision apparatus with telephone. Either one will enable millions of people to see and hear the same performance simultaneously, . . . or successively from kinetoscope and phonographic records. . . . These inventions will become cheap enough to be . . . in every home. . . . You will have the home theatre of 1930, oh ye of little faith.

Gilfillan believed that both the "CATV" and over-the-air broadcast form of video would coexist by 1930, and also that there would be a television of abundance with libraries of material from which one could choose. He thought great art would drive out bad; he described an evening program of Tchaikovsky, ballet, Shakespeare, educational lectures, and a speech by a presidential candidate on "The Management of Monopolies"; he thought the moral tone of the home theater would be excellent. And he pointed out that the difficulties in having all this were not technical but human. Let the reader draw his own conclusions!

By 1938, when the Walker investigation of the telephone industry published its report, television already existed but had not yet reached the general public. The Walker Report projected two ways it could develop:

Television offers the possibility of a nation-wide visual and auditory communication service, and this service might be developed under either of two broad methods. The first is by the eventual establishment of a series of local television broadcasting stations similar to the present local radio broadcasting stations . . . or conceivably it may develop into some form of wire plant transmission utilizing the present basic distributing network of the Bell System, with the addition of coaxial cable or carrier techniques now available or likely to be developed out of the Bell System's present research on new methods of broadband wire transmission. [24]

The one forecast the report did not make, and which now seems the most plausible, is that over time television would be delivered first one way and then the other. With time unspecified both forecasts may be realized.

**Crime**

From these forecasts—all of which concerned the telephone system's development—we turn now to forecasts regarding its impact on society. From the first decade of the century, there has been much discussion of the telephone's relationship to crime with diametrically opposite predictions. The telephone is portrayed as both the promoter and the conquerer of crime. A villainous anarchist in a 1902 *Chicago Tribune* short story wires a bomb to a phone to be detonated by his call. But in the happy ending, the bomb is detached seconds before he rings; the police trace his call and catch him.

The telephone is portrayed as part of a process of urbanization, with decay of traditional moral values and social controls. The "call girl" was the new form of prostitute; obscene callers took advantage of the replacement of operators by dial phones. The phone company was repeatedly berated by reformers for not policing the uses to which the phone was put. As early as 1907, *Cosmopolitan Magazine* had a muckraking story by Josiah Flynt entitled "The Telegraph and Telephone Companies as Allies of the Criminal Pool-Rooms."[25] Flynt charges:

Because they are among the country's great "business interests," because the stock in them is owned by eminent respectables in business, and because they can hide behind the impersonality of their corporate existence, they have not been compelled to bear their just share of the terrific burden of guilt. But they have been drawing from five to ten million dollars a year as their "rakeoff" from the pool rooms. . . . Every one of the estimated four thousand pool rooms throughout the United States is equipped with telephones used for gambling purposes and for nothing else.

Flynt charges that 2 percent of the New York Telephone company revenues, or a million dollars a year, was derived from gambling in pool rooms. He rejects the argument that the phone company should not attend to what subscribers say on their lines; the company, he says, knows full well who the criminal users are but simply does not wish to forgo the profits of sin.

Prohibition coincided with the telephone system's years of growth to a national network and total penetration. The bootleggers and the rackets made full use of whatever was available to run their operations; it is hard, however, to take seriously the argument of causality—that somehow there would have been less crime without the telephone.

Side by side with that accusation, one can also find in the 1920s and 1930s the reverse argument. The telephone, it was said, gives the police such an enormous advantage over criminals that law and security will come to prevail in American cities.

When a girl operator in the exchange hears a cry for help— "Quick!" "The Police!"—she seldom waits to hear the number. She knows it. She is trained to save half seconds. And it is at such moments, if ever, that the users of a telephone can appreciate its insurance value.[26]

There were forecasts that crime would decrease, for the criminal would have little chance of escape once telephones were everywhere and the police could be notified ahead about the fleeing culprit. "Police officials feel that the scarcity of dramatic crimes may be due somewhat to the preventive factor present in modern police communication systems."[27]

In fact, enforcement agencies gradually adopted new communications technologies, but usually only slowly and after their usefulness was well demonstrated. Telegraphy was first used in law enforcement to connect police stations and headquarters, but it was not of great importance until encouraged by the International Association of Chiefs of Police, formed in 1893. Recognition of the need for a complex communication system for crime prevention followed civil service reform and the recognition of police work as a specialized profession.[28]

The problem of communication between the police station and the patrolman on the beat received little attention until the 1880s. Before the telephone, the technology available was that used since 1851 for fire alarms. (Between 1852 and 1881, 106 electric fire alarm systems were installed in American cities.) The first electric police-communication system of record was installed in 1867. Between 1867 and 1882 only seven more systems were put in operation. . . . 56 systems were installed from 1882 to 1891, 76 systems in the next decade, and 84 . . . from 1892 to 1902.[29]

A survey in 1902 found that of 148 systems, 125 were telegraphic, 19 telephonic, and 3 mixed. Although police departments had subscribed to telephones ever since the Washington department took 15 in 1878, they did not deploy them to the beats. In 1886, the *New York Tribune* in an editorial criticized the New York police for not connecting the stations with the central office by telephone as the Brooklyn police had. The *Tribune* remarked that "doubtless the time may come when every patrolman's beat will be furnished with one of these instruments."[30]

The Chicago department had been the first to move in that direction. Between 1880 and 1893, over 1,000 street boxes were installed. The popularity of such systems received a boost in 1889 when a murderer was caught at the railroad station a few hours after all police in the city had been notified of his description by the phone network. Telephone boxes began replacing signal boxes; yet by 1917 there were still only 8,094 telephone boxes to 86,759 of the latter in police and fire service. In short, ideas for police use of communications technology were prevalent but their adoption came slowly. One idea which was never adopted, but was discussed by 1910, was that each individual should have a number by which he could be reached telephonically wherever he might be.[31]

Hindsight induces a jaundiced view of the forecasts that saw the telephone defeating crime. Such prognoses, we should note, were not made by developers of the telephone system, nor by law enforcement experts, but rather by journalists and reformers. Yet let us not be too complacent about our hindsight. Why were these forecasts wrong? Even with all the advantages of hindsight, it is hard to say. A priori it seems sensible that an instrument permitting well-organized, dispersed police agents to contact and warn each other about suspects much faster than the suspects could move should make things harder for lawbreakers. Yet crime increased in the same years the telephone became available to the authorities; this tells something about the limits of social forecasting based on assessment of one isolated technology. To understand the anomaly of growing crime in the same period as improved technologies of law enforcement, one must understand such matters as the public's attitude toward minor crime, the judges' behavior in sentencing, the organizational incentives in the legal process, the social structure of migrant and ethnic groups in the society, and the nature and reliability of crime statistics.

We do not understand those matters very well, even with hindsight. The forecast that the telephone would help the police was not wrong; it has helped them. The impact on the amount of crime, however, depended primarily on what people wanted to do.

There is substantial evidence that whatever the net trend in criminal activity, the telephone has added to the citizen's sense of security. Alan Wurtzel's and Colin Turner's study in Chapter 11 on how people were affected by the New York City exchange fire in 1975 supports this. The ability to call for help is an important security, yet the relationship of communications to law enforcement is many-sided and complex.

The telephone became part of the pattern of both crime and law enforcement, affecting both. Criminals and policemen alike came to use the telephone, and it changed the way they did things. There even came to be special telephone crimes and telephone methods of enforcement; tapping is an example of both. Yet it is hard to argue that the level of crime or the overall success or failure of law enforcement had any obvious or single-valued relationship with the development of the telephone network.

**The Structure of Cities**

One of our working hypotheses as we began this study was that the automobile and the telephone—between them—were responsible for the vast growth of American suburbia and exurbia, and for the phenomenon of urban sprawl. There is some truth to that, but there is also truth to the reverse proposition that the telephone made possible the skyscraper and increased the congestion downtown.

The movement out to residential suburbs began in the decade before the telephone and long before the automobile. As Alan Moyer describes it in Chapter 16, the streetcar was the key at the beginning. Today streetcars have vanished, and the automobile and the telephone do help make it possible for metropolitan regions to spread over thousands of square miles. But the impact of the phone today and its net impact seventy years ago are almost reverse. As John J. Carty tells it:

It may sound ridiculous to say that Bell and his successors were the fathers of modern commercial architecture—of the skyscraper. But wait a minute. Take the Singer Building, the Flatiron, the Broad Exchange, the Trinity, or any of the giant office buildings. How many messages do you suppose go in and out of those buildings every day? Suppose there was no telephone and every message had to be carried by a personal messenger. How much room do you think the necessary elevators would leave for offices? Such structures would be an economic impossibility.[32]

The prehistory of the skyscraper begins with the elevator in the 1850s; the first Otis elevator was installed in a New York City store in 1857, and with adaptation to electric power in the 1880s, it came into general use.[33] "The need to rebuild Chicago after the 1871 fire, rapid growth, and rising land values encouraged experimentation in construction." In 1884, Jenney erected a ten-story building with a steel skeleton as a frame. The Woolworth Building with fifty-seven stories was opened in 1913. "By 1929 American cities had 377 skyscrapers of more than twenty stories."[34]

The telephone contributed to that development in several ways. We have already noted that human messengers would have required too many elevators at the core of the building to make it economic. Furthermore, telephones were useful in skyscraper construction; phones allowed the superintendent on the ground to keep in touch with the workers on the scaffolding. As the building went up, a line was dropped from the upper girders to the ground.

As the telephone broke down old business neighborhoods and made it possible to move to cheaper quarters, the telephone/tall-building combination offered an option of moving up instead of moving out. Before the telephone, businessmen needed to locate close to their business contacts. Every city had a furrier's neighborhood, a hatter's neighborhood, a wool neighborhood, a fishmarket, an egg market, a financial district, a shipper's district, and many others. Businessmen would pay mightily for an office within the few blocks of their trade center; they did business by walking up and down the block and dropping in on the places where they might buy or sell. For lunch or coffee, they might stop by the corner restaurant or tavern where their colleagues congregated.

Once the telephone was available, business could move to cheaper quarters and still keep in touch. A firm could move outward, as many businesses did, or move up to the tenth or twentieth story of one of the new tall buildings. Being up there without a telephone would have put an intolerable burden on communication.

The option of moving out from the core city and the resulting urban sprawl has been much discussed, but most observers have lost sight of the duality of the movement; the skyscraper slowed the spread. It helped keep many people downtown and intensified the downtown congestion. Contemporary observers noted this, but in recent decades we have tended to forget it. Burlingame, for example, said:

It is evident that the skyscraper and all the vertical congestion of city business centers would have been impossible without the telephone. Whether, in the future, with its new capacities, it will move to destroy the city it helped to build is a question for prophets rather than historians.[35]

Burlingame, before World War II, already sensed that things were changing. The flight from downtown was already perceptible enough to be noted as a qualification to his description of the process of concentration; both processes have taken place at once throughout the era of the telephone. The telephone is a facilitator used by people with opposite purposes; so we saw it with crime, and so it is here, too. It served communication needs despite either the obstacle of congested verticality or the obstacle of distance; the magnitude of the opposed effects may differ from time to time, and with it the net effect. At an early stage the telephone helped dissolve the solid knots of traditional business neighborhoods and helped create the great new downtowns; but at a later stage, it helped disperse those downtowns to new suburban business and shopping centers.

The telephone contributed in some further ways to downtown concentration in the early years—we have forgotten how bad urban mail service was. The interurban mails worked reasonably well, but a letter across town might take a week to arrive. Given the miserable state of intracity communication, the telephone met a genuine need for those who conducted business within the city.[36]

The telephone also contributed to urban concentration in the early days because the company was a supporter of zoning. The reasons were similar to ones motivating cablecasters today. Cablecasters are inclined to string CATV through comfortable middle-class neighborhoods where houses are fairly close together but in which utility lines do not have to go underground; they get their highest rate of penetration at the lowest price that way. Only under the impact of regulation do they cover a city completely.

The situation of early telephone systems was in some ways similar, though in some ways different. The Bell System strongly adhered to universal service as the goal, yet economics also favored first pushing into neighborhoods where there would be most businesses. At the beginning, when telephone systems first graduated from the renting of private lines between a factory and its office into providing a community system on a switchboard, many persons in a particular business or profession tended to be signed up fairly simultaneously. In some of the New England towns, the physicians made up a large proportion of the early subscribers. In London, there were few physicians but many solicitors. Eventually the subscribers became more diversified, yet there was still a tendency for customers to be drawn from certain segments of the population until penetration became more or less universal. Telephone companies, therefore, found it in their interest to have stable, well-defined neighborhoods in cities in which they were laying trunks and locating central offices. Shifting and deteriorating neighborhoods were not good for business.

Zoning of a city helped in planning for future services, so the phone companies (along with other utilities) became supporters of the zoning movement.[37] The Department of Commerce's zoning primer of 1923 stated that "expensive public services are maintained at great waste in order to get through the blighted districts to the more distant and fashionable locations." In the initial development of phone service it was economic to avoid blighted neighborhoods. With a bluntness that reflects the times and would be unthinkable today, Smith and Campbell said:

It should not be taken for granted that this satisfies the requirements unless there is at least one telephone for every eight inhabitants in an average American city, in which practically everyone is white. Where a large portion of the population belongs to the negro race, or a considerable portion of the population is made up of very poor workers in factories, the requirements will be less. In some cities one telephone to fifteen inhabitants is all that can be expected.[38]

Zoning, along with other efforts at urban planning, became popular around the turn of the century. After the Chicago fire in 1871, building codes were enacted (around 1890) with explicit provisions for fireproofing.[39] Codes dealt with allowable building heights and the location of tall buildings in the city. The idea of a planned city was contained in such books as Robinson's *Town and City* (1901) and Ebenezer Howard's *The Garden City*; zoning actually began in New York in 1916. In the intervening years the phone companies were one of the main sources of information fed into the new urban plans. We have already noted that AT&T urged each local phone company to do a development study to arrive at a Fundamental Plan. To do this they collected large amounts of neighborhood data on the population trends in the city, its businesses, and its neighborhoods; the telephone was used as a device for conducting the research. "The most direct means of approaching citizens on the planning issue was reported in Los Angeles where a battery of phone girls called everyone in the city to secure reactions, while mailing an explanatory folder."[40]

Part of a telephone development study was a "house count" in which the classes of buildings and their uses were determined.[41] In general, the urban reformers and zoning planners received good cooperation from the phone companies and derived much of their data from phone company research.

On a few points, however, city authorities and telephone interests were sometimes at odds. Zoning was often used to prevent the construction of the tall buildings that were heavy users of telephones. Cities also tried to prevent the growth of suburbs outside their boundaries by prohibiting utilities from extending their services beyond city limits. That, if enforceable, would have been a particularly severe restriction on the telephone system because its whole function required that it be interconnected.[42] Robinson noted in 1901 the connection between telephone communications and family cohesion when a family moves to the suburbs.

Thus we find many relationships between the development of the telephone system and the quality of urban life; strikingly, the relationships change with time and with the level of telephone penetration. The same device at one stage contributed to the growth of the great downtowns and at a later stage to suburban migration. The same device, when it was scarce, served to accentuate the structure of differentiated neighborhoods. When it became a facility available to all, however, it reduced the role of the geographic neighborhood. A technology assessment of the device would be misleading unless it included an assessment of the device available in specified numbers.

## THE RECORD OF FORESIGHT AND HINDSIGHT

We may distinguish between those who made technical or business predictions and those who made predictions about the telephone's social impact. Some men did both (Gilfillan, for example), but a separation according to principle emphasis will be useful.

### Technical and Business Forecasters
In our research we have looked at the business predictions of Bell, Vail, Western Electric, Elisha Grey, Edison, Hammond Hayes, chief engineer from 1886–1907, independent phone companies, John Carty, chief engineer after 1907, J. E. Otterson, general commercial manager of Western Electric and later president of ERPI, AT&T's motion picture subsidiary, and S. C. Gilfillan, a sociologist and historian of technology.

Bell's excellent predictions about the evolution of the switchboard and central office, automatic switching, the development of long-distance service, underground cables, and universal penetration have already been noted. Vail, too, was remarkably prescient. While Bell made his predictions through either scientific logic or visionary insight, Vail predicted goals to be fulfilled and made his dream of universality happen. While on most issues his sense of strategy was keen, he made some errors, as we judge with benefit of hindsight. When Vail and Carty took over in 1907, they shelved the handset (or French phone) which the company had begun to install. In addition, AT&T moved slowly on the automatic exchange because of its massive investment in manual systems. Independents, such as the Home Telephone Company, and European systems (because they were growing more slowly) could adopt automatic switching more easily and earlier. Finally, Vail underestimated the potential of wireless telegraphy, but Carty convinced him of its importance by 1911. Yet none of these items gainsay his remarkable insight into how to build the business so that it worked.

Western Union and Elisha Grey, on the other hand, demonstrate the drawbacks of perseveration on established perspectives. Grey could probably have invented the telephone at least a year before Bell had he realized its commercial value.[43] In 1875 in a letter to his patent lawyer Grey wrote: "Bell seems to be spending all his energies on [the] talking telegraph. While this is very interesting scientifically, it has no commercial value at present, for they can do more business over a line by methods already in use than by that system."[44] In another letter after the Philadelphia Centennial Exposition, he added: "Of course it may, if perfected, have a certain value as a speaking tube. . . . This is the verdict of a practical telegraph man."[45]

Hounshell points to a number of reasons for Grey's misjudgment: his extensive experience in telegraphy, his association with and his respect for the leaders of the telegraph industry. He committed the fallacy of historical analogy. He and Western Union thought of what the telephone could do to extend the existing telegraph system. Bell was a speech expert who approached the problems of telecommunication from the outside.

Hounshell illustrates the point by the story of one Western Union officer who anticipated the phone's use for transmitting speech between telegraphers. He could not visualize the elimination of the traditional telegrapher.[46]

Daniel Boorstin argues that Thomas Edison committed the same fallacy with the phonograph: he invented it as a repeater because he believed that few people would be able to afford their own telephone. His notion was that offices (such as telegraph offices) would use it to record spoken messages that would be transmitted by phone to a recorder at another office where the addressee could hear it.[47] Partly as a result of this misperception, it took Edison fifteen years to realize the entertainment potential of the phonograph.

Clearly capitalist investors or other market-oriented technical men can forecast badly, too. The combination of technical understanding and appreciation of the market may be a necessary condition for good assessment of a new technology. But capitalist investors fail more often than they succeed. History focuses on successes; failure is treated as a kind of environmental wasteland too dreary to gaze upon. And so innovators who followed the wrong path lost their money and have been forgotten.

Yet the melding of technical and economic considerations has been a key to whatever outstanding successes of forecasting have occurred. Where those requisites have been present but foresight has still failed, a common and easily identified reason has been a lack of imagination about the range of possible change and perseverance in an established way of doing things.[48]

### Forecasts of Social Consequences

The record of social forecasting is far less impressive. To evaluate the forecast record on the telephone's social effects, we have looked at writings by journalists, historians, and sociologists. The first conclusion is that they have had very little to say on the subject.

We reviewed the indexes of some histories of technology. The result presented in Table 1 shows that attention to transportation has been much greater than attention to communication. Among the references to communications, attention to the telephone is salient only in histories of American technology.

Table 2 presents a similar analysis of the number of paragraphs devoted to different technologies in some sociological
treatises, mostly from the 1930s. The distribution reflects the
bias of social scientists toward the present in contrast to the historians' focus on the past. The technologies the sociologists
attended to most were the ones that were relatively new at the
period when they wrote: radio, television, automobile, and aviation; the telegraph had receded into the past. But note that the
telephone gets less attention than railroads, even though it is
younger. Figure 1 illustrates the greater focus on transportation
than communication, age being held constant.

Quantity, however, is not the key thing; more disturbing is
that historians, social scientists, journalists, and current commentators have given us very few significant forecasts or analy-
ses on the telephone's social effects. What few there are tend to
add the telephone to a list of forces that are all asserted to work
in the same way. Sociological writing of the 1920s and 1930s
was heavily shaped by a grand conception (which came to
America from Germany about 1908) of the decline of "gemein-
schaft" and the rise of "gesellschaft": the decline of the traditional primary group and the growth of a complex society
dominated by impersonal relations. In writing about any technology, the impulse was to make it fit that model. Writers discussed the growth of the city, the breakdown of the family, or
some other aspect of this grand historical process, noted how
the automobile or some other technical change led to it, and
added (as an aside) "along with such other innovations as the
telephone, the telegraph, or you-name-it."

Such metatheories without detailed cold-blooded study of the
historical facts have obscured the telephone's real history. Its effect has not necessarily been in any single direction, nor in the
same direction as other devices such as the mass media, the
telegraph, or the automobile.[49] The telephone is a device with
subtle and manifold effects which cannot be well guessed a
priori. There are, in our society, significant problems of privacy,
alienation, crime, and urban environment. In no instance is it
clear what the telephone's net effect has been on any of these.

**Table 1**
Relative Concentration on Various Technologies in History of Technology Literature

| Reference | Number of References to Different Technologies:* | | | | | |
| --- | --- | --- | --- | --- | --- | --- |
| | Telephone | Telegraph | Radio | Television | Railroad | Automobile |
| **I Early Technology (1600–1900):** | (9) | (27) | (—) | (—) | (123) | (40) |
| Kranzburg Technology in Western-Civilization Vol. 1 (1976) (1600–1900) | 7 | 14 | — | — | 62 | 6 |
| Singer, ed. A History of Technology (1968) (1850–1900) | 2 | 13 | — | — | 61 | 34 |
| **II General Technology Histories (Ancient to 1950 +):** | (8) | (13) | (17) | (6) | (71) | (56) |
| Ferguson Bibliography of the History of Technology (1968)** | 3 | 6 | 3 | — | 33 | 16 |
| Lilley Men, Machines, and History (1965) | 4 | 3 | 13 | 5 | 18 | 18 |
| Armytage A Social History of Engineering (1961) | 1 | 4 | 1 | 1 | 20 | 22 |
| **III American Technology (1700 + to 1935 +):** | (77) | (60) | (96) | (47) | (231) | (152) |
| Boorstin The Americans: The Democratic Experience (1973) (1860–1970) | 19 | 10 | 41 | 36 | 64 | 50 |
| Oliver History of American Technology (1956) (1730–1950) | 20 | 24 | 26 | 7 | 85 | 28 |
| Allen The Big Change (1952) (1900–1950) | 1 | 1 | 3 | 1 | 10 | 15 |
| Burlingame Engines of Democracy (1940) (1865–1935) | 37 | 25 | 26 | 3 | 72 | 59 |

| | | | | | | |
|---|---|---|---|---|---|---|
| Grand Totals | 94 | 100 | 113 | 53 | 425 | 258 |
| Totals II & III only | 85 | 73 | 113 | 53 | 302 | 218 |

*Numbers indicate number of pages on which technology is mentioned plus number of different subtopics mentioned in index.

**Numbers for this book refer only to the number of documents on each technology cited in this bibliography.

**Table 2**
Relative Concentration on Various Technologies in Social Impact and Trend Literature of 1930s and 1940s*

| Reference | Telegraph | Telephone | Radio | Television | Railroad | Automobile | Aviation |
|---|---|---|---|---|---|---|---|
| Mumford *Technics and Civilization* (1934) | 2 | 4 | 2 | 1 | 7 | 6 | 2 |
| Leonard *Tools for Tomorrow*** (1935) | 5 | 5 | 6 | 3 | 3 | 6 | 10 |
| Gilfillan "Social Effects of Invention"*** (1937) | 1 | 3 | 17 | 19 | 2 | 6 | 15 |
| Ogburn *Machines and Tomorrow's World* (1938) | 1 | 6 | 7 | 4 | 7 | 19 | 2 |
| Roger *Technology and Society*** (1941) | 1 | 1 | 31 | 4 | 10 | 4 | 6 |
| Ogburn *Technology and the Changing Family* (1955) | — | 2 | 6 | 10 | — | 1 | 1 |
| Totals | 10 | 21 | 69 | 41 | 29 | 42 | 36 |

* Number of paragraphs related to technology except where noted.
** Number of references to technology in index, for these books only.
*** From Subcommittee on Technology of the National Resources Committee report, *Technological Trends and National Policy*, Ogburn Chairman, 1938.

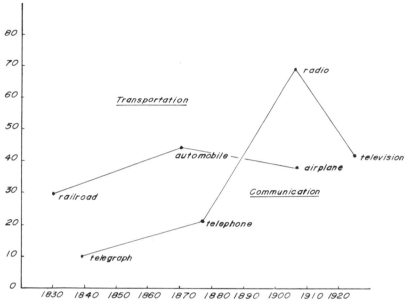

**Figure 1**
Relationship between date invented and attention in the social impact literature of the 30s for transportation and communication technology.

We should not be too harsh about forecasters, however; we have the benefit of hindsight now, yet it is not much easier to answer the questions about the past than about the present or future. Postdiction is almost as hard as prediction; aftcasts almost as hard as forecasts. What would have happened to American cities if the telephone had never been invented? Would there, perhaps, have been an enormous proliferation of teletypewriters to serve some of the same purposes? If so, would our cities be bigger or smaller than they are today, more densely settled or more dispersed? It is not much easier to answer the "what if" questions of history than the "what if" questions about the future.

## NOTES

1. Marion May Dilts, *The Telephone in a Changing World* (New York: Longman's Green, 1941), p. 11.
2. Ibid., p. 16.

3. Ibid., pp. 188ff.

4. Thomas P. Hughes (according to *Science*, Nov. 21, 1975, p. 763) notes that Edison was not only a "consummate inventor" but a "complete capitalist." His notebooks "while devising the nations first public power system . . . show on every other page calculations of the system's market potential, the price charged for competing gas illumination, the cost of copper wiring, and other entrepreneurial concerns."

5. That would have been particularly so, as Cherry has noted, without the switchboard. As Arthur Vaughan Abbott calculated in 1894 (*Bulletin of the University of Wisconsin*, Engineering Series, vol. 1, no. 4, p. 70), a system for 10,000 subscribers (the approximate number than in New York or Chicago) with a separate line between each pair of subscribers would have taken an underground conduit a yard square, or a pole above ground 1,000 feet high.

6. Dilts, *The Telephone in a Changing World*, p. 28.

7. In 1904, the AT&T *Annual Report* asserted: "As a general principle, it seems perfectly certain that it will always be the case that the larger and more densely populated the community, the higher must be the standard of rates for the comprehensive service required for that community. Not only the investment and the cost of operation, but the general difficulty of doing business which can only be overcome by enlarged expenditure, increase in passing from smaller towns and cities to the larger."

8. A further incentive for this change was the growing percentage of AT&T ownership of local phone companies in large cities such as New York. As long as AT&T's revenue was largely from franchise fees for the use of Bell equipment, the company stood to gain from the growth of local companies but not directly from their profitability. As their equity position grew, their direct concern for profitability did, too. Faced with competition from price-cutting independents (some using message rates), it became important to meet the competition at the low end while keeping revenues up. A pricing system that segmented the market maintained earnings. A message charge also served to discourage neighbors' using a telephone and thus increased the likelihood of their getting their own.

9. This figure, which was suggested to local companies for use in the Fundamental Plan, is a rather crude average. More detailed planning in AT&T, of course, used a variety of figures for different types of equipment.

10. Arthur B. Smith and William L. Campbell, *Automatic Telephony*, (New York: McGraw-Hill, 1915), p. 379.

11. In 1912, Arnold Bennett, in a series in *Harper's Monthly*, gave his impressions of the United States. He begins the fourth (vol. 125, July, pp. 191–192) commenting: "What strikes and frightens the backward European almost as much as anything in the United States is the efficiency and fearful universality of the telephone. Just as I think of the big cities as agglomerations pierced everywhere by elevator-shafts full

of movement, so I think of them as being threaded under pavements and over roofs and between floors and ceilings and between walls, by millions upon millions of live filaments that unite all the privacies of the organism—and destroy them in order to make one immense publicity. I do not mean that Europe has failed to adopt the telephone, nor that in Europe there are no hotels with the dreadful curse of an active telephone in every room. But I do mean that the European telephone is a toy, and a somewhat clumsy one, compared to the seriousness of the American telephone. Many otherwise highly civilized Europeans are as timid in addressing a telephone as they would be in addressing a royal sovereign. The average European middle-class householder still speaks of his telephone, if he has one, in the same falsely casual tone as the corresponding American is liable to speak of his motor car. . . . Is it possible that you have been in the United States a month without understanding that the United States is primarily nothing but a vast congeries of telephone cabins?''

12. Dilts, *The Telephone in a Changing World*, p. 4.

13. Herbert N. Casson, "The Telephone As It Is Today," *World's Work*, vol. 19 (1910), p. 12775. Vail understood the strategic advantage of controlling the long lines. From 1881 until 1897 the company issued a national telephone directory, like the national index directory today. Eventually it had to be abandoned for it became too big.

14. Roger Burlingame, *Engines of Democracy* (New York: Charles Scribner, 1940), pp. 118ff.

15. Casson, in *World's Work*, p. 12776. Casson's tale simplifies slightly in that the Boston to New York line was not an instant success. It was initially noisy and had the severe problems of all long-distance lines until adequate repeaters were developed. However, its very existence was a triumph, and quality of service gradually improved.

16. An article in *Current Literature*, vol. 50, May 1911, p. 504, on "The Immediate Future of the Long-Distance Telephone," reports the laying of an experimental submarine telephone cable between France and England. Conversations between London and the whole European continent are, it says, now possible, and conversations are now possible over up to 1700 miles of cable. The article looks forward to the time, far in the future, when the whole globe will be linked together.

17. Chapter 15 gives the data on the decreasing sensitivity of phone charges to distance.

18. It is worth noting that the same points can be made about Marconi and the history of wireless telegraphy. After Herzian waves were discovered in 1888, a number of scientists recognized that they could be used, as electrical transmissions over wires were already being used, for communications devices. Marconi's important insight, aside from his entrepreneurial ones, was his successfully demonstrated conviction that those waves could go long distances. His great triumph was his trans-Atlantic transmission.

19. Alexander Bain outlined the principles of telegraphic transmission of pictures in a British patent in 1843.

20. Robert V. Bruce, *Bell* (Boston: Little Brown), p. 242.

21. Dilts, *The Telephone in a Changing World*, p. 22.

22. *World's Work*, vol. 20, May 1910, pp. 12,916–12,917.

23. *The Independent*, vol. 73, October 17, 1912, pp. 886–891.

24. Proposed Report, Telephone Investigation, FCC, USGPO, 1938, pp. 238–239.

25. Vol. 43, May 1907, pp. 50–57.

26. Herbert Casson, "The Social Value of the Telephone," *The Independent*, vol. 71, October 26, 1911, p. 903. See also Burlingame, *Engines of Democracy*, p. 125, and Dilts, *The Telephone in a Changing World*, pp. 82, 95–96.

27. Dilts, *The Telephone in a Changing World*, p. 177. She quotes the *Worcester Telegram* on the capture of a gunman: "A bullet killed him, but radio and teletype and telephone had already doomed him" (p. 178).

28. V. A. Leonard, *Police Communication Systems*, (Berkeley, Calif.: U. of Calif. Press, 1938).

29. Ibid., pp. 6–7.

30. Ibid., p. 10.

31. Herbert Casson, "The Future of the Telephone," *World's Work*, May 1910, pp. 1908–1913.

32. In John Kimberly Mumford, "This Land of Opportunity, The Nerve Center of Business," *Harper's Weekly*, vol. 52, August 1, 1908, p. 23. The same point was made in the trade journal *Telephony*, vol. 4, no. 2, 1902.

33. Charles N. Glaab and A. Theodore Brown, *A History of Urban America* (New York: Macmillan, 1967), pp. 144–145.

34. Ibid., p. 280.

35. *Engines of Democracy*, p. 96.

36. We have noted several times how much worse European phone service was than American. Conversely, the local mail service was often very much better. In many big European cities express letters could be sent rapidly and reliably. There may have been some tradeoff between alternative communication devices for achieving the same goals.

37. Telephone, electric light, gas, and trolley companies report that zoning is making it possible for them to eliminate much of their guesswork as to what services they must provide, said John Noland's book *City Planning* shortly after zoning was enacted in New York. Hubbard and Hubbard wrote: "The utility companies as a rule may be counted favorable to zoning. The general attitude of the telephone companies has been expressed in favor of the stability brought about by zoning."

38. *Automatic Telephony*, p. 379.

39. In the early years there was considerable concern about telephone wires and safety. The *American Architect and Building News* had a section in each issue on new inventions. It did not note the telephone in

1876, but by 1881 there had been ten references to the telephone in the magazine; four of these concerned safety. There was particular concern about the proliferation of overhead wires.

40. H. V. and T. K. Hubbard, *Your Cities Today and Tomorrow* (Cambridge, Mass: Harvard Univ. Press, 1929), p. 93.

41. Smith and Campbell, *Automatic Telephony*, p. 379.

42. However, occasionally phone companies took refuge in such restrictions. In *Young vs. Southwestern Telegraph and Telephone Co.* 192 F. 200, 1912, the Arkansas Circuit Court ruled that the phone company was not discriminating when it refused to construct a line beyond city limits.

43. David A. Hounshell, "Elisha Grey and the Telephone: On the Disadvantages of Being an Expert," *Technology and Culture*, vol. 16, no. 2, p. 159.

44. Ibid., p. 152.

45. Ibid., p. 157.

46. Ibid., p. 145.

47. Daniel Boorstin, *The Americans: The Democratic Experience* (New York: Random House, 1973), p. 379.

48. When Vail left the company in 1887, the organization of engineering research fell to Hammond Hayes. Between 1887 and the management reorganization of 1907, the Bell company conducted little fundamental research. Rather, research activities focused upon improvements of long-distance service and minor incremental improvements in apparatus. Sources outside the company contributed the significant breakthroughs in telephone: the automatic exchange system by Strowger in 1889, the loaded coil by Pupin in 1900, and the vacuum tube by De Forest in 1907.

Vail's return in 1907 signaled a change in that research philosophy. The obvious frontier was "wireless" communication, made possible by the De Forest patents. Research in that area "would not only react most favorably upon our service where wires are used, but might put us in a position of control with respect to the art of wireless telephony should it turn out to be a factor of importance." (Quoted in N. R. Danielson, *AT&T: The Story of Industrial Conquest* [New York: Vanguard, 1939], pp. 104–105.) The Bell interests rapidly acquired and developed wireless technology and moved on, in Vail's phrase, "to occupy the field" of telephonic research. That is, by 1910 the company was committed to the internal development of both fundamental research and of incremental engineering improvements. Today, the Bell Labs still follow the basic mandate and organization that followed from Vail's management.

49. See Chapters 5 and 8 for a strong case for the difference between the social effects of the telephone and the mass media.

**APPENDIX**

Kensington, March 25, 1878.

To the capitalists of the Electric Telephone Company:

Gentlemen—It has been suggested that at this, our first meeting, I should lay before you a few ideas, concerning the future of the electric telephone, together with any suggestions that occur to me in regard to the best mode of introducing the instrument to the public.

The telephone may be briefly described as an electrical contrivance for reproducing, in distant places, the tones and articulations of a speaker's voice, so that conversation can be carried on by word of mouth between persons in different rooms, in different streets, or in different towns.

The great advantage it possesses over every other form of electrical apparatus consists in the fact that it requires no skill to operate the instrument. All other telegraphic machines produce signals which require to be translated by experts, and such instruments are therefore extremely limited in their application, but the telephone actually speaks, and for this reason it can be utilized for nearly every purpose for which speech is employed.

At the present time we have a perfect network of gas pipes and water pipes throughout our large cities. We have main pipes laid under the streets communicating by side pipes with the various dwellings, enabling the members to draw their supplies of gas and water from a common source.

In a similar manner it is conceivable that cables of telephone wires could be laid under ground, or suspended overhead, communicating by branch wires with private dwellings, counting houses, shops, manufactories, etc., uniting them through the main cable with a central office where the wire could be connected as desired, establishing direct communication between any two places in the city. Such a plan as this, though impracticable at the present moment, will, I firmly believe, be the outcome of the introduction of the telephone to the public. Not only so, but I believe in the future wires will unite the head offices of telephone companies in different cities, and a man in one part of the country may communicate by word of mouth with another in a distant place.

In regard to other present uses for the telephone, the instrument can be supplied so cheaply as to compete on favorable terms with speaking tubes, bells and annunciators, as a means of communication between different parts of the house. This seems to be a very favorable application of the telephone, not only on account of the large number of telephones that would be wanted, but because it would lead eventually to the plan of intercommunication referred to above. I would therefore recommend that special arrangements be made for the introduction of the telephone into hotels and private buildings in place of the speaking tubes and annunciators, at present employed. Telephones sold for this purpose could be stamped or numbered in such a way as to distinguish them from those employed for business purposes, and an agreement could be signed by the purchaser that the telephones should become forfeited to the company if used for other purposes than those specified in the agreement.

It is probable that such a use of the telephone would speedily become popular, and that as the public became accustomed to the telephone in their houses they would recognize the advantage of a system of intercommunication.

In conclusion, I would say that it seems to me that the telephone should immediately be brought prominently before the public, as a means of communication between bankers, merchants, manufacturers, wholesale and retail dealers, dock companies, water companies, police offices, fire stations, newspaper offices, hospitals and public buildings and for use in railway offices, in mines and other operations.

Although there is a great field for the telephone in the immediate present, I believe there is still greater in the future.

By bearing in mind the great object to be ultimately achieved, I believe that the telephone company cannot only secure for itself a business of the most remunerative kind, but also benefit the public in a way that has never been previously attempted.

I am, gentlemen, your obedient servant,

Alexander Graham Bell.

# Editor's Comment

The chapters thus far have looked at the earliest days of the telephone. They have asked what early imprinting imposed itself on the system we know today, what presumptions the founders initially had, what choices the technology itself offered, and what decisions society eventually made.

The author of the next chapter has devoted his life to telephone research, most of it in Bell Labs. It scans the whole 100 years and gives an overview of the vast, complex system existing now. Pierce presents the basics of telephone growth and usage, the characteristics of the present system's functions, and the system's sensitivity to abuse. As he notes, the telephone system is the most complex machine yet created by man; it is remarkable how well it works. There are places in the world, experience reminds us, where it does not work well. The telephone can be degraded by a variety of misuses; Pierce presents his conclusions about the conditions of its successful or degraded operation.

# 7

## The Telephone and Society in the Past 100 Years

### John R. Pierce

## INTRODUCTION

In the 100 years that the telephone has been with us, it has directly and profoundly affected man's daily life. It has also had a large impact on the arts and on the sciences. In rendering a new sort of service, it has given rise to a new sort of industry—a complex, high-technology system whose actual operation, in the sense of the origination and completion of calls, is carried out directly by the users.

In supplying telephone service, different nations have taken various paths with varied success and impact on their citizens; all telephone systems, however, interconnect. The new service and the complex system that the telephone's invention has brought into being are resources that can be and have been exploited in many ways unforeseen by its originators.

I propose to say something about all these aspects of the telephone and society: its novel nature, its exploitation, the influence of telephone systems, and the impact of telephony in our individual lives.

In 1876, when Bell invented the telephone, there was no need for it. Society did very well without it, just as it did without electric power or automobiles. Today the telephone has become more than a luxury or a convenience; it has become a basic part of man's world.

The telephone's power is not that of an idea, a creed or an ideology; it is the power of science and technology to enlarge man's life. This enlargement began 100 years ago through one man, Alexander Graham Bell. Like many Americans, Bell was born in another country, Scotland. A scientist of speech, his interest in electricity was pragmatic. He tried to improve telegraphy and he invented telephony. James Clerk Maxwell's Rede lecture of 1878 put this very well:

Now, Prof. Graham Bell, the inventor of the telephone, is not an electrician who has found out how to make a tin plate speak, but a speaker who, to gain his private ends, has become an electrician. He is the son of a very remarkable man, Alexander Melville Bell, author of a book called *Visible Speech*. . . .

The inventor of the telephone was thus prepared, by early training in the practical analysis of the elements of speech, to associate whatever scientific knowledge he might afterwards acquire with those elementary sensations and actions, which each of us must learn from himself, because they lie too deep within us to be described to others. . . .

I shall, therefore, consider the telephone as a material symbol of the widely separated departments of human knowledge, the cultivation of which has led, by as many converging paths, to the invention of this instrument by Professor Graham Bell.

For whatever may be said about the importance of aiming at depth rather than width in our studies, and however strong the demand of the present age may be for specialists, there will always be work, not only for those who build up particular sciences and write monographs on them, but for those who open up such communications between the different groups of builders as will facilitate a healthy interaction between them.

Bell brought together different parts of knowledge and produced a workable telephone; once this was accomplished, he sought to exploit it. After he patented it, he showed it at the Philadelphia Centennial Exposition; a year later in association with others, he formed a telephone company.

From this man and these early acts has grown all that I write about.

## STATISTICS AND THEIR INTERPRETATION

Happily, many statistics concerning the telephone and its usage are available for the entire history of telephony. In his chapter "The Telephone and the Uses of Time," Martin Mayer has made excellent use of statistical data to elucidate many details of telephone usage. The statistics I use are largely aggregated; they explain overall usage in times and at places where usage may vary widely among the population (many having no telephones). The message conveyed by the statistics may not be entirely clear, but our considerations should begin with incontrovertible numerical data rather than preconceived ideas.

When was telephony introduced and how rapidly did it grow? Figure 1 shows telephones per 100,000 people as a function of time. We see that telephony grew rapidly from the year of its invention. This is characteristic of good, useful, inexpensive new products or services that work well from the start. The history of Xerography and hand calculators must be similar.

The rate of growth of telephony increased abruptly in the mid 1890s since Bell's original patent expired in 1893, and it was possible for anyone to make telephone equipment and sell service. Many competing companies were formed. Some served cities that already had Bell exchanges, but many provided service where none was available before. Partially, the expansion of telephony near the turn of the century must also have been associated with the new potential of commercial long-distance service (Table 1).

Telephony's spurt of growth in the 1890s is clearly associated with non-Bell capital and enterprise. Figure 2 shows the fraction

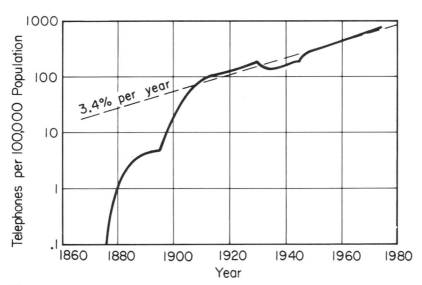

**Figure 1**

Telephones per 100,000 population. (Data from *Historical Statistics of the United States: Colonial Times to 1957*, U.S. Department of Commerce.)

**Table 1**

Initiation of Commercial Long-Distance Service

| Date | Cities |
|------|--------|
| 1881 | Boston–Salem |
| 1884 | New York–Boston |
| 1892 | New York–Chicago |
| 1893 | Boston–Chicago |
| 1893 | New York–Cincinnati |
| 1895 | Chicago–Nashville |
| 1896 | Kansas City–Omaha |
| 1896 | New York–St. Louis |
| 1897 | New York–Charleston |
| 1897 | New York–Minneapolis |
| 1897 | New York–Norfolk, Virginia |
| 1898 | New York–Kansas City |

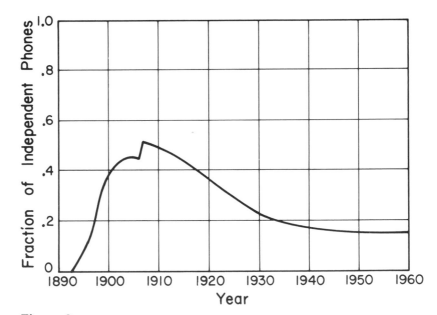

**Figure 2**

Fraction of independent phones. (Data from *Historical Statistics; Statistical Abstracts, 1974*, U.S. Department of Commerce.)

of independent (non-Bell) phones as a function of time. The initial growth of independents was rapid, peaked around 1910, slowly fell off, and rose again somewhat after 1960.

We have seen that long-distance service developed gradually after 1881. When the independents began operation in 1894, many phone systems were isolated. Independents (Figure 3) first interconnected with Bell System phones in 1898; this fraction increased gradually until 1907. Initially, Bell sometimes refused to interconnect and give competitive independents the advantage of serving Bell customers. In 1908, Theodore N. Vail, president of AT&T, announced a policy of eliminating dual service in one city by either buying the competing service or abandoning the field to it. A rapid rise in interconnection followed, and also in the fraction of independent phones between 1907 and 1908 (Figure 2).

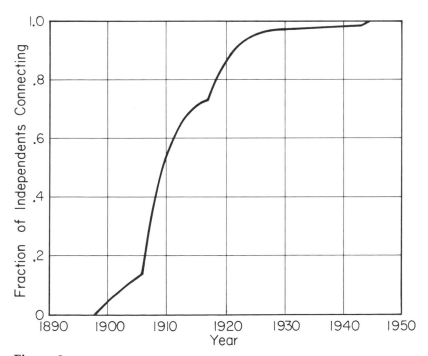

**Figure 3**
Fraction of independent phone companies that connect with Bell. (Data from *Historical Statistics; Statistical Abstracts*.)

In 1913, AT&T agreed to provide long-distance connection to any independent and not to purchase any more independents (except when approved by the Interstate Commerce Commission), but competition of independents not interconnected with Bell in the same service areas nevertheless persisted for some years. However, the next spurt of interconnections did not come until 1918, and it may have resulted more from World War I than from phone company policies. From July 14, 1918, to July 30, 1919, the country's telephone systems were operated under the Post Office Department by direction of President Wilson; after 1918, interconnection of independents rose smoothly.

The chief growth of independents preceded the interconnection agreement of 1913. It seems clear that the provision of telephone service did not prove to be a get-rich-quick business; rather, telephony became an increasingly challenging task that required much capital and technical know-how and was beset by many problems. Today, independents supply a minority of service at a price no less than Bell's, but the service is sometimes inferior. The fraction of growth in independent and Bell phones is now chiefly dictated by growth in the geographical areas served by each; while Bell serves most of the population, the independents control over half the geographical area of the country; this area's population is currently growing at a more rapid rate than that of the Bell area.

Let us return to Figure 1, which gives the number of phones as a function of time. As indicated by the dashed line, telephones per person have grown exponentially at a rate of about 3.4 percent a year since 1910. This growth was interrupted by the depression of 1929 and perturbed by World War II, but it has been remarkably steady. Most homes and all businesses now have telephones. Today the growth in number of telephones per person must be attributed to more phones per home and per employee.

We have seen that telephony has grown steadily since its inception. What has this done to other modes of communication?

Is telephony replacing travel? No. Very roughly, in recent years the number of telephone calls and the number of air miles flown have increased at about the same rate, and the number of car miles traveled about half as fast. Undoubtedly, a telephone call sometimes substitutes for a trip, but more and faster communication tends to engender widespread associations and activities that result in trips.

What about books as a means of communication? Is the telephone somehow affecting publication? No. Figure 4 shows this clearly.

Has the telephone call displaced the letter? The number of telephone calls per person per year (Figure 5) is over three times as great as the number of first-class plus airmail letters per person per year, and the recent rate of growth of telephone calls is over twice as great as the rate of growth of letters. Letters per person still appear to be growing, but this may change slowly or abruptly; it seems plausible that telephony together with data transmission will eventually displace much mail.

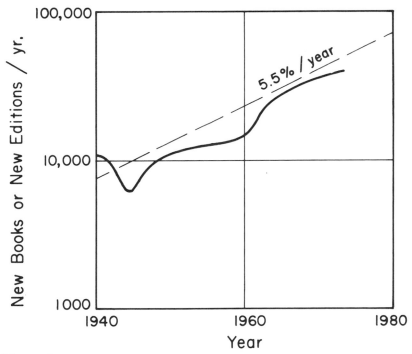

**Figure 4**
New books or new editions published in various years. (Data from *The World's Telephones*[AT&T]; *The American Almanac, 1972.*)

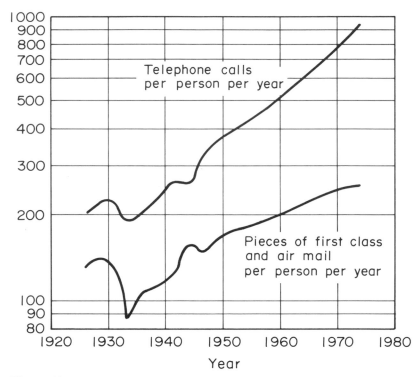

**Figure 5**
Domestic telegraph messages per year. (Data from *The American Almanac, 1972.*)

The telephone call has definitely displaced the telegram (Figure 6); and the yearly number of telegraph messages has declined steadily since 1945, while telephone calls have steadily increased.

Telephone calls are immediate and convenient, but text is advantageous for many purposes; indeed, data communication is growing. Telegraph service is falling in volume partly because it no longer delivers text (messages are telephoned) and partly because it is costly and increasingly inconvenient. I can remember when telegraph messengers did deliver text, and when office switches summoned telegraph messengers to pick up text to be telegraphed.

Telegraphy has failed primarily because it is labor-intensive in a world where the cost of labor rises continually; telephony is

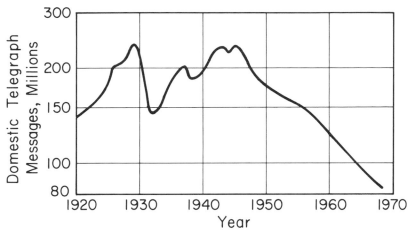

**Figure 6**
Domestic telegraph messages per year. (Data from *The American Almanac, 1972.*)

a combination of automation and do-it-yourself. The user dials, picks up the phone and replaces it. Except in credit-card and person-to-person calls, machinery does almost everything. Credit-card calls and person-to-person calls are necessarily expensive, and directory assistance is a great financial burden to telephone companies.

Unless our society changes fundamentally, labor-intensive services will decline in comparison with do-it-yourself and automated services. The burden of delivery continually increases postal cost; telegraphy has abandoned physical delivery, but an employee still has to transcribe the message for transmission and read the message to the recipient. Economic forces have worked to the disadvantage of mails and telegraphy, but data transmission (such as teletypewriter service) may increase when terminals become cheap enough. The traditional telegram is on the way out.

We have noted the importance of automation to the survival of telephony. In Figure 7 the number of telephone calls per Bell System employee per day is shown as a function of time. Up to about 1920 there is no consistent rise or fall; this is surprising. As the number of subscribers increases, it becomes more complex for one subscriber to reach another; thus, we might expect

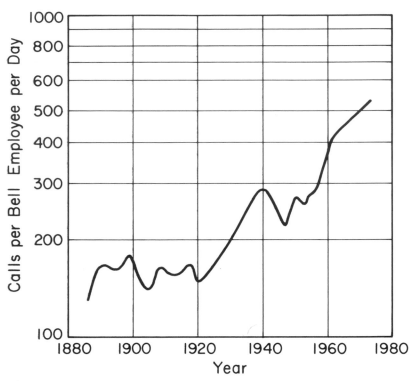

**Figure 7**
Calls per Bell employee per day. (Data from *Historical Statistics; Statistical Abstracts, 1974.*)

a decrease in calls per employee.

Automatic switching was invented in 1879 and tried in the Bell System in 1902, but the first large-scale switching system was not put into use in the Bell System until November of 1919. The introduction of automatic switching convincingly explains why the average rise in calls per employee has been about 2.5 percent per year and shows no sign of falling.

While rising manufacturing productivity is the source of America's affluence, steadily rising service productivity is atypical. A rise in productivity has been attained through an intense program of automation, not only of calling but of testing and maintenance as well.

While many telephone statistics are aggregated, separate figures are available on local, toll, and overseas calls. Calls per person per year in these categories are shown for recent years in Figure 8. Toll calls grow more rapidly than local calls, and overseas calls grow more rapidly than toll calls. This reflects a changing pattern of life and work. Families tend to disperse and children go to far universities, yet friends and relatives keep in touch by telephone. Commercial, scientific, and cultural relations are increasingly dispersed and increasingly international. Despite the intense nationalism characteristic of present governments, the scope of man's activities does not halt at national boundaries. This is reflected in the rapid rise of overseas calls, which in turn reflects improved technology.

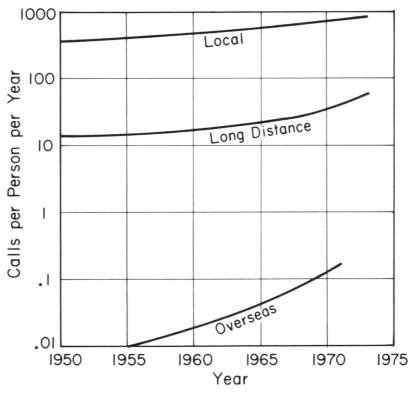

**Figure 8**
Overseas, long distance, and local calls per person per year. (Data from *The World's Telephones*.)

There was no commercial overseas telephony until radio
service was established in 1926, and consistently high-quality
service came only with the first trans-Atlantic cable in 1956;
trans-Atlantic calls rose rapidly thereafter. Since the first commercial telephone service by communication satellite in 1965,
the number of nations that can be reached by telephone has increased remarkably. The bulk of international telephone traffic,
however, is still among the highly developed nations, and the
part carried by satellite is dictated by government regulation
rather than by cost or quality of service.

The discussion of international telephone traffic leads naturally to the matter of telephony in various countries. In Figure 9,
the horizontal axis denotes gross national product in dollars per
year per person and the vertical axis denotes telephone calls per
person per year.

It is plausible that telephone usage in the United States,
Canada, and Sweden should be similar, because these three
countries have a high gross national product and very good telephone service. We also note that these countries have the only
telephone systems that own manufacturing as well as research
and operating facilities.

It is easy to see why telephone usage in France is low: service
is abominable. It is difficult to get a telephone, hard to get dial
tone, and hard to get the called party—even if one has a phone.
Telephone service in Italy is poor, too, as it is in many other
countries.

There must be other reasons, however, that calls per person
are so low in such nations as Switzerland, West Germany, and
the Netherlands. This may partly reflect a lack of aggressive promotion of service, but it also may be due to language problems.
Teletypewriter service for businesses is much more common in
continental Europe than it is in the United States; many people
who have an adequate reading knowledge of another language
cannot communicate satisfactorily in that language over the telephone. Perhaps the reliance on text in business carries over into
private life, but this is mere speculation. Whatever the cause,
telephony is not as prominent in the lives of many advanced
nations as it is in our own.

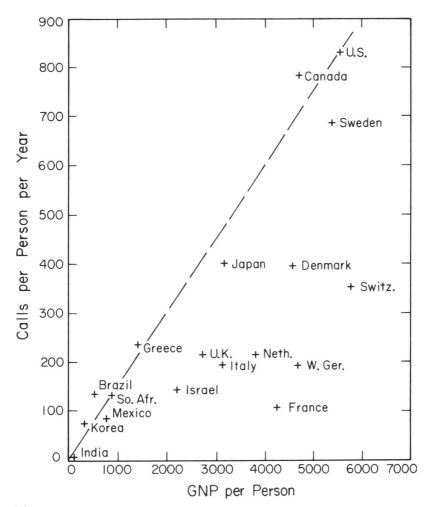

**Figure 9**
Comparison of per capita calls and GNP. (Data from *The World's Telephones; The World Almanac, 1975.*)

In terms of GNP (Figure 9), telephone usage is high in Korea, South Africa, Brazil, Mexico, and Greece. In Korea we may suspect American influence, but in all the countries cited, telephone usage is probably intense among a small segment of the population who consume most of the GNP. That is, average GNP and average telephone usage are both diluted by the same body of nonconsumers and nonusers. In the case of South Africa, the nonconsumers and nonusers are presumably black; in Greece, Brazil, Mexico, and Korea, they are presumably merely poor.

Here we conclude our examination of telephone statistics. What have we learned?

Telephony was attractive enough to find an immediate place in society (Figure 1); telephony expanded rapidly after the expiration of Bell's first patent and has had a remarkably steady growth since 1910. Independent companies played a great role (Figure 2) after 1895 and a lesser role in later years; the fraction of the market served by independents declined (Figures 2 and 3) as interconnection increased. A surge in interconnection followed (Figure 3), not Bell's 1913 agreement to interconnect with any independent, but rather America's entry into World War I when the government took over operation of telephones.

Telephony has grown with, rather than superseded, transportation. It has not cut into the publication of books (Figure 4), it is replacing first class and airmail only gradually (Figure 5), and has largely replaced the traditional telegram (Figure 6) although not the teletypewriter. The increase in calls per employee (Figure 7), attained through automation, has allowed telephony to flourish despite rising labor costs. Long-distance calls (Figure 8) are increasing more rapidly than local calls, and overseas calls more rapidly than either. Telephony does not play as large a part in the lives of peoples of some other nations (Figure 9), even technologically advanced nations, as it does in our lives.

We should keep these facts in mind as we consider various social aspects of telephony.

**TELEPHONE USAGE**

Our knowledge of telephony's role in our lives and work has largely been common and anecdotal. In their papers, Martin

Mayer, A. A. L. Reid, John Brooks, and Jean Gottmann have added considerably to our knowledge of telephone usage. It may be worthwhile, however, to note a few aspects of the telephone and of telephone usage that sometimes escape attention.

The telephone provides quick and private access to others at a time of one's choice. We may note that it seems quicker and easier to reach individuals in the Soviet Union by phone, from outside or inside the borders, than in any other way. The telephone is used widely among friends for gossiping, verbal amusement, or reassurance. The picture of the housewife, phone propped on shoulder, gossiping while she irons, is drawn from life, but many people gossip by telephone while they are not working.

Such gossiping might seem superficially akin to watching TV or listening to the radio since both are electrical communication and both provide amusement. In all of their uses, however, the telephone and mass communication are poles apart; mass communication distracts people from their neighbors and community and focuses attention on national soap operas or national or international affairs. The telephone draws neighbors and communities together (or splits them apart) by providing people with shared knowledge. And, unlike mass communications, telephone service might be described as *discretionary*. The telephone, unlike TV or radio, allows the user to choose whom or what service he shall call and when.

In a broader context, mass communication tells lots of people at one time what is happening, while the telephone enables them to choose their own time and associates to protest against it. Mass communication is from the few to the many; the telephone is always from person to person. Mass communication fits people to schedules; the telephone fits people. It is no wonder that politicians subsidize the delivery of periodicals, encourage television, and tax telephone calls.

The discretionary person-to-person characteristics of the telephone take many forms; they may be used for entertaining gossip or friendly greetings or for arranging dates or parties, often on the spur of the moment. This sort of social interaction is rare in France where the telephone does not really work.

One can place orders by phone (if one can get them delivered). A special cut of meat may be ordered and then picked

up. The Yellow Pages, which are very profitable, are both convenient and informative; they may be considered the successor
to the Sears Roebuck catalog. Although they do not list the exact
item or its price, they do give us a start; with the telephone, we
can do the rest ourselves.

The telephone is useful in making reservations for dinner, a
play, or concert, for air travel, for a hotel, motel, or rental
car; many businesses have a toll-free number through which
reservations can be made.

Do you want to know something about Social Security, municipal services, or how hard the local water is? The telephone
will tell you. If misfortune strikes, you call the fire department,
the police, or a doctor; in this last case, it is said that the telephone once responded, "Take two aspirins and go to bed. This
is a recording."

While the role of the telephone in our personal lives is extremely broad and very difficult to particularize, its role in business is perhaps narrower and clearer. Compared with the mails
or the teletypewriter, the telephone is faster and cheaper if we
take the cost of typing into account. For many purposes telephony is better than text, but not for examining legal documents, including patent specifications. Legalese goes in one ear
and out the other with no intervening understanding; detailed
contracts and specifications must be read line by line at leisure.
When text is necessary, it is often mailed separately, preceded
by a phone call or not, and the text is discussed during a subsequent phone call.

Internally in business and among close associates, a phone
call is a quick way to get information or to ask someone to do
something. As Alex Reid notes in Chapter 18, such a business
phone call is different from a face-to-face encounter because it
needs no social element; one does not have to ask about health,
family, or the weather. A phone call can thus be more economical than and preferable to a face-to-face meeting. It is significant
that data cited by Reid show that 53 percent of business phone
calls give or receive orders, instructions, or information; only 24
percent of face-to-face encounters do this.

The all-pervasive use of telephones in business involves priority and etiquette. In stores, the telephone takes precedence
over the live customer; the presumption must be that a tele-

phone customer might get away, but one waiting is hooked.

Busy people who are harried by callers often have their secretaries answer the phone; they accept the call only if they are free and willing. Some have their secretaries make calls as well. Except in the case of person-to-person calls, or when one does not know the number, this can be slow and wasteful if one has direct dialing. Going through a PBX for outside calls means misery for someone. The insistent priority of the telephone call is reasonable when there is no way for the caller to leave a message. I do not see how telephony could function without it.

The telephone allows businessmen, scholars, and ordinary people to search out information and individuals. I once organized a meeting of economists to discuss our declining balance of trade (this was in 1970 when people were not interested). At the beginning of the day I knew one economist, but by the end of that day I knew a good many. The telephone, together with people who know, can be an information source of great convenience and power.

## IMPACT ON OTHER FIELDS

In "The First and Only Century of Telephone Literature," John Brooks exhibits the telephone and its place in our lives as depicted in stories and plays. Here I address briefly another matter on which Brooks touches: how the very existence of telephony and the development of the technology associated with telephony have influenced the arts, inspired and made feasible new work in the sciences.

In his book *The Summing Up*, Somerset Maugham writes:

The drama pictures the manners and customs of the day, and in its turn affects them, and as these change minor changes follow both in the trappings and in the themes. The invention of the telephone, for instance, has made many scenes redundant, has quickened the pace of plays and has made it possible to avoid certain improbabilities.

The messenger need no longer arrive coincidentally as in Greek tragedies; we are spared both this nonentity and his salary. But the exploitation of the telephone has gone further than one might guess from Maugham's words. A sung or spoken soliloquy is unnatural; a telephone conversation, even a long and varied one is not. And like face-to-face dialogue, such a conversation differs profoundly from communication by text.

Some study of communication preceded the invention of the telephone, but the telephone's existence has inspired studies of vocal communication in absence of sight. Such studies have told us new things and confirmed old impressions.

Written communication differs profoundly from spontaneous spoken communication. Indeed, transcriptions of talks often surprise the speaker by seeming nearly meaningless; a text is written to stand on its own. Once a speaker senses understanding on the part of a listener or an audience, he can exploit this very freely. The recording of actual utterances has been essential in revealing the nature of spoken language.

The study of speech and hearing became "scientific" with Helmholtz's work, which largely preceded the invention of the telephone, but telephony inspired studies of speech and hearing and provided apparatus which made new approaches possible. Much of this new work was done by Fletcher and his colleagues at the Bell Laboratories. The work included a major survey of hearing acuity conducted at World's Fairs in San Francisco (1939) and New York (1939–1940), the devising of hearing aids, investigations of the masking of one tone or sound by another, investigations of articulation (intelligibility of speech as affected by bandwidth and noise), the measurement of limens or just noticeable differences in frequency and sound level, and a host of other studies of speech and hearing.

The orthophonic phonograph and early talking pictures came from work in the Bell System. Homer Dudley at Bell Laboratories invented the vocoder (which has given us much insight into the nature of speech), and R. K. Potter and his associates devised the speech spectrogram, a running short-term record of spectrum vs. time which has become variously known as the sonogram or voiceprint.

The development of automatic telephone switching led to developments of another sort. Regarded as a whole, the switched telephone network (which now includes electronic as well as electromechanical switching systems) is certainly the most complicated machine system in existence. It was certainly the first multiprocessor computer and the first large computer devised to do "nonnumerical" tasks. The first electromechanical computer for arithmetic calculations was the Complex Computer made at

Bell Laboratories by George Stibitz in 1940. Other areas of wide importance which came directly from Bell work include the transistor, with its many manifestations, and Shannon's information theory.

To close, as we began, with the influence of telephony on the arts, we can cite the use of computers in the study and production of musical sounds. This has spread from the work of Max Mathews at Bell Laboratories to Boulez's Institute for Research and Coordination of Acoustics/Music (I.R.C.A.M.), a center of art and culture in Paris, to the work of Chowning and his associates at Stanford, and to a host of other places.

## TELEPHONY AS A HIGH-TECHNOLOGY SYSTEM

It is a commonplace notion that the goods of life are a product of science and technology. Many practical efforts have been made to exploit technology systematically for human use. Edison's laboratory was an early example. The laboratories and factories of General Electric and chemical and pharmaceutical concerns are other examples. In telephony we see a high-technology system that changes very rapidly through research and invention.

Superficially, telephony may seem to be a slowly changing service. Such innovations as new types of phones, automatic vs. manual switching, transoceanic service, direct distance dialing, wide area telephone service (WATS), and new forms of PBX service (including direct dialing) have been spaced years apart. The provision of TV network transmission and of data transmission goes almost unnoticed by the average user.

Internally, the telephone plant changes rapidly. The problem of providing an ever more extensive and complicated service in the face of increased costs of labor and material has led to a continual search for new materials, devices, techniques, and methods. New plastics have replaced the wood, metal, and hard rubber of older telephone sets. The expensive lead of cable sheathing has been replaced by plastic. The first submarine telephone cables looked much like telegraph cables; in newer cables the steel for strength is in the center, not on the outside. Materials have also come to include highly purified semiconductors for integrated circuits and light-emitting diodes, artificial magnetic materials, artificial quartz for frequency-selective networks, and highly transparent glass fibres for optical communication.

Techniques have advanced from transmission on single wires with a ground return to transmission on pairs of overhead wires, to transmission on pairs of wires in a cable with "loading coils" to decrease loss, to amplification with vacuum tubes, to putting many conversations on one pair of wires by frequency-division multiplex (and later time-division multiplex in which voice signals are transmitted as coded pulse trains), to more conversations on coaxial cables, to microwave radio for countrywide telephony and TV, to communication satellites (the first demonstration of transoceanic telephone and TV via satellite was done and paid for by AT&T), to waveguides and optical fibres.

Amplification and control have progressed from vacuum tubes and relays to transistors and integrated circuits. Switching has progressed from "step-by-step" in which control is spread through the switching network, to "common control" systems such as crossbar which can handle more complicated types of calls, to electronically controlled switches, and to all-electronic switching. As a product or byproduct of this advance, the telephone network has provided facilities for the transmission of TV and data.

The technological challenges in providing universal telephone service have been unique in that they involved integration of all sorts of new technologies into a stable and reliable system.

Let us compare telephony with the mails, for instance. Pen, ink, and paper existed long before there was a unified postal system; so did modes of conveyance. The postal system brings together current technical facilities for the collection, transportation, and delivery of the mails.

In travel by road and by canal, the path of transportation (road or canal) is supplied by one source and the vehicle (car or barge) by another; fuel comes from still another source. Problems of compatibility are fairly simple. Compatibility of motor fuel and oil is perhaps the most complex and is worked out between oil companies and car manufacturers. Tire problems are handled in a similar manner, and complex interrelations govern safety of vehicles.

The distribution of electric power and the interconnection of power systems pose problems of considerable complexity, but less than those of telephone systems. Air travel (which came after telephony) is complex and high technology, but the number and variety of operations are small compared with telephony.

Railroads with their tracks, rolling stock, schedules, and service are perhaps closer in nature and complexity to telephony than any other service; still, they are simpler, and they have scarcely maintained a high technology.

Telephony is characterized by an unusual combination of aspects:

1. In providing telephone service to many people, there is an inherent diseconomy of scale. It is more complicated to have access to 100 million other people than to have access to one other person. Electric power and cable TV can be distributed by means of a tree-like network which changes only slowly with time, but in telephony an individual pair of wires goes to each subscriber, and the network is reconfigured from call to call. Only high technology combined with economies of scale in research, design, manufacture, and operation can keep the cost of telephony down as service expands.

2. In carrying out its functions, telephone equipment must be operated by users without formal training (in this it is similar to auto travel). The function of telephone employees is to plan, install, maintain, and bill the customers.

3. The customer's equipment (the telephone set) is an almost negligible fraction of the telephone plant; most of the cost and complexity lies in wires to the central offices, switching, and transmission between offices.

4. While the purpose of the telephone system is to serve individual users, maintenance costs for the complex telephone network and for the user's terminal fall on the telephone company. And, unlike the burned-out light bulb or smoking power tool, there can be no clear distinction in the user's mind between different modes of failure.

The telephone company must maintain reasonably stable but ever expanding and improving services while continually improving the economy and internal functioning of the system. This involves research, exploratory development, development, manufacture, installation, and operation. It involves planning and systems engineering.

Because telephony makes use of rapidly changing technologies, planning must be very general. Interconnection of the nation's and then the world's telephones, conversion from manual to dial switches, direct distance dialing, a transition from vacuum tubes to solid-state devices, a general expansion of digital transmission as opposed to analog transmission, automation of testing functions, and automated (computer) aids to installation, maintenance, and operation are broad, sound goals; they constitute planning of the highest order. Such goals cannot be implemented from the top down, however. They must be reached from the bottom up by those who are aware of the goals and adept in science and technology.

Systems engineering is necessary in reaching the overall goals of a general plan. Systems engineers try to take three things into account: the current state of the technological art, the current nature and (unfilled) needs of the telephone system, and the economics of development, manufacture, and operation. On these bases, systems engineers propose new systems that may or may not be accepted for manufacture. During development, systems engineering considerations are invoked when (as always) changes must be made because of misjudgments about the state or cost of technology or the cost or value of various operating features.

Systems engineering is a very chancy business. It must go hand in hand with development, and when systems engineers become isolated from development they misjudge the actual state of technology and, hence, both cost and performance.

Through science, technology, planning, and systems engineering, new devices, techniques, and principles must be incorporated into a complex, reliable, maintainable, yet ever-changing operating system. This complex system must not "crash" (in the computer sense) through obscure and unforeseen internal interactions. It must be maintained and operated by available personnel. Next to the armed forces, the Bell System runs the largest teaching program in the country. But training is not the whole answer to people problems.

Whatever machine fallibility may be, human fallibility is more varied, unpredictable, and harder to control. Hence, there is a

tremendous incentive in telephony to automate all the operations (or parts of operations) that can be automated and to use
human beings only for functions requiring human qualities.
Human beings intervene in person-to-person, credit card, and
coin calls, but they have been supplied with computerized aids.
Human skill and labor are still required in maintenance, but
automated testing is more dependable than scheduled testing by
maintenance personnel.

Automation is also furthered by the sort of personnel available to telephone companies in this country today and especially in large cities. Telephone personnel do not stay on the job as
long as they used to, and their level of education is generally
lower. For better or worse, it seems unlikely that anyone will
ever again begin a career digging holes for telephone poles and
finally become Chairman of the Board of AT&T, as Frederick R.
Kappel did.

At one time the ability to plan, engineer, and operate complex
reliable systems was largely confined to telephone companies.
The Nike antiaircraft missile systems were designed by Bell
Laboratories and manufactured by Western Electric. Today there
is widespread competence in complex, high-technology systems,
but the telephone systems remain the largest and most complex
systems in the world.

## OPERATING TELEPHONE SYSTEMS

We have seen something of the history and nature of the telephone system, the broad impact of telephony on the arts and
sciences, and the difficult technical problems of building, expanding, improving, and maintaining a telephone system.

Telephone systems express high technology and serve users.
As entities in themselves, they are part of the social and economic structure of any country. Although the services telephone
systems render are similar in different nations, their organization varies from country to country. We can recognize four general patterns for supplying telephone service:
1. National telephone systems have been operated by a private
organization such as the International Telephone and Telegraph
Company, either directly or under contract for the national government. During this period of growing nationalism and grow-

ing "governmentalism," such systems have been nationalized as
a matter of public (or political) policy.

2. Some telephone systems, including that of France, are operated as departments of the government (as the U.S. Post Office
used to be), revenues go into the treasury, and both expenses
and capital needs are met (or not met) through legislative appro-
priations.

3. The telephone system may be operated as a public corporation (as in Japan, and more recently in England), with public financing as well as sale of bonds. Sometimes there is a separate
organization which handles international traffic.

4. The telephone system may be operated by one or more private companies, as in the United States and Canada (domestic
service only in Canada). Such companies are always regulated
and of course pay taxes; in the United States, federal, state, and
local taxes are a large item of expense, second only to labor. In
1974 the total taxes shown on Bell System books were nearly 30
percent of operating expenses; however, because of deferred
taxes (accelerated depreciation) and investment tax credits, the
taxes actually paid in that year were a little less than 20 percent
of total operating expenses. This does not include local sales
taxes and federal excise taxes levied on telephone users; these
taxes, for which the telephone companies serve as collection
agencies, totaled about $2 billion in 1974. All such taxes should
be taken into account in comparing private and government
operation.

Whatever the nature of any telephone system, it primarily
supplies a technological service but faces financial problems in
doing so. When a telephone system is operated by a government department or administration and revenues go directly
into the treasury, money for operating expenses, expansion of
plant, and research and development must all come through
legislative appropriation. Commonly, money for telephony is
practically last in legislative interest and enthusiasm. Communication is less pressing than defense, pensions, welfare, or medical service. Telephone service is starved for decades, until it is
so bad it scarcely seems to merit support.

The public corporation is an attempt to solve this problem of
support; the telephone company is allowed to retain its revenues and may borrow money. In principle, this is much better.

However, it is sometimes difficult to sell bonds. The Nippon Telephone Public Corporation ingeniously insists that a new subscriber buy enough bonds to cover the capital cost the company incurs in serving him. Some public corporations, like the U.S. Post Office, do poorly because they struggle with the consequences of many years of bad practices and policies during operation as a government department. Past mischief seems almost irreparable.

Private telephone companies retain earnings, sell bonds, and sell stock when they can. The total Bell System construction expense in 1975 was about $9.5 billion; some came from depreciation, some from tax deferrals (accelerated depreciation), some from investment tax credits, and some from retained earnings. About $2.9 billion had to be raised through debt or equity. This seems a small amount compared with a GNP of $1,400 billion or even the $36 billion of auto sales in 1973; however, $2.9 billion is a very large amount to raise in competition with government borrowing as well as borrowing by and investment in industry.

Whatever their nature, telephone companies or administrations have tough financial problems; adequate money is a necessary but not sufficient condition for success. The success or failure of a telephone company or administration lies in its ability to generate and apply rapidly changing technology in its constant struggle with increasing volume of service, increasing costs of materials and labor, and the provision of new communication services.

How do various telephone systems cope with technological change? This is easiest for small companies, whether they be independent companies in the U.S. or companies or administrations in small independent nations; small companies or administrations cannot afford extensive research and development. What they can do is buy good, modern equipment that has been developed for a larger market. A certain amount of engineering is necessary to fit the equipment to particular needs, but this is far simpler than creating new technology and embodying it in new systems. In essence, small telephone companies or administrations exploit the state of the international telephone art. Because they can do it very well, some small companies or administrations provide excellent and economical service, but others spend too little on equipment and give poor service.

We might also note that small nations with good governments are more adept at managing telephone systems than large nations are or can be. In a country like Sweden, public policies can be established and adhered to in a way that is impossible in a large and unhomogeneous country like the United States. Moreover, the Swedish telephone administration is very aggressive in promoting both good service and good technology. It owns a research and development company (Ellemtel) jointly with L. M. Ericsson. It also owns a company that manufactures telephone equipment.

All national telephone companies or administrations have research and development laboratories of some sort, but all companies except the Bell System, Bell Canada, and the Swedish telephone administration must rely largely on outside suppliers. Thus, in general there is difficulty in relating the problems, ideas, and proposals of the telephone administrations to the actual production of telephone equipment. Further, the suppliers themselves have research laboratories, and there is a problem of access in the telephone administration's laboratories to the real state of the art as understood by the suppliers.

These problems are usually approached through what would be regarded in this country as a collusive collaboration between the telephone administration or public company and several chosen suppliers. The telephone laboratory itself either develops a prototype system or components for such a system; because the laboratory (except in Sweden) has no manufacturing facilities with which to interact, the system or components are not manufacturable. The telephone administration, company, or laboratory also supports the development of prototype systems by one or, preferably, more than one supplier. By keeping in close touch during such work, the telephone administration or company keeps reasonably informed concerning current technology and practical problems, and the suppliers are kept informed concerning the telephone system's problems and goals. This procedure somewhat resembles the prototype concept David Packard introduced into defense procurement; several contractors undertake to produce prototype planes designed to meet a need rather than rigid specifications.

The only alternative in the relations between a telephone administration or company and suppliers in the development of new systems (as opposed to the purchase of existing systems) would be the antiseptic procedure of bidding to rigid specifications and awarding the contract to the lowest qualified bidder. This cannot really succeed, because both technology and goals must be modified during the course of development to cope with costs and technological reality. Such antiseptic contracting results in cost overruns and technological monstrosities and may create such barriers between the procurer and supplier that neither understands what the other knows, needs, or can do. It is understandable that government administrations and government-owned companies have uniformly pursued a policy of collusive collaboration with suppliers in which each learns something of the work of the other.

In the United States, the creation and application of new technology in telephony is pursued quite differently. The American Telephone and Telegraph Company owns the telephone operating companies and the Western Electric Company, which manufactures much of the telephone equipment used by the Bell operating companies. Together, AT&T and Western Electric own the Bell Telephone laboratories.

The work of the majority of Bell Laboratories' employees is the development (right up to drawings for manufacture) of systems and devices to be manufactured by Western Electric; they also do the systems engineering necessary for such systems and device development. This work is done in Bell Laboratories rather than in Western Electric to give the users more influence on the features and performance of the systems they will use and to give a more stable support to exploratory development than that afforded by manufacture. The sales of Western Electric, like those of other manufacturing companies, change much more drastically through the economic cycle than do the revenues derived from telephone service. While Bell Laboratories derives support from Western Electric for specific development projects, it also receives support for longer-term exploratory and research work from the telephone operating companies through AT&T. Thus, support from operating revenues gives Bell Laboratories stability, and design for manufacture by Western Electric keeps Bell Laboratories in touch with reality.

Among the 16,000 Bell Laboratory employees, a little more
than 1,000 are in a research division; research work is supported
almost wholly by funds derived from the telephone operating
companies. Research work is long-term exploration relevant to
the problems of communication. It includes fundamental work
in physics and chemistry. Four men have received the Nobel
Prize while working for Bell Laboratories, and one ex-employee
received that prize. Research also includes the fabrication of
new devices and the construction and testing of novel transmission and switching systems; sometimes work continues for decades (as in the case of waveguides) before a practical application
is found. The strength of research at Bell Laboratories derives
from the long-term support from the relatively stable operating
revenues, close association with development, manufacture, and
operation, and from an overall Bell management that appreciates
the benefits of science and technology.

We have seen that telephone companies or administrations
and their operations differ greatly in different countries. What
can we say of their effectiveness? Some small telephone systems, such as that of Sweden, are admirable. Among large telephone systems, it is universally accepted that the Bell System is
best, both in standard of service and in innovation. Despite a
heavy burden of taxation that government systems do not have,
its rates are low.

Can or does the success of telephone systems change with
time? A strong effort has been made to improve telephone service in the United Kingdom by transferring operation from a
government department to a public corporation, but the outcome remains to be seen. Telephone systems embody large
amounts of equipment built in the past that must be made to
operate in the present. Large organizations are lethargic; it is
hard to improve telephone service in a short time.

It would be far easier for telephone service to degenerate rapidly. We have seen an example of such degeneration of service
in the American railroads. Through regulation or blindness, the
railroads became tied to a technology rather than a service. Others exploited buses, trucks, and planes. Overregulation and bad
management added to the catastrophe.

The Bell System has fought for the right to evolve and exploit
new technologies of communication. So far it has failed to

maintain this freedom only in international satellite communication. By and large, the Bell System has been well managed, though there have been instances of bad management (as in New York) and the consequences have been disastrous and dramatic. So far, the Bell System has always recovered.

It is quite conceivable, however, that telephone and other communication services in this country could degenerate dramatically in a very few years to the level of many other countries. The likeliest causes would be drastic government actions aimed at bringing the Bell System in line with current ideology. Inability to raise sufficient funds for renewal and expansion or bad management are other possibilities.

In general, telephone systems are wonders of the world. It is remarkable that such complex systems can operate, grow, and change to provide new services. It is remarkable that any human organization can operate such systems, yet several forms of organization do operate telephone systems. The Bell System is the clearly acknowledged greatest wonder of telephony, and has grown into what it is over a century. Past wonders have not been eternal. Who knows what the future may hold?

## INTERNATIONAL RELATIONS IN TELEPHONY

Is it not remarkable that the telephone systems of various nations interconnect, so that we can call from one nation to another? Yes and no. The problems of interconnection are twofold. One aspect of interconnection is financial and procedural. How do telephone companies of different nations agree to operate together? How do they arrange financial matters? Here the experience of postal service and telegraphy service provided precedents.

The other aspect of interconnection is technological. How is it technologically possible to interconnect telephone systems? Here the answer is very simple. Before World War II (and for some years thereafter), world telephony copied American standards almost slavishly. In recent years, conformity to international standards has become a two-way street. Common standards, however, are built on a firm foundation of past technological standards that were largely American or copies of American standards. Only in one other field—airplanes and their operation—has American leadership so dominated world practice. That domination came through technological leadership, not political intervention. Indeed, the first major political intervention into telephony (America's approach to international satellite communication) first introduced a nationalistic and competitive element into what had been an area of technological and operational cooperation.

Formally, international cooperation in telephony takes place through the International Telecommunications Union, which grew out of the International Telegraph Union created in 1865. It was an intergovernmental organization for countries to work out uniform agreements on rates, equipment, and operating techniques. Today's basic charge unit for telephone calls—three minutes—was laid down at an ITU Conference in Budapest in 1896.

The International Telecommunications Union now finds itself in a strange environment of rampant nationalism and rivalry, for it has become an agency of the United Nations. Nonetheless, it retains much of its character of international technological cooperation.

The ITU has a General Secretariat responsible for preparing ITU conferences and publishing ITU recommendations. The International Consultative Committee for Radio (CCIR) studies technical and operating matters concerning radio and makes recommendations. The International Consultative Committee for Telephone and Telegraph (CCITT) has similar responsibilities for telephony. The International Frequency Registration Board maintains a registration of all radio frequencies used for all purposes except military.

Through the CCITT, the ITU is at its best in providing a means for technical people from various countries to consult together and arrive at operating standards enabling communication systems in various nations to interconnect effectively. Recommendations are just recommendations, but they are usually followed in whole or in part.

The CCITT is weakest in trying to plan for the future in the absence of firm technological input. Much early futile planning concerning data communication was done when data networks were just emerging and no sound information concerning the practical advantages of various approaches was available. The effectiveness of the ITU must be in promoting international exchange and cooperation; grandiose and detailed plans for the future are either futile or inhibit progress, whatever their source.

Many problems of international communication have traditionally been worked out directly between the telephone companies of different countries. The first telephone cable across the Atlantic was brought into being and owned jointly by the American Telephone and Telegraph Company and the British Post Office. Later cables have been owned and operated between AT&T, German, and French telephone administrations, and other groups. The greatest telephone traffic has been between particular nations or among particular groups of nations, and these nations have worked together in pairs or groups to provide for such traffic.

Satellite communication has been a revolutionary departure from this general pattern because of the approach adopted by the American government. The nature and history of this departure are complex. The final result has been Intelsat, an international organization serving over 100 countries, territories, and possessions. In principle, nothing can be done without the consent of all members. The leading and American member, Comsat (the Communications Satellite Corporation), acts as manager.

American policy has been not to launch regional satellites to link groups of nations outside of Intelsat, but it will launch national satellites such as Canada's domestic satellites.

Communication satellites have been a resounding technical success. They have also been a subject of much national rivalry and international and domestic confusion; an extended discussion would be difficult and probably unrewarding. Perhaps it is best to defer any further discussion of communication satellites until 2076.

## EXPLOITING TELEPHONE SYSTEMS

Telephone companies and the widespread use of telephony have provided opportunities for various types of exploitation not initially contemplated. One of these forms of exploitation is wiretapping. People appear to trust that telephone conversations are private. Few phones are tapped, legally or illegally by gov-

ernments or illegally by others, but telephony has provided a
new mode of surveillance, akin to skulking behind bushes or
opening letters.

Tapping phone lines and listening to conversations is more
difficult and time consuming than intercepting and reading letters. If we accept common report, telephony within and to the
Soviet Union is relatively reliable and secure compared with the
mails. Recent news reports in this country indicate a considerable government interception and reading of letters, but the
public impact of talk about wiretapping has been greater.

Wiretapping strikes at the individual by exploiting the telephone system; other forms of exploitation strike at users by increasing the cost of service. Telephone companies cannot print
money, and only government telephone companies can pass
costs on to the general taxpayer. Added costs in telephony must
be paid for by telephone users.

One of the best-known forms of exploitation is the "blue
box." A subscriber is allowed to exercise control over the telephone network in dialing, but other forms of control are in theory reserved for the telephone company. In many cases, both
sorts of control are accomplished by the transmission of audible
tones. The blue box generates tones that control functions of the
telephone network ordinarily controlled by the telephone company, so a successful blue box enables the user to make long-
distance calls without being billed. The cost of these calls is
spread over paying users. Newer telephone equipment makes
use of common channel signaling, where control signals are not
sent over the talking path, but over separate signaling channels.
This speeds the setting up of connections and also renders the
blue box ineffective.

Codes are an alternative to the blue box. In one code, the distant party (a husband on a trip, for instance) calls at a prearranged time, lets the phone ring a prearranged number of
times, and hangs up. By counting the rings, the called party
gets the message without picking up the phone. Again, the
costs are spread over other users. In a more elaborate code, the
caller puts in a person-to-person call to a given number. The
person called is never there, for the name is fictitious, and the
name itself conveys the message. This is a very powerful form
of communication; just three initials can be used to specify

17,576 messages, and the number of names is of course far greater. The cost to the caller is zero and again the cost is spread over other users.

When the prices of various services are not closely related to the costs of the services (through regulation, "social policy" on the part of the telephone company, historical accident, or inadvertence), the difference can be exploited through a procedure called "cream skimming." As an example, the price of leased circuits is based on airline miles only, yet the costs of circuits is less over high-volume routes served by advanced equipment than it is over low-volume routes. In this country, a number of companies have engaged to provide data and voice service between large cities where the volume of traffic is high and the cost of transmission is low. They sell this service for less than the Bell System price. Where the cost of transmission is high (as in local connections), they rent circuits from the Bell System.

If prices were closely related to costs, cream skimming would be impossible. Through a long history of relating prices to average cost per mile rather than to cost for a particular route, cream skimming has become possible. The Bell System has proposed to go part way toward relating price to cost, but the Federal Communications Commission, which regulates common carriers, is reluctant. Of course, the costs of cream skimming are passed on to Bell System users.

Governments find taxes on businesses a convenient way of raising money for general use. As has been pointed out, a considerable part of the user's telephone bills goes to pay federal, state, and local taxes. In addition, the telephone companies serve as a collection agency for the $2 billion in sales taxes and federal excise taxes that telephone users pay each year. This exploitation of telephony as a way of getting money from users for general government purposes is not, of course, unique to telephony, but telephony is a very lucrative source.

Another form of exploitation may be either advantageous or disadvantageous to the user. Telephone companies provide circuits between remote points, and a variety of terminal devices can be connected to these circuits, whether they are dialed-up connections or private lines. Connectible terminals include second-hand telephones, aids for the deaf, dialing and answering devices, teletypewriters, various data terminals, computers, mobile telephone systems, and private telephone services (PBX's—private branch exchanges serving various organizations).

Why shouldn't anyone connect any old thing to the telephone network? Careless interconnection *can* have several bothersome consequences. Accidental connection of electric power to telephone lines can certainly startle and might conceivably injure or kill telephone maintenance men and can wreak havoc with telephone equipment. Milder problems include electrically unbalancing telephone lines (causing cross-talk) and dialing wrong numbers or false numbers, which ties up telephone equipment.

For technical and economic reasons, a telephone company would like to avoid interconnection as much as possible. In general, telephone companies cordially interconnect only with other telephone companies, but in an era of mushrooming new uses (especially data), interconnection has advantages for the person who interconnects if not for other users.

Interconnection is a troubled field. Unbridled or poorly planned and controlled interconnection can harm communication networks, yet no interconnection is too little. However advantageous it is, interconnection essentially exploits an existing resource, the telephone network. This is not to say that it is an undesirable exploitation.

Another form of exploitation of telephone companies and their customers is the imposition of social goals that the company would not ordinarily work for and the customers would not willingly pay for.

Like electric, gas, and water distribution, telephone service is a "natural monopoly." Provision of service in a given area by two or more competing companies would be costly to the user, and it seems unlikely that in free competition more than one company would survive. In these circumstances a single company in a single area is tolerated by the government, provided that the government regulates rates and some aspects of the service supplied.

A common explanation of regulation has been that it substitutes for competition in assuring a low price and good service. The effect of regulation on price has been studied most carefully, not in a natural monopoly, such as telephony, but in a case where regulation limits competition. Recently, economists have compared prices in the unregulated intrastate segment of the airline industry and in the interstate regulated segment. It seems clear that an unregulated airline forces the prices down.

An examination of regulation where there is no natural monopoly shows that regulators impose expensive practices and restraints on businesses, for which consumers would not pay if given a choice. Some economists have thus come to view regulation not as a means for reducing prices or providing the service for which a consumer is willing to pay, but rather as a means for attaining social goals desired by the regulator but not the consumer, at the consumer's expense.

This leads us to ask whether such exploitive regulatory practices are to be found in telephony. We find that in telephony regulation may be used to promote satellite communication (if that indeed needs promoting) by keeping transoceanic rates up and insisting that some quota of calls go by satellite. Regulation may be used to cause new entry into the common carrier field by promoting cream skimming; the regulators merely have to deny the existing carriers rates based on cost or deny any lowering of rates when advanced technology lowers costs.

Beyond regulation lie laws, new and old, including antitrust laws. Here there is no very serious pretense that the goal is benefit to the users of the telephone; the goal sought is abstract right or wrong, as deduced from history and enacted into law.

A final form of technological exploitation appears to benefit a great many people other than telephone users. Telephony draws from a large pool of evolving science and technology, and through Bell Laboratories it contributes to that pool. From some of the contributions, such as the transistor, huge and profitable industries have grown both here and abroad. In the consent decree of 1956, AT&T agreed to engage in no activities but common carrier communications and government projects. Thus, while technological fruits of Bell Laboratories' inventions have accrued to telephone users, the largest profits have gone to companies licensed under Bell patents.

An acute Soviet observer remarked: "In the United States, man is exploited by man. With us it is just the other way around." Exploitation is a universal feature of society, but universals have their particulars; the exploitation of telephone service and telephone companies is a little different from the exploitation of mineral resources, gullible investors, or slaves.

## IN CONCLUSION

A story tells of the little old lady whose son asked her to fly across the country to visit him. "Me, get into one of those things?" she replied. "No, I'll sit by my fireside and watch TV, as the good Lord intended."

Although TV and telephones are acts of men, not acts of God, we tend to regard them as a part of nature. We encounter telephones so early in our lives that we seldom think of them, any more than we meditate on speech, hearing, vision, or the functioning of our internal organs—unless any of these give us trouble. Then we are annoyed.

If the reader who has followed me this far asks what this has all been about, my reply must be that it has been about a very uncommon commonplace of our lives. Like an iceberg, most of the telephone system is hidden from us, but all of it affects us directly. All of it is part of and affects our society.

Plenty of people look forward to glories 100 years hence, but I have rather looked backward over 100 years of progress. I have said something about the inventor of the telephone and something about the exploiters of the telephone. I have tried to give a picture of the history, nature, and impact of telephony through statistics, through usage, and through its impacts on art and science. I have pointed to telephony as a high-technology system and have depicted the differences among telephone systems in different countries, the problems they have in common, and international organization and relations in telephony.

What I have said is central but limited. The reader can find further information in the other papers prepared in connection with the centenary of the telephone and in earlier papers and books. In a number of articles and books he will encounter much ill-founded prophecy and polemics. On the whole, perhaps the wisest approach is to observe telephony, but with wonder rather than complacence.

## REFERENCES

*Events in Telephone History*, AT&T Public Relations Department.

Michael E. Levine, "Is Regulation Necessary? California Air Transportation and National Regulatory Policy," *Yale Law Journal* 74 (1965).

Brenda Maddox, *Beyond Babel* (New York: Simon and Schuster, 1972).

Richard Posner, "Taxation by Regulation," *Bell Journal of Economics and Management Science* 2 (1971).

## ADDITIONAL REFERENCES

John Brooks, *Telephone, The First Hundred Years* (New York: Harper and Row, 1976).

Arthur C. Clarke, *Voices Across the Sea* (William Luscombe, 1974).

M. D. Fagan (ed.), *A History of Engineering and Science in the Bell System. The Early Years (1875–1925)* (Bell Telephone Laboratories, 1975).

Harald T. Friis, *Seventy-five Years in an Exciting World* (San Francisco: San Francisco Press, 1971).

Frederick R. Kappel, *Business Purpose and Performance* (Dull, Sloan and Pearce, 1964).

Prescott C. Mabon, *Mission Communications, The Story of Bell Laboratories* (Bell Telephone Laboratories, 1975).

J. R. Pierce, *Symbols, Signals and Noise* (Harper and Brothers, 1961).

_____, *The Beginnings of Satellite Communication* (San Francisco: San Francisco Press, 1968).

_____, "Communications," *Scientific American,* September 1972.

**II**    The Telephone in Life

# Editor's Comment

Between 1890 and 1920, there was more commentary on the telephone's impact on daily life than there is today. By now, as Henry Boettinger says, "The telephone is so much a part of our lives that use of it is habitual rather than conscious." Then it was a novelty, a startling innovation that changed both the routines of living and man's sense of the universe. Magazines ran cartoons and short stories about such curiosities as a suitor proposing on the telephone, an inquisitive old lady eavesdropping on the party line, or the embarrassment of an error about whom one was conversing with. There was also some awareness of the more momentous changes taking place, as psychic life was expanded to include customers, colleagues, informants, and friends who were outside one's physical neighborhood. Corporate organizations were extended, "invisible colleges" of specialists grew [the phrase is from Derek Price, *Little Science, Big Science* (New York: Columbia University Press, 1963)], and the scope of available candidates for personal relations expanded.

The chapters in the coming section explore these human impacts of the telephone on daily life. Boettinger provides a survey of those common telephone experiences all of us know from habit but do not think about because we take the phone for granted.

Brooks uses fiction and drama as the mirror of life in which to observe what the telephone means to people. Starting with Mark Twain, he finds a rich telephone imagery (particularly in the early years) which wanes as the phone becomes so pervasive that it no longer evokes images of mystery, amusement, fear, and magic.

Martin Mayer's mirror of life is the statistics of telephone use.

The simple facts of who uses the phone most, how often, and for what purposes have rarely been collected by sociologists, but the phone company has compiled some of the facts for its own operational needs. Mayer obtained studies from the phone company and reports on these.

Wurtzel and Turner use one of the classic devices of sociologists to ascertain the social function of a familiar and ubiquitous phenomenon: looking for a pathological situation where it is suddenly absent. One learns the significance of the newspaper or elevator by observing what happens in a strike. Wurtzel and Turner's study occurred when a fire knocked out an exchange in New York; the authors then interviewed people about what they used for substitutes and what its absence had meant to their sense of security and community.

Brenda Maddox deals with the telephone system's impact on women, particularly on the role of the switchboard in creating a career for many thousands of young women as telephone operators. With the arrival of automatic switching, operators have begun to vanish: that happens precisely because there were so many of them. As the size of the network grows, the necessary number of operators grows more than proportionately (Chapter 6 explains why).

It is said that someone in the phone company once projected the need for operators from the system's growth curve and found that eventually there would be more operators than young women in the United States; the requirement to go automatic was obvious. Yet while the era of the operator was a short one, the existence of a burgeoning demand and respectable path for young women to leave home for white collar work played a significant part in the process of liberation.

Suzanne Keller's chapter deals with the ways in which the telephone has changed patterns of social interaction. Until the telephone, virtually all human interactions were governed by contiguity. Most people related only to those they saw face to face. The question posed by Keller and later (in the section on the telephone and the city) by Gottman, Abler, and Thorngren is how the telephone has changed that. How far are today's human interactions relieved of the constraints of location? How far has the telephone made possible community without contiguity?

# 8

## Our Sixth-and-a-Half Sense

Henry M. Boettinger

The telephone is so much a part of our lives that use of it is habitual rather than conscious. In the history of invention and technology, no other device can be used (in safety) with such total disregard of the thing itself. Nearly every invention of that flood whose tide rose in the Industrial Revolution amplified and extended man's muscles, or enhanced a single sense. Steam engines, textile machinery, railroads, ships, machine tools, telegraphs, chemicals, mining equipment, telescopes, microscopes, metallurgy, automobiles, or the household appliances now considered essential for life (even at poverty levels) all require some transmutation of power or special training in their use. All, if used ineptly, can cause accidents, sometimes fatal, to their operators or bystanders. The telephone is an outstanding exception; its uniquely phenomenal growth and pervasiveness in our lives can in large measure be ascribed to its ease and safety of use.

Most inventions alter the linear, reciprocating motions of the human body (based on extension and contraction of muscles) into more mechanically efficient rotary motions, such as circular saws, trench-digging machines, concrete mixers, or grindstones. (The classic reciprocating-to-rotary motion converter is the *crank*—one of the basic inventions of our culture.) But the telephone accepts the native, natural sounds of a human voice, silently translates them to forms suitable for electrical processes, and delivers a faithful replica of the original voice directly to a listener's ear. In early demonstrations of the telephone, audi-

This paper is adapted from a chapter in my book on the social history of the telephone's first century, *The Telephone Book*, Riverwood Press, Croton-on-Hudson, 1976.

ences gave standing ovations to the transmission of a conversation between two Choctaw Indians who were introduced to the instruments for the first time. While we may smile at the scientific naiveté of those frock-coated and crinolined crowds, were we not so accustomed to it today we could still wonder at how the expressions of *any* human voice were sped with the velocity of light. The name itself, derived from the two Greek words for *far* (tele) and *voice* (phone), is found in the vocabulary of nearly every language possessing a written literature. Exceptions are like the German *Fernsprecher* (far speaker) the result of those rashes of linguistic insularity that sweep over even mature nations from time to time.

About 300 million phones are in daily use throughout the world, from Pitcairn Island's 31 to the United States' 150 million; they have altered in a thousand subtle ways the daily lives of those who use them. Telephones sit in alcoves of luxurious penthouses and the corners of rural hovels, on the desks of heads of state and the walls of shanty offices. Each day in unnumbered interactions, different people struggle through their labors or converse with acquaintances through the silent servant that has become so unobtrusively and totally blended into the background of living.

And it all began in the strange, driven behavior of two improbable inventors, Bell and Watson. They came of a creative era, but seldom have two less likely types created a silent revolution of such social impact. What they gave us is now so taken for qranted that we become aware of the telephone only when it does not work or when we are in regions without it. In fact, that army of scientists, engineers, operators, craftsmen, and managers dedicated to the telephone's continuous operation gauge the excellence of any improvement by whether "no one noticed what we did!" This peculiar motivation betokens a triumph of duty, but creates large challenges and situations when public understanding of problems is essential.

Over the years many variations and adaptations of design in telephone apparatus have allowed handicapped persons— whether blind, deaf, without larynx, or paralyzed—to enjoy near-normal contacts with others. In fact, certain forms of stuttering disappear when conversing by telephone and no one really knows why. Infants barely able to crawl are taught to

twirl a dial marked "Operator" in cases of emergency or their
need for help. In kitchens and bedrooms, useful or vital numbers are placed nearby, ready to summon every kind of aid.
Lonely people, smothered by a city's indifference, sometimes
randomly dial numbers hoping for a moment's comfort from a
sympathetic voice.

One eminent politician, when asked recently how he graded
his responses to problems, said, "If it's routine, I write an official letter; if really important, I phone him; only for averting disaster do I visit him." Notice the social change implied, almost
an inversion from nineteenth-century practice.

The return to oral discourse (older than Neolithic villages) and
neglect of written modes has caused a melancholy decline in
personal letter-writing skills and volume. Schoolchildren and
debutantes are taught "telephone etiquette," yet often they cannot frame a paragraph. Histories and biographies of the future
will mark the change, for only in perverse cases will private
conversations of value and insight be available to scholars.
Novels that use long letters in their plots, like those of Trollope,
Jane Austen, Smollett, and Richardson, appear antique to modern readers.

Another behavioral characteristic peculiar to our era is that
produced by the insistence of a ringing telephone bell. Chapters
5 and 7 above have noted how a sales clerk will turn from a customer to reply to a telephone inquiry. Most persons will answer
the phone's ring, almost compulsively. The mysterious charm of
the unknown call, the impulse to know the news, good or bad,
which spices and poisons our lives offers modern dramatists a
splendid device to introduce discontinuities and developments
in a plot that Brooks has analyzed in his chapter. Novels of sus-
pense and espionage cannot do without the instrument.

When the plans for rebuilding the bombed-out House of
Commons to a modern, efficient design came before Churchill,
he asked the members how many of them wished to alter their
procedures and traditions. Not a voice was heard. He then said,
"Very well. Then we will restore the House exactly as it stood,
for first we shape our buildings, and then our buildings shape
us." So also are our lives shaped by this innocent and unobtrusive invention, which began as a toy, acquired status as a luxury, became a comfort, and is now such a necessity that welfare

families can have their telephones paid for by the state. We shaped it to us, and now it shapes our lives in ways to which we give no thought.

With a phone we can call nearly anyone we wish. In business, education, armies, government, and the church we can, as Cherry has noted, call powerful persons with little anxiety, even though both would be uncomfortable with a visit in person. Thus, the element of awe in power relationships is subtly and progressively undermined. Leadership must then be founded on real knowledge rather than conferred status.

Another aspect of social change triggered by this invention is the presence of women in offices, who now outnumber men. In Chapter 12 Brenda Maddox describes how, together with the typewriter, the telephone created an entering wedge which leveled barriers to women's employment in clerical domains. In advertisements at the turn of the century, "typewriter" and "telephonist" were prestige positions, and fashion responded by producing appropriate attire (like the shirtwaist and blouse) for those new waves of women "going to business."

The imagination of early moralists was so fevered by prurient thoughts that one suggested laws prohibiting telephone installation in bedrooms. A blizzard of popular music, beginning with lines in a *Pinafore* quartet, developed the opportunities lurking in love-at-a-distance, and titles like *Hello, Central, Give Me Heaven*, and *All Alone by the Telephone* exploited new dimensions in romance.

Such events underscore the central characteristic of *telephone* conversation—a direct person-to-person conversation conducted in privacy, with no record of the conversation available to others. This freedom of expression caused Stalin to veto Trotsky's plans for developing a modern telephone system soon after the Russian Revolution. "It will unmake our work. No greater instrument for counterrevolution and conspiracy can be imagined" was his comment, reported in Trotsky's *Life of Stalin*. Controlled broadcasting and a subservient press are the natural allies of a totalitarian regime, since they are one-way techniques. The telephone, as Cherry has argued in Chapter 5, is the ally of democratic societies where individuals share ideas, opinions, and reactions with one another.

This free expression in oral discourse is the bedrock of telephone utility and explains the legitimate outrage of users at perversions like wiretapping and seizure of call records no matter what the reason. The telephone companies profoundly understand this and have opposed all efforts to undermine the integrity of their service. However difficult that may make law enforcement, it is simply too precious a right to imperil. In countries where wiretapping and surveillance are common and expected, use and growth of telephones are inhibited. A President of France, eager to signify new relationships between his government and citizens, abolished the official wiretapping headquarters as one of his first acts, to great popular acclaim. Deep trust is thus involved when a society embraces telephone conversation as a habit, and all concerned with the service know it.

One measure of value for anything we possess is the sense of loss accompanying its deprivation. We take health for granted until disease strikes, become newly sensitive to the beauty of voices and music at the onset of deafness, are unaware of muscles until we overstrain them, and appreciate our friends' virtues only when death takes them from us. So, too, with the telephone. Chapter 11, by Wurtzel and Turner, describes how people feel when it is taken out of service, as may happen in a thunderstorm or if peaks in usage cause system breakdowns. "Community" and "communications" share the same etymological root, and one cannot exist without the other. Hence, a communications breakdown isolates us from the security, involvement, and mutual dependence of our community. Awareness of this has produced in those who provide communications an attitude (inexplicable to cynics) known as the "spirit of service." This spirit is real and calls forth responses ranging from acts of true heroism in disasters, to doing dirty jobs deep in the muck of flooded manholes, or dealing with a person in difficulty who needs assistance.

Before the advent of electrical communications, older economics textbooks used "communications" to mean modes of transportation—railroads, canals, roads, ships, and bridges—and "messages" were confined to either oral reports by weary travelers or written letters carried from place to place. Even the telegraph (Greek for "far-writing") tried to emulate the copy of a real letter. The evolution of every society depends on such means of interactions among its members, facilitated by their *physical* movement; the central streets of towns and villages still carry the idea of congregations of people meeting on Main Street, High Street, Market Street, Broad Street, and Broadway. The tendency since Biblical times for all trades, crafts, dealers, and professions to settle into common districts is still with us, diminishing only now under explosions of population and fracturing of old communities. *The telephone was the first device to allow the spirit of a person expressed in his own voice to carry its message directly without transporting his body.* This was the quiet social revolution inherent in the telephone, so little remarked in its origins, yet fundamentally responsible for its phenomenal growth. In this, it was a liberating force of the first magnitude. At least, it is hard to see how the United States population could have grown the way it did, from 50 million in 1876 to 220 million in 1976, without the enhanced "nervous system" provided by this invention.

Throughout this book, several authors note how the combination of automobile and telephone created the possibilities for suburban and exurban living by providing mobility, safety, and community access. In fact, the telephone system is a form of societal insurance; no one knows how many lives are saved or injuries prevented by its reliability and speed of access to every kind of service and assistance. People moving to a new home unconsciously assume the service will be there on the day they settle in.

New patterns of commerce, trade, and government have been built on the system: finance and travel, decentralized manufacturing plants, regional offices, libraries, retail and wholesale purchasing. "Let Your Fingers Do the Walking," a slogan of Yellow Pages classified advertising, encapsulated a profound social shift in the way shopping had been done for countless generations. To many people, that thick compendium of specialized help or services is their most-used reference work, and listings like *AAAAAA Atomic Television Repair Shop* (named to be at the head of its alphabetized columns) show how canny proprietors calculate the instinctive behavior of customers faced with equipment breakdowns.

The rise of international travel has seen another alliance blossom: jet planes for movement, telephones for maintaining contact with home. Overseas calling is the fastest growing sector of telephone usage, doubling every three years. Political campaigns, selling by telephone calls, polling opinion on public questions, and even Christmas card mailing all depend on the directories furnished without charge to all users. They represent a symbolic map of a community's members. We tend to take the directory's existence for granted, but in post–World War II Japan, easily remembered telephone numbers were sold at auction because a directory service was lacking. Even when West Berlin was the center of bitter controversies, France, Britain, Russia, and the United States cooperated to keep the city's telephone system functioning. Thus, what was just a scientific toy a century ago has grown into something we cannot do without, yet take completely for granted. We love it and hate it at certain moments in our lives—when awaiting news of the safe arrival of a dear one or repulsing the interruption of a bumptious, fast-talking salesman.

Few inventions that founded great businesses have been able to sustain their importance for a century without decline or collapse. The telephone probably holds the record in this respect, and its congruence with *personal* use of an expanding population may be the key to its history. Erected upon the basic network of personal, private conversations have been the advanced technologies of data, broadcast, and other specialized transmissions; the overall criterion of success applied to any telephone system, however, is how well it enables one member of a society to be part of a larger community by allowing him to talk with neighbors, family, friends, associates in business or profession, or people whose help or services he needs. And he wants to do this at will, any time of night or day to any place in the nation or the world, at prices he can afford and thinks reasonable. It is a challenge fit for scientific genius and noble character, for cool managerial judgment, and the warmest of individual service; each has a place and is found in every nook and cranny of that human and technical system we call "the phone."

Even increase in social status—one of the foibles of American life—has been affected by this device and its appurtenances.

The life story of a Hollywood star has been described in three sentences:

| | |
|---|---|
| Early Days | "You can get me on the coin box at the druggist next door." |
| First Job | "Call me. I'm in the book." |
| Established | "Call my agent. Only he has my unlisted number." |

Thus, a quiet revolution took place in a historical context where this device rapidly and subtly became an extension of the senses of human beings. Few had any awareness of the power of a new liberating force, set in motion a century ago.

# 9

## The First and Only Century of Telephone Literature

John Brooks

One way of approaching the subject of the effects of an important product of technology, such as the telephone, on human society is to look at the reflections of the product as they appear in the mirror of society, literature. The social effects of the telephone are, of course, not all measurable or subject to rational analysis. As Professor Gottmann shows in Chapter 14, it has not changed human nature much; but it has had far-reaching effects on people's perceptions and attitudes as well as their actions and may therefore be better understood through society's collective imagination—as expressed in the works of its creative writers.

The telephone has appeared frequently in the imaginative literature of the United States and to a lesser extent that of Europe for a good part of the century of its existence. Everyone who has more than a nodding acquaintance with American literature can without the slightest effort call to mind an appearance of the telephone in a favorite work. Yet close inspection in most literary uses shows that the telephone is only a conduit, a stage prop. Surely it must be the single most convenient prop ever introduced into the warehouse of the imaginative writer. If the novel and the drama are basically about people talking to each other, think of how the scope of those forms is increased by the existence of a device that enables people to talk without being in the same place. Think of the now-vanished problems faced by the storyteller before 1876; if John were in London and Mary in Bath, nothing of words could pass between them, short of telepathy, except by that slow-moving and temporarily one-way form of intercourse, the letter. Somehow the author had to bring them together, or let them brood on each other separately, or forget each other for the time apart. It seems likely that if Aristotle were writing in the twentieth century, he would modify his famous edict calling for unity of time, place, and action in drama. For time and action, the situation remains more or less unchanged; but the problem of place has been all but defeated, not by a literary device, but by an electrical one.

But the telephone as a literary prop, however interesting in itself, is not the subject here. The following literary references to the telephone have been selected not because of the ways the device has helped solve the writer's craft problems, but because each one contains a comment, stated or implied, by the writer on the telephone and its role in society at the moment being described. Through these references, we may perhaps catch a glimpse of the nature of human—and especially American—apprehension of the telephone and of how that apprehension has changed, step by step, over a century.

In approaching the first telephone literature, we must know how the idea of the telephone was looked upon just before its invention, and how the device itself was looked upon just after. In sum, both were widely viewed with suspicion bordering on terror. Hearing voices when no one was present was after all historically considered evidence of either mystical communion or insanity: Joan of Arc or Bedlam. Physicists took it as an axiom that electricity, whatever its other miraculous qualities, could not convey or produce the human voice, a skill presumably under patent grant to God.

It is a reasonable speculation that this climate of opinion actually delayed the invention of the telephone. In 1860, the German Philipp Reis built what modern telephonists say could have been made into a workable telephone. But Reis did not make the adjustment. His "telephone" therefore could transmit the pitch but not the quality of a sound, and he never claimed to have invented the telephone. His failure was one of faith rather than technology—he failed because he did not really believe that the telephone was possible, a shortfall not present sixteen years later in the cool Scottish-born elocutionist Alexander Graham Bell. Again, in 1877 when Bell and Thomas Watson were giving public demonstrations of the telephone in New England and New York, the newspaper reports were full of forebodings of witchcraft. "It is difficult," said the *Providence Press*, "to really resist the notion that the powers of darkness are in league with it." The Boston *Advertiser* spoke of a "weirdness" never before felt in that city; and the New York *Herald* found the telephone "almost supernatural."

Yet, what may be the first piece of telephone literature was not gothic in tone. ("Literature" is used here in the fairly strict sense of the serious work of writers of wide reputation.) As early as October 1877, the London *Telegraphic Journal* commented that the telephone had already established "a literature of its own" and went on to explain that by "literature" it meant jokes in comic papers and sermons from pulpits—forms of social comment excluded from consideration here. Perhaps it may be noted in passing, too, that George Bernard Shaw in 1879, at the age of 23, was briefly an employee of the Edison Telephone Company of London, in a capacity mysteriously described as "wayleave manager"; but never in his subsequent writings did he deal substantively with the telephone as a social phenomenon. Rather, the first discussion was cool, rational, sophisticated, and knowing, as if the telephone were already long familiar and accepted. It was produced by the cool, rational, and magically prescient Mark Twain and first published in 1880 when Twain—always a devotee of new technology, sometimes with disastrous financial consequences to himself—had been a telephone user for two years. In 1878, Twain had a telephone installed in his house and used the occasion to exercise his wit, saying to the installers, "If Bell had invented a muffler or gag, he would have done a real service . . . . Now you fellows come along and seek to complicate matters." His 1880 sketch, "A Telephone Conversation," is a three-page account of the then-novel experience of listening to just one end of a telephone conversation being conducted by someone else. The person Mark Twain overhears is a female member of his household, and what he overhears are references—necessarily cryptic—to such topics of contemporary and universal concern as cooking, the Bible, beauty care, children, church, and a husband. There is a section suggesting gossip, even scandal, that is totally incomprehensible (and thus presumably infuriating) to the overhearer, and the whole thing ends with the prolonged, repetitious good-byes, suggestive of the coda of a romantic symphony, that had apparently become a feature of telephone conversations among women right at the start. This slight, charming sketch introduced the persistent hero of subsequent telephone literature, the woman user, and its dominant theme, her special love of the instrument and special way of using it. The piece might almost have been written in the 1970s.

Why something so rational and workaday, without overtones of the sinister or the supernatural? Certainly because this first telephone fiction happened to be written by Mark Twain; but just as much, perhaps, because the themes of "weirdness" and "powers of darkness" disappeared the moment people began hearing the familiar voices of friends and relatives on their own telephones rather than strange voices in public demonstrations. In becoming an infant public utility, the telephone had demoted itself one giant step from a fearsome and possibly religious object to a useful if irritating one. But the sense of official wonder survived for a time. The great century of invention was near its climax—the phonograph came in 1878, the electric lamp in 1879, and the skyscraper in the 1880s, while the automobile and the airplane were waiting in the wings—and there were plenty of workaday dithyrambists to celebrate each new marvel of technology in heroic couplets. The telephone was no exception. Here are a few ringing (perhaps a bit tinnily ringing) lines from "The Wonders of Forty Years," written by Benjamin Franklin Taylor in 1886:

The far is near. Our feeblest whispers fly
Where cannon falter, thunders faint and die.
Your little song the telephone can float
As free of fetters as a bluebird's note,
Quick as a player ascending into Heaven.
Quick as the answer, "all thy sins forgiven" . . .
The Lightning writes it, God's electric clerk;
The engine bears it, buckling to the work
Till miles are minutes and the minutes breaths . . .

This epic strain in American apprehension of the telephone faded almost as quickly as had the initial feelings of fear and trembling. By the 1880s, literary men here and abroad followed in the steps of the British Luddites earlier in the century and

developed a marked bias against technology in most of its
manifestations; the telephone in particular seemed to threaten
one of their most cherished possessions, their privacy. Thus, we
find Robert Louis Stevenson in 1889 complaining in a letter to a
newspaper editor that the telephone invaded "our bed and
board . . . our business and bosoms . . . bleating like a desert-
ed infant." Mark Twain himself, in a Christmas 1890 piece for
the New York *World*, sent his good wishes to one and all "except the inventor of the telephone." (He and Gardiner Hubbard,
Bell's father-in-law and principal backer, subsequently had a
good-humored exchange of letters in which Twain explained
that he had meant nothing personal about Bell, but rather had
intended his wrath for the Hartford telephone system, which he
pronounced "the very worst on the face of the earth.") This
new, querulous note was to become a subdominant theme in
telephone literature for years to come.

The next note to be struck in the United States was the result
of structural change in the national telephone industry and telephone service. After the original Bell patents expired in 1893
and 1894, telephony was open to competition. Thousands of
new independent companies moved in to fill the breach, so that
for a short time early in the new century, they had more telephones in service than the Bell System had. Although many of
these telephones were in cities where Bell already had service—
bringing about for a time the memorable confusion of two competing telephone systems in a single place—the majority of
them were in rural areas where Bell had never reached; and so
rapid and pervasive was the growth of rural telephony that by
1907 the states with the densest concentrations of telephones per
population were not the Eastern ones where telephony had begun but Iowa, Nebraska, Washington, and California. (Today
several cities, all in the East, have more working telephones
than people.) The explosion of rural telephony brought into being a hero custom-made for the optimistic popular literature of
the time: the rural operator. Invariably female (male operators
were used briefly in the very first years of service and then
abandoned because of their egregious inefficiency), she quickly
came to have informal functions as a community message center

and informal secretary, and in particular, as an alarm giver in case of fire or flood.

A particular circumstance of rural telephony that caught the fancy of writers was that it was at first always organized by party lines, making it a natural carrier of gossip. A popular short story about rural telephony early in the century was Harriet Prescott Spofford's "A Rural Telephone." The community discovers a tyrannical old woman's eavesdropping on the party line because the ticking of her grandfather clock can be heard on the line and also because she sometimes cannot resist breaking in to offer advice. "Fact is, Nancy, it's like a continnered story in the papers," the old lady explains to her daughter. When a neighbor complains, the old lady is astonished: "Do you mean to say that my neighbors grudge me gittin' what plaisure I can out o' this telephone?" So, as we learn from such stories, voices rang over wires across the lonely plains in those years, providing perhaps the chief entertainment available to many of the country's most characteristic citizens. Thus, party-line telephony, which permitted eavesdropping, may also have helped set the stage for a national disregard for privacy that has led us to FBI files and the gossip-filled dossiers, often irresponsibly collected and handled, of private credit agencies. No wonder wire-tapping is an inherent part of modern telephony; it began with listening in on the party line.

Meanwhile, the telephone as it existed early in this century entrenched itself firmly in the consciousness of Europe and, in reflection, in two of the modern masterpieces of literature dealing with that period, James Joyce's *Ulysses* and Marcel Proust's *Remembrance of Things Past*. *Ulysses* was first published in 1922, but the events it describes take place in Dublin on June 16, 1904, and to judge from the prominence of the telephone in its action and in the ruminations of its characters—even allowing for possible anachronism attributable to the author—the place of the device in the Irish consciousness of 1904 was considerable and complex. The telephone appears eight times in *Ulysses*, and not as something new and strange but as an accepted (and morally equivocal) part of life. It seems to have graduated from the status of a technical marvel or curiosity to something with a personality of its own. Indeed, it is in *Ulysses* that we first see this personification of the telephone, later to become a commonplace of telephone literature. Twice its ring is described as an actively aggressive "whir"; once it rings "rudely" in a woman's ear. Once it is treated as a potential necessity for the dead: "Of course he is dead. Monday he died. They ought to have some law to . . . make sure or an electric clock or a telephone in the coffin . . . ." At another place, it becomes a sort of pander in the flirtation of the young rogue Blazes Boylan with a flower girl. It figures twice in the famous Nighttown scene, first when a drunk uses it for a ribald joke, then, more interestingly, in a fantasy of the guilt-ridden hero Leopold Bloom, who imagines a "medley of voices" cataloguing the sins of his past: "He went through a form of clandestine marriage with at least one woman in the shadow of the Black Church. Unspeakable messages he telephoned mentally to Miss Dunn at an address in d'Olier Street while he presented himself indecently to the instrument in the callbox." Obviously the word "indecently" is the key here: a man who presents himself naked to an inanimate object like a mirror (or a telephone) has done nothing indecent and has certainly not committed a sin. In Bloom's imagination, and Joyce's, the instrument in the callbox has become the person he supposes to be on the other end of the line. That Bloom considers his exposure a "sin" even though he realizes his call to Miss Dunn is only "mental" suggests that for him the telephone has become a real and sternly censorious presence on its own.

This strikingly early view of the telephone as a threatening moral presence is set off by a more down-to-earth and less apocalyptic view of it simultaneously as a social amenity and a nuisance in Proust's *The Captive*, the fifth volume of *Remembrance of Things Past*. As Attali has told us in Chapter 4, the French have always been and continue to be among the more resistant of peoples in developed countries to the habitual use of the telephone for social or business relations. French telephony in 1889 had been virtually confiscated by the government, which had proceeded to inflict upon it the worst evils of bureaucracy; as a result, by the turn of the century—when there were hardly more than 30,000 working telephones in France— the quality of service was widely considered a national disgrace, although not one serious enough to call for drastic remedial action. Even today, according to Bell Laboratories' employees in charge of liaison with foreign telephone companies, French businessmen are often reluctant to use the telephone to do business when a face-to-face meeting is possible, and the count of telephones per population lags behind that in otherwise comparable nations. Where Americans immediately took to the telephone like ducks to water and Englishmen came to find it useful enough, Frenchmen apparently still regard it with a suspicion that may arise out of temperament or out of the particularly bizarre service difficulties that were formerly prevalent in their country.

The telephone passage in *The Captive* is perhaps the most famous in literature. The hero, Marcel, has a telephone that apparently worked better than most telephones in France at the time (about 1905). His aged housemaid Françoise epitomizes French resistance to using the instrument, an act that she finds "as unpleasant as vaccination or as dangerous as the aeroplane." She simply refuses to learn to use it. Marcel uses it enthusiastically, but of course has trouble with the notorious French operators. Calling Andrée, a friend of his flirtatious mistress, Albertine, to get information about Albertine's comings and goings, Marcel reports:

I took hold of the receiver, invoked the implacable deities, but succeeded only in arousing their fury which expressed itself in the single word "Engaged!" Andrée was indeed engaged in talking to someone else. As I waited for her to finish her conversation, I asked myself how it was—now that so many of our painters are seeking to revive the feminine portraits of the eighteenth century, in which the cleverly devised setting is a pretext for portraying expressions of expectation, spleen, interest, distraction—how it was that none of our modern Bouchers or Fragonards had yet painted, instead of "The Letter" or "The Harpsichord," this scene which might be entitled "At the Telephone," in which there would come spontaneously to the lips of the listener a smile all the more genuine in that it is conscious of being unobserved.

At last Marcel gets Andrée on the line, but the telephonic troubles have only begun. As soon as he falls silent for a moment, the operator—now called an "irascible" deity—comes on the line and says, "Come along, I've been holding the line for you all this time; I shall cut you off." He hastily resumes the conversation and continues it without further pauses, placating the irascible deity who might cut him off by thinking of her as a "great poet" able not only to evoke "Andrée's presence" but to envelop it "in the atmosphere peculiar to the home, the district, the very life itself of Albertine's friend."

Where Joyce's Irish hero thought of the technological product itself as a demoniacal personage, Proust's French hero—although explicitly expressing the idea that supernatural power is somehow involved, through his repeated use of the word "deity"—more matter of factly treats the instrument itself as a mere convenience, attributing the supernatural power to a human operator and finding interest in the telephone purely through the reactions it calls forth from himself and others. There is a hint that Proust's is a more sophisticated way of saying much the same thing that Joyce says; the fact remains that the two authors set forth the two contrasting strains of subsequent telephone literature—the practical and rational on the one hand, the hysterical and supernatural on the other.

The coming of long-distance service, marked by the opening of the first coast-to-coast line in 1915, changed the American artistic perception of telephony as it changed the nature of telephony itself. What had been a local phenomenon, its localness underlined by the ever-present "Central" and her de facto position as the neighborhood message center, now became a national one, awesome rather than cozy. The change gave rise to a new wave of epic celebration in telephone literature; this time it was Carl Sandburg who struck the keynote, in his 1916 poem "Under a Telephone Pole":

I am a copper wire slung in the air,
Slim against the sun I make not even a clear line of shadow.
Night and day I keep singing—humming and thrumming;
It is love and war and money; it is the fighting and the tears,
    the work and want,
Death and laughter of men and women passing through me,
    carrier of your speech.

"God's electric clerk" had come to seem more like God's executive vice president. But again, as in the early days, the epic strain was quick to fade. The new wonder, transforming the conduct of our national commerce, became accepted as a part of life; poems like Sandburg's came to sound naive and boyish, as they do now; people assimilated national telephony into their minds as if into their bodies, as if it were the result of a new step in human evolution that increased the range of their voices to the limits of the national map. This second-cycle blasé attitude is at the heart of the greatest era of telephone literature, the decade after World War I.

By the 1920s, a telephone exchange name could be used to describe and by implication characterize the neighborhood where it was used—a fact recognized by John O'Hara when he named his second novel after the famous silk-stocking-district exchange *Butterfield 8*. If the 1920s were the first age of mass sophistication in America, the telephone was the symbol of that sophistication. By then, almost everybody who was anybody had a telephone and used it constantly; a negligent style of using it (lying in bed, say, or on the floor, or in the bathtub) became a mark of sophistication as soon as the long-delayed introduction of the French handset by the Bell System in 1927 made such use of it anatomically practical. (Movie stars, the gods and goddesses of American society in those days, were allowed to differentiate themselves by using white instruments that were obtainable by anyone else only with great difficulty and extra expense.) The bootlegger and party-giver Jay Gatsby in F. Scott Fitzgerald's famous novel leaves his guests only to answer the telephone; an attendant says to him "Chicago calling" or "Detroit calling," he says a few cryptic words into the instrument, and the business is done. So early an industrialist as E. H. Harriman had been a compulsive telephone user, but that had been considered freakish at the time; now it seemed natural for a fictional figure emblematic of his nation's moral status to conduct his business affairs entirely on the telephone.

As a now necessary and ever-present artifact of human intercourse, the telephone came to loom so large in stage plays that it was a character in them—not as a dangerous presence as in *Ulysses*, but as a reliable and predictable one for others to measure themselves against, a straight man. The telephone onstage became the leading cliché of Broadway. The curtain would go up on an empty set—a well-furnished living room with a single telephone so placed as to claim the audience's attention immediately. The telephone would ring. The lady of the house would rush to answer it; the ensuing conversation, of which the audi-

ence would hear only one end, would announce the beginning
of the plot. Thus, through the device of the telephone, the play-
wright was able to create a sense of reality, start the action,
accomplish some useful exposition, and arrange a dramatic entrance for a star actor, all at a single stroke. Finally, with plays
like Ben Hecht's and Charles MacArthur's 1928 Broadway hit,
*The Front Page*—in which seven telephones are onstage and constantly ringing, and the plot is unfolded largely through conversations on them—the telephone has taken over: it dwarfs the
characters and makes them seem subservient to it. It raises the
spectre of a monster that threatens to direct and dominate life
rather than being a convenience to serve it.

Meanwhile, the theme introduced by Mark Twain in 1880—
the special relationships between the telephone and the woman
user—is seen to be moving from Twain's light satire through
darker hues to tragedy. Noel Coward's 1923 sketch, *Sorry You've
Been Troubled*, introduced in saucy yet foreboding fashion what
was to become a staple of stage and later screen, the telephone
monologue or one-sided dialogue by a woman. Poppy Baker, a
young Londoner of fashion, is lying in bed one morning when
a call from the police informs her that her husband has committed suicide by jumping off Waterloo Bridge. In the course of
several subsequent gossipy conversations of Poppy's with her
friends, we learn that she regrets her husband's death scarcely
at all. The twist comes when she learns, to her open disgust,
that a mistake has been made; the man who jumped was not
her husband but that of the woman who lives upstairs. Violent
death and murderous marital hostility played for knowing
smiles, and the telephone as a mettlesome toy: the height of
Jazz-Age sophistication. But now the note of gaiety fades; the
real truth, it appears, is that the telephone is the external expression of female desperation of a very unfunny kind. The
"alone by the telephone" theme, sentimentally announced in Ir-
ving Berlin's 1924 popular song "All Alone," is carried to heart-
rending lengths in two literary productions of 1930. Dorothy
Parker's famous story "A Telephone Call" is a monologue addressed to God by a woman waiting in vain for a call from her
lover. The telephone she waits beside is a palpable presence
and a hostile one because of its refusal to ring. The woman fi-

nally longs even for a call in which the lover would reject her, because it would release her from her torment. In a few pages, the author creates a spinning pattern around the telephone, an object that keeps the heroine sane with hope while it drives her insane with frustration. This minor classic ends with the telephone still not having rung, and the woman left counting to five hundred by fives.

A darker and more florid version of the same theme is the subject of Jean Cocteau's one-act play *The Human Voice* (later set to music by Poulenc), in which "a woman in a long nightdress lies as if murdered" on the floor of her bedroom waiting for a promised call from her ex-lover. This time the telephone does ring, repeatedly. First there are two wrong numbers; then the ex-lover calls, and there follows a long conversation, interrupted several times by broken connections or by one or the other of the callers hanging up, in which we learn from what the woman says that her hopes of resuming the affair are in vain and that she has recently attempted suicide. In her hysteria, she makes the telephone instrument an explicit symbol of her feelings; she takes it to bed with her as if it were the lover himself, she imagines it as a diver's air tube without which she would drown, and finally she winds the cord around her neck in a symbolic suicide. Then, mustering her courage, she says to the ex-lover over the telephone, "I'm brave. Be quick. Break off. Quick. Break."

The curiously crucial role of the telephone is in modern love— its ability to make the breaking off of love relationships both easier and harder than it would otherwise be, and its ability to both relieve and intensify the tensions and rejections of life—is only one of the implications of these two works. In them, the telephone is both a giver and a taker of life, an air supply and a suicide noose; it fills the symbolic role of a god. With them— and perhaps we may add Robert Frost's poem "The Telephone," in which a flower serves as a telepathic "telephone" over which is conveyed a lover's one-word message, "Come"—telephone literature reaches its emotional climax: at the end of an age of sophistication, we are back to our original perception of the telephone as magic or a part of nature and the supernatural. The demons that people associated with the telephone when they first heard voices over it had survived a half century of rational use of it in commercial and social life.

If it may appear (as it does to the writer) that creative writing about the telephone has been in decline since 1930, the reason seems pretty clear: the telephone's ubiquitousness and the special forms of human conduct it sometimes gives rise to have since that date ceased to be new and fresh. But some new themes and subthemes in the telephone literature of the past generation are worth noting.

In the 1930s, a time of worldwide economic depression and rising fear of world war, we find the telephone for the first time being associated with feelings of nonsexual frustration and being used as a means of combating those feelings. The British novelist Norman Douglas is nostalgic for the lost era before its invention: "We can hardly realize now," he writes, "the blissful quietude of the pre-telephone epoch." H. G. Wells, bothered by the ringing telephone's intrusion on his concentration and privacy (a theme that strangely has scarcely appeared previously in telephone literature), longs for "a one-way telephone, so that when we wanted news we could ask for it, and when we were not in a state to receive and digest news, we should not have it forced upon us." Wells has, in effect, fantasized the telephone into an all-news radio station. In America, Robert Benchley sees the telephones as a means by which one may feel more important than one is: "There is something about saying 'OK' and hanging up the receiver with a bang that kids a man into feeling that he has pulled off a big deal, even if he has only called up Central to find out the correct time." In the same spirit, in the late 1930s a man named Abe Pickens in Cleveland, Ohio, placed more or less completed long-distance calls to Franco, Mussolini, Chamberlain, Hirohito, and Hitler, at a cost to himself of ten thousand dollars. The telephone had, in Benchley's words, become a way of kidding oneself.

These themes dominate telephone writing since World War II: powerlessness against technology, association with violence and horror, and emotional association with the telephone. We may pass quickly over Gian-Carlo Menotti's 1947 opera *The Telephone*; although it was the first important appearance of the telephone in opera, it was merely a replay of the old joke about women's telephone long-windedness introduced by Mark

Twain. We can get on to the first of these themes. In Robert M. Coates' 1947 *New Yorker* story "Will You Wait?" surrogates for the called party—an office switchboard, a lady's housemaid, a Minister of Finance's secretary—are all unable to put the calls through immediately and ask the callers to wait, with the result that the circuits become clogged and nobody reaches anybody. "After all," the author comments, "such things happen in New York almost every day, and it's a tribute to our strength of character that no one, really, seems to mind." As a practical matter, the story appears in retrospect as an argument for Centrex equipment and direct dialing, innovations that did not appear until some years later. But the intended point, of course, is that technology meant to increase communication is now impeding and even paralyzing it. ("Mrs. Belling, after holding the phone so long, decided that suicide would be silly, and started making plans for a career in the ballet instead.") Many stories along similar lines followed that one.

In Lucille Fletcher's one-act play of 1948, *Sorry, Wrong Number*, an invalid, neurotic, petulant, high-strung woman is the telephone monologuist who, in the tradition set by Dorothy Parker and Cocteau, advances the subject of that genre from lost love to violent death. Here the lady gradually learns through a series of telephone conversations about a plot to murder her but is powerless—partly because of technical failures, but more because of bureaucratic indifference—to prevent the murder. Bad news, including the precise hour when the talker will be murdered, can come in via the telephone, but her cries for help cannot go out. This hair-raising account of the telephone's ability to accentuate paranoia and horror set the style for a whole series of dramatic works on similar themes. Thus, for the first time the telephone in literature came to be associated, as a paradigm for all evil technology, with fear and death—interestingly enough, before the epidemic of threatening and obscene anonymous calls that occurred in the 1960s.

We have seen that in the works previously cited, with the exception of *Ulysses* and perhaps to a lesser extent *The Captive*, the people with the strongest emotional relationships to the telephone have been women rather than men. Women have sighed, cried, and screamed on it; men have treated it practically or with cool wit, irony, or exasperation. In the postwar world,

with traditional sexual role definitions gradually fading and the
telephone more than ever in the thick of everyone's life, this
seems to have changed. "What really knocks me out," says
Holden Caulfield in perhaps the best-known passage of J. D.
Salinger's *The Catcher in the Rye* (1951), "is a book that, when
you're done reading it, you wish the author that wrote it was a
terrific friend of yours and you could call him up on the phone
whenever you felt like it." For the youth of a new generation,
friendship has been defined as freedom with someone else's
telephone time. Abe Pickens in the 1930s had *done* it (although
his callees had not been writers but politicians); in the 1950s
with hundreds rather than tens of millions of phones in use
around the world, the telephone as a potential means of communication with anyone anywhere—a realization of the global
village—had become a fantasy, and one that, Marshall McLuhan
would soon argue, could be satisfied best not by the telephone,
but by television. Again, E. B. White in his 1951 obituary of his
friend and colleague, Harold Ross, spoke of how immediately
after Ross's death, "when our phone rang just now . . . we
thought, 'Good! Here it comes!' But this old connection is
broken beyond fixing. The phone has lost its power to explode
at the right moment and in the right way." In Daniel Fuchs'
1955 story, "Twilight in Southern California," a fading and failing West Coast businessman, fearful of losing his swimming
pool and his beautiful young wife along with his business, desperately telephones a well-known moneyman for a loan, and
succeeds only in having the moneyman's butler hang up on
him—a final rejection that he tries in vain to forestall with the
*cri de coeur* "Don't hang up!"

Worldwide soul contact imagined by telephone; the death of
a beloved friend described by the metaphor of a broken telephone connection; life's ultimate rejection epitomized as a telephone hang-up—and all by men. Long treated as almost entirely
the province of women, the telephone in the postwar world becomes the agent of hope, anguish, and despair for everybody.

In the 1970s the telephone has virtually disappeared from the
serious theater, which has adopted a grotesquely simplified social landscape and substituted existential brooding for sophisticated social observation. Poets are thoroughly tired of it. The
novel, literature's tenuous remaining connection with realism,

remains the telephone's principal literary arena. One can hardly find a novel about contemporary society that does not comment on the telephone, usually sardonically. And of course there are still telephone movies—like "The President's Analyst," which in presenting a telephone system automated to what seems a fantastic degree may in fact be showing a preview of the real future. In fiction, what is most striking is what has not been done. The "novel of letters" was once common; indeed, a novel consisting entirely of letters, Richardson's *Pamela*, is often credited with having been the first novel. Why not then—now that the telephone call has gone so far toward replacing the letter—a telephone novel, consisting from start to finish of telephone talk and nothing else? I have not found such a novel. To tell the truth, I don't really want to, because I am afraid it would be a boring tour de force. Perhaps a closer approach to the ultimate telephone fiction is Carol Emshwiller's story "Peninsula," published in 1974, in which an abandoned woman taking comfort in the thought that "we are all connected by telephone wires," suits the action to the words by walking away from a roof on a telephone wire and using a long piece of molding as a balancing pole.

My impression is that the great days of telephone literature are over: no matter how striking the changes in telephone use that future technology may bring, the telephone as a subject for the creative imagination has been exhausted. The fact remains that over the telephone's first century—when the emotional collision of it and its users was new and repeatedly renewed by advances in its technology—telephone literature faithfully reported on the dramatic effects of that first collision; at its best it went beyond reporting to help create the relationship of people with their nerve-end to society.

## REFERENCES

Mark Twain, "A Telephone Conversation," reprinted in Charles Neider (ed.), *Complete Humorous Sketches and Tales of Mark Twain* (1961).

Benjamin Franklin Taylor, "The Telephone," in *Complete Poetical Works* (1886).

Harriet Prescott Spofford, "A Rural Telephone," in *The Elder's People* (1920).

James Joyce, *Ulysses*, Modern Library Edition (1933).

Marcel Proust, *The Captive*, Modern Library Edition (1929).

Carl Sandburg, "Under a Telephone Pole," in *Chicago Poems* (1916).

Noel Coward, "Sorry You've Been Troubled," in *Collected Sketches and Lyrics* (1932).

F. Scott Fitzgerald, *The Great Gatsby* (1925).

Ben Hecht and Charles MacArthur, *The Front Page*, in John Gassner (ed.), *25 Best Plays of the Modern American Theatre* (1949).

Robert Frost, "The Telephone," in Virginia Shortridge (ed.), *Songs of Science* (1930).

Dorothy Parker, "A Telephone Call," in *The Viking Portable Dorothy Parker* (1944).

Jean Cocteau, *The Human Voice*, translated by Carl Wildman (1930).

H. G. Wells, *Experiment in Autobiography* (1934).

Robert M. Coates, "Will You Wait?" in *The Hour After Westerly and Other Stories* (1957).

Lucille Fletcher, *Sorry, Wrong Number*, in Bennett Cerf and Van H. Cartmell (eds.), *24 Favorite One-Act Plays* (1958).

J. D. Salinger, *The Catcher in the Rye* (1951).

E. B. White on Harold Ross: "The Talk of the Town," *The New Yorker*, Dec. 15, 1951.

Daniel Fuchs, "Twilight in Southern California," in Paul Engel and Hansford Martin (eds.), *Prize Stories 1955: The O. Henry Awards (1955)*.

Peter Marks, *Hang-Ups* (1973).

Carol Emshwiller, *Joy in Our Cause* (1974).

All the references to telephone history are to be found, along with their sources, in my book, *Telephone: The First Hundred Years* (1976), from which portions of this paper are adapted. I am indebted for research help to Nancy Hallinan.

# 10            The Telephone and the Uses of Time

## Martin Mayer

In the last paragraph of his admirable book on the telephone company, John Brooks suggests that the reason for its continuing vitality is its constant contact with customers.[1] I think that is essentially right, but the nature of the relationship is an even more skeletonized key. For interacting technical and social reasons the telephone company has a forever *changing* relationship with its customers. The fundamental nature of that relationship is that the telephone, while obviously a service produced and paid for, is to an ever-changing degree an extension of its user's self. Moreover, it is an extension that in both manageable and unmanageable ways confronts the self with the great economizing choice of a developed society: the uses of time.

Charles Ramond argues in *The Art of Using Science in Marketing* that "the telephone has changed the behavior of Western man more than any technology in history." This goes farther than most of us would, but Ramond also explains our reluctance, urging that by now the changes wrought by "the speed of communication . . . can be seen only in comparison with times so distant as to seem irrelevant."[2] Given the obvious importance of the subject, whether it is as important as Ramond thinks, I find it astonishing that so little has been done by sociologists (or indeed by anybody else) to investigate what the telephone has meant and means to daily life in this country. The statistical information, all of it proprietary within the telephone company, remains fragmentary; and a certain amount of what follows in these pages is necessarily impressionistic—suggestions for work to be done, rather than reports on work that

has been done. On the other side, much of what is here now
appears in print for the first time.[3]

The telephone became an extension of self in America more
than anywhere else because Vail perceived that its utility was a
function of its ubiquity: the more people had the telephone, the
more valuable the service would be. Almost from the beginning
then, the Bell System companies have arranged their pricing to
encourage people to have the service. The costs of installation
(now averaging about $70 per instrument) are capitalized rather
than expensed. They are paid off in the householder's bill over
a number of years—more years, indeed, than he is likely to be
in this house and using this connection and this instrument.
The costs of providing a local service have historically been pro-
rated to give all households the statistical quality of a flat rate
for any number of calls of any duration within a given area. In
places like Atlanta and Denver, for traditional reasons, the area
covered by the flat rate is more than fifty miles across, but that's
panache: within a metropolitan area, about 70 percent of the
calls from residences are made to places within a five-mile radius, well within the flat-rate zone almost everywhere.

Figure 1 shows the distance of local calls originating at a
representative central office in a large state. The symbol "FR"
stands for "Flat Rate Residential." Undoubtedly graphs very
similar to this one can be drawn for almost any residential
central office. (A graph from another central office, showing
roughly the same distribution of calls, appears in Lawrence Garfinkle's article in the February 18, 1976, issue of *Telephony*.) All
studies show that between 40 and 50 percent of the telephone
calls originating from a household are made within a two-mile
radius. In short, people make most of their telephone calls within the neighborhood in which they live.

Today, almost 95 percent of American households have a telephone, and about 90 percent of the telephone households enjoy
a flat-rate service. The average price is between $7 and $8 a
month. Having paid that $7 or $8, the ordinary American now
feels he has a *right* to unlimited service, like the right to the
pursuit of happiness—only more enforceable. "Taking away
what a man considers his free access to the telephone," says a
Bell System researcher, "is like taking away his mother, or the
flag. You just can't do that to him." Because the monthly rate is

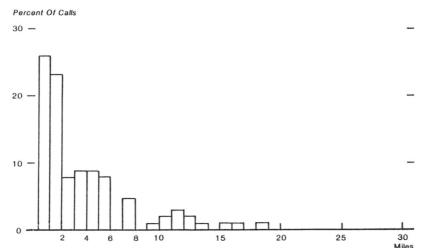

**Distance Of Local Calls**
1FR Class Of Service

**Figure 1**

flat, the average price per call varies hugely from household to household, depending on the number of calls made. The average household makes about 120 calls a month, just under four a day. (The median would be considerably lower, under ninety calls, because households that make a large number of calls pull up the averages.)

Figure 2 shows that the mode of telephone usage lies at two calls per day, that the dropoff after four calls per day is quite steep, and that there are households on this exchange that make more than twenty calls a day. Some of these are undoubtedly businesses conducted from the home, and they pay a residential rather than a business rate.

Researchers find that people invariably overestimate the number of calls made each day from their home. The reason is that people do not distinguish between their outgoing and their incoming calls; if the telephone is an extension of personality, it may not matter much to you who spoke the first word in a conversation. When the researcher reminds the respondent to count only outgoing calls, the estimate goes below the actual number of calls made; people do not count calls that were not completed to the person intended. If Mrs. Jones' daughter says she's gone

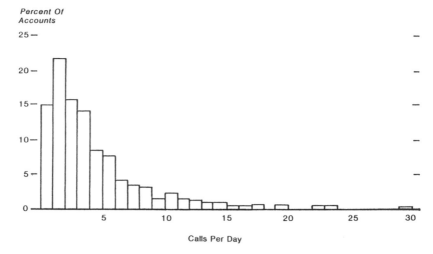

Originating Local Calls Per Customer Per Day
1FR

Figure 2

out, you do not feel you've really made a call to the Jones house and if there is a flat rate so that the call is not paid for, there is no particular reason why you *should* think so.

About one-fifth of all residential calls (these data come from a different, more urban study) go to the same receiving number; and the next four most frequently called numbers account for another 30 to 40 percent of all calls. About half the calls from the median urban household, in other words, go to only five different numbers; over a month, the median household will dial about twenty-five different numbers, half of them served by switching systems in the same central office building. Most people, in fact, have a rather limited acquaintance, and one does not meet new friends on the telephone.

The average *length* of a call from a residential telephone is just over 4¼ minutes in one statewide study; here again, the average is highly deceptive, dragged up by the people who hang on the phone. Figure 3 describes this situation. The x-axis is marked at intervals of thirty seconds. Note that almost 30 percent of the calls last less than thirty seconds, and almost 50 percent last less than a minute. Note also the size of the cell representing calls that last more than twenty minutes. Some of these calls are made by garrulous people or moonstruck adoles-

**1FR Service**
**Local Area Message Holding Time**

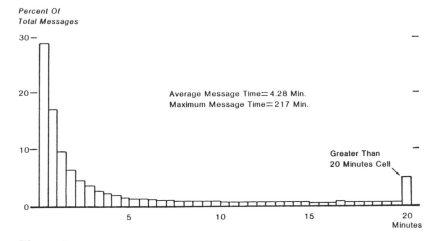

*Percent Of*
*Total Messages*

Average Message Time= 4.28 Min.
Maximum Message Time= 217 Min.

Greater Than
20 Minutes Cell

**Figure 3**

cents, but the telephone company—which is *not* listening in—
believes there are other causes, too.

On one line, for example, may be an engineer who has a
dataset at home and is interacting with a computer, hour after
hour. He has a particularly good deal because he has bought the
minimum two-party-line service, which he enjoys in solitary oc-
cupancy. Those assigned to share it with him quickly demand a
new line from the phone company because the line they have
gives them a lot of miscellaneous squeakings rather than a dial
tone. Another of these long calls, if tapped, would present the
sound of a baby's easy breathing—the parents have called the
neighbor they are about to visit, then left their receiver off the
hook beside the baby's crib and the neighbor's receiver off its
hook beside the bridge table so that they can monitor developments in the nursery while they play cards.

These very long calls are also to be found, incidentally, in the
few places (mostly New York, Boston, and Chicago) where people are billed per call and per minute, rather than on a flat-rate
basis, because even there a call within one's immediate neighborhood is billed at a single unit for its day-part, regardless of
how long it continues. In fact, a study done in one of these
cities showed an average call length of 5.4 minutes, 25 percent
higher than the average length in the flat-rate study. Total

monthly usage in this urban study varied spectacularly. Tabulated by the originating phone, the median local call usage was
400 minutes a month; the household at the bottom of the top
quartile was on the phone on its own local calls for 800 minutes
a month; the household at the top of the bottom quartile originated only 175 minutes of phone calls per month.

We have been looking at the telephone's influence on the allocation of time entirely from the viewpoint of the machine itself;
these data are the fundament. It should also be noted that until
recently, this much was not known in any formal sense, even
within the telephone company. When billing is done on a flat-
rate basis, there is little reason to undertake the expense of me-
tering the calls from individual telephones. What has made a
difference is the company's desire to move toward usage-
sensitive pricing, which cannot be proposed to the public, let
alone implemented, without precise knowledge of whose ox
would be gored, whose bread would be buttered. The arrival of
the Electronic Switching System (ESS) has made it cheaper to
gather the data. (The trick, for the technically minded, is to fool
the ESS into believing that all calls from the numbers in the
sample are toll calls and must therefore be registered in the
accounting machinery.)

For usage-sensitive pricing purposes, average origination and
duration data are not enough, and a few of the Bell System companies have asked their researchers to learn more precisely who
is making the calls. The available conclusions are tentative,
based on thin data from a handful of relatively small studies.
Moreover, the studies are not entirely consistent with each other, especially on the influence of household income on telephone use. Still, most of the patterns are stable from different
surveys conducted by different people in different places, and
are probably entitled to some degree of confidence. Mostly they
show people behaving with considerable good sense.

For example, what household would be the heaviest telephone user? The answer, confirmed in two studies, is the
household with teenaged kids that recently moved to a new
neighborhood in the same metropolitan area. What better use
for an extension of self than the maintenance of cherished relations now severed by space? People who move continue to live
their daydreams in the neighborhood they left, and it is the

sign of adjustment to new surroundings when the concerns of new friends become more absorbing than the concerns of friends left behind. Now some of the daydreaming can be avoided, and the pains of transition—especially for teenagers— can be soothed by keeping in touch on the telephone. The findings here are not confirmed in a study conducted in a center city, but it seems probable that the failure of this study to show increased telephone use by transients is merely another demonstration of the pathological causes of transiency in the center city.

Controlling for income and transiency, the most important single factor in determining how many calls a household will make, is, as it should be, the number of people in the household who are old enough to use the telephone.

Figure 4 graphs this information. Note that single people use the telephone more than couples do. Apart from the simple numbers, the most important single factor is the presence of a woman between the ages of 19 and 64. The second most important boost to telephone originations is the presence of a man over 65 who is not the head of the household. Then comes the presence of girls between the ages of 13 and 18, then the presence of girls or boys under 10. When the head of the household is over 55, the number of calls made diminishes fairly rapidly, and the length of time per call diminishes even more rapidly.

The age factor should be lingered over a little, because it al-

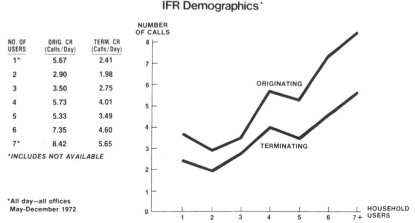

### IFR Demographics*

| NO. OF USERS | ORIG. CR (Calls/Day) | TERM. CR (Calls/Day) |
|---|---|---|
| 1* | 5.67 | 2.41 |
| 2 | 2.90 | 1.98 |
| 3 | 3.50 | 2.75 |
| 4 | 5.73 | 4.01 |
| 5 | 5.33 | 3.49 |
| 6 | 7.35 | 4.60 |
| 7* | 8.42 | 5.65 |

*INCLUDES NOT AVAILABLE

*All day—all offices
May-December 1972

Figure 4

most certainly relates to people's attitudes toward the instrument. The telephone company has done an excellent job of teaching boys and girls in kindergarten and first grade how to use the telephone; it is unusual today for a six-year-old to be afraid to pick up the telephone and fairly common for a seven-year-old to make his own calls. (They tend, fortunately, to be brief: "Hey, Joe, can I come over?" "Sure." "See ya.") Especially during the last decade, the instrument has become an extension of a lot of very small new selves.

But for those born before 1920, the telephone is still and always will be an intrusion. A Bell System study about five years ago indicated that the sound of the telephone bell was most frequently an unpleasant experience for older people, who feared it was bringing bad news—though partly because of this association of the telephone with bad news older people feel deeply dependent on it. The same ringing sound was a stimulating experience to most people in early middle age and younger, for whom it promised relief from boredom. Among the causes of steady growth in telephone usage—at the rate of about 2 percent per household for some years—must be the gradual dying off of the community for whom the telephone seemed a threat and its replacement by a community for whom the telephone is a means of keeping what one researcher calls "multilevel commitments" that could not be maintained otherwise.

The influence of income on telephone usage is more difficult to pin down. Everywhere, the total telephone bill tends to rise with the income level of the household; it rises rather slowly up to about $15,000 in 1972 dollars and then more steeply. People apparently will ration the uses of the telephone that cost money until the total money at their disposal passes a discretionary trigger point; thereafter the toll call and the urban-suburban call become highly appreciated and often purchased luxuries. In the big-city area where customers are billed by usage, the number of local calls made by a household seemed also to be a function of income; but in the more broad-based study of flat-rate customers, including suburban exchanges as well as center cities, the heaviest users of the telephone were people with the lowest incomes.

Nobody likes the look of the graphs in Figures 5a and 5b. The easiest explanation is that racial factors contaminate the result—

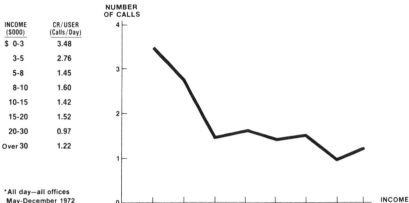

| INCOME ($000) | CR/USER (Calls/Day) |
|---|---|
| $ 0-3 | 3.48 |
| 3-5 | 2.76 |
| 5-8 | 1.45 |
| 8-10 | 1.60 |
| 10-15 | 1.42 |
| 15-20 | 1.52 |
| 20-30 | 0.97 |
| Over 30 | 1.22 |

*All day—all offices
May-December 1972

**Figure 5a**

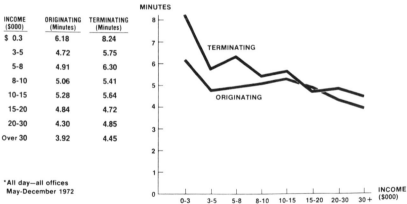

| INCOME ($000) | ORIGINATING (Minutes) | TERMINATING (Minutes) |
|---|---|---|
| $ 0.3 | 6.18 | 8.24 |
| 3-5 | 4.72 | 5.75 |
| 5-8 | 4.91 | 6.30 |
| 8-10 | 5.06 | 5.41 |
| 10-15 | 5.28 | 5.64 |
| 15-20 | 4.84 | 4.72 |
| 20-30 | 4.30 | 4.85 |
| Over 30 | 3.92 | 4.45 |

*All day—all offices
May-December 1972

**Figure 5b**

blacks make more telephone calls than whites, other things being equal, and they take more time on each call—but that would argue that in the central city where the telephone is billed by usage, blacks are exceptionally sensitive to cost factors, which is not true in purchases of services other than the telephone.

Until now we have dealt with gross averages and medians on a monthly basis, but people do not make calls randomly with the passage of the day; the network is most heavily used on certain days, Mothers' Day being the most notorious, and during certain parts of the day. Volume rises dramatically on wet days; people economize on the time they might otherwise spend in the rain. Workers in the bowels of a telephone company central office in a northern climate, entombed though they are between windowless walls, are among the first to know that it is snowing, because the relays start clacking. Before anyone had carried the news into their areas, these workers also knew that something terrible had happened on the afternoon of Jack Kennedy's assassination, because all the networks went busy at once.

Calls from business telephones peak around 11 A.M., drop off to only about half that peak volume between noon and 1 P.M., then rise to a secondary peak, not quite so high as the morning totals, around 3 P.M.. In a center city, the peak calling rate from residential telephones comes in early evening (except for households with heads over 65, for whom the peak calling time is in the morning). In the suburbs, to cite a different study from a different state, the peak calling rate from residential telephones may come between 3:30 and 4:30 P.M. when the high-school kids are disgorged from the school buses and get on the telephone to make sense of the day that has passed. One central office produced an unprecedented peak between 5 and 6 P.M., which stimulated a little on-the-spot research. This office, it turned out, served a large university, and that 5 to 6 P.M. was the favorite telephoning time of college students. Elsewhere in the country, of course, that hour is a bottom time, because the offices and businesses close down and people are not yet home.

As shown in Figure 6, calls tend to get longer as the night passes; after 10:30 the average duration approaches eight minutes. The graph suggests that something is going on here. The sharp rise in duration on the business side as midnight approaches is the result of computers beginning to talk to each

**Average Holding Time By Time Of Day
Weekdays**

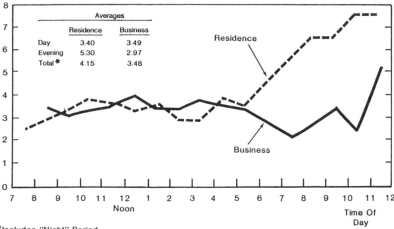

*Includes "Night" Period

**Figure 6**

other. There is not yet any convincing explanation of why the average length of residential calls rises after 9 P.M.—just evidence someone should explore. Teenagers after their parents go to bed? A loneliness factor? Perhaps it is just the elimination of short, purposive, or abortive calls that during the day counterbalance long calls, which may in fact be no more frequent at night than they were at other hours. Can someone spare a Ph.D. candidate to work on this problem?

Anyone who believes that the power of broadcasting has diminished in this country can learn otherwise by a glance at a chart of the minute-by-minute totals of incoming and outgoing calls in the residential central offices of New York, Boston, and Chicago, where the machine normally generates local usage data for the billing department. If we had that chart we would see that the greatest single cause of random peaks in a sampling of the graphs of incoming calls is the presence of a radio station on the exchange, because there are always either commercials offering something to somebody who calls in or station promotions that urge a call-in. Where there is no radio station, the regular evening peaks in telephone usage—the graph looks like a schematic of a mountain range—come in the period a minute and

a half before to a minute and a half after the hour, when the
television stations broadcast commercials. Usage doubles and
triples from the minute five minutes before the hour to the
striking of the clock, then drops again in the same proportion
before the minute hand gets five minutes past the hour. The
peak on the hour itself is very sharp. There are a very few,
mostly upper-income neighborhoods where this phenomenon is
not observed. In the Midwest, incidentally, the generally declining trend of calls from 8 P.M. on is arrested at 10:30, when the
people who have been watching the late news pick up their
telephones.

Because toll calls have always been metered, time-of-day data
are much more readily available on the long lines; and because
rates on toll calls have always been a function of the time of
day, the data reveal in a way most satisfying to an economist
the pervasiveness and power of prices as an information system. Sometimes the story can be spectacular. Michigan Bell, for
example, recently launched a special promotion of "a nickel a
minute" for calls between any two points within the state, after
11 P.M.. A few days later, the telephone engineers learned of two
well-regarded mining and metallurgy schools in northern Michigan, populated by boys who left girls behind in the southern
part of the state. In northern Michigan, the network was most
heavily used after 11 P.M. as the boys and girls made contact
with each other.

Figure 7 is perhaps the most interesting of all. It shows toll
calls of over 500 miles as a percentage of evening traffic by
evening hour in different years, when different discount schedules were in effect. The lines are drawn between points, each of
which is the average for the full hour indicated on the x-axis. In
1955, the evening discount took effect at 6 P.M.; calls between 6
and 7 P.M. were more than double those between 5 and 6 P.M.
and rose a little more between 7 and 8 P.M. (when the local networks are busiest) before trailing off.

In 1963, the evening discount was not effective until 9 P.M.; as
a result, traffic between 6 and 8 P.M. dropped by about 50 percent while traffic between 9 and 10 P.M. actually *tripled*. In telephone company language, this bombed the networks: people
who put in long-distance calls between 9 and 10 P.M. were waiting quite a while to get through. To handle a peak of that size

# Residence Toll Distribution
# (Calls Over 500 Miles)

**Percentage Of**
**Evening Traffic**

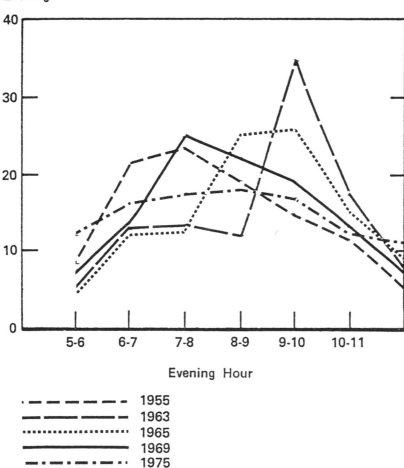

Figure 7

properly would have required extensive additional investment in long-lines facilities. The discount time was finally moved back an hour.

In 1965, the bargain calls started about 8 P.M.. This flattened the peak; calls between 9 and 10 P.M. were reduced by about one-third and calls between 8 and 9 P.M. doubled. In 1969, the discount time was moved back another hour, returning the peak time to the 7–8 P.M. slot; but it was still a fairly high peak.

Finally in 1975, the telephone company pushed the discount on long-distance calls all the way back to 5 P.M., with the result that there is no significant peak, and the smooth demand for long-distance service can be more economically handled.

The conclusion of this long experiment with the clock significantly stimulated the movement toward usage-sensitive pricing in the local companies, which presently runs smack into public attitudes deriving from the view of the telephone as an extension of self. The startling fact seems to be that poor people want flat-rate telephone service, no counting of the number of calls, and no timing of their length, even if they know that a flat-rate service will cost them more.

In their original analysis of the problem, the telephone company marketing experts felt that the company's first obligation was to provide access to the lifeline, a "crisis service" at a rock bottom price with a charge per call and per minute over a small base usage. When offered, this service sold well; and people familiar with the realities of the consumer movement will not be surprised that the average income of the people who bought it was between $30,000 and $40,000 per year.

An analysis of purchasing of the different rate packages offered in one jurisdiction (where the company has authority to employ usage-sensitive pricing) showed that 40 percent of all subscribers are not taking the package that would be most economical for them. But all educational efforts failed.

Poor people must ration so much of their consumption by price that they seem unwilling to add the telephone to the list. They will pay for the service, and after that they wish to be able to use it "free." The telephone and all the contacts made through it are conceived of as necessities that must not be governed by the vulgar calculus of cost. About the only concession the opinion researchers have been able to wring from the lower-

income fraction of the using community is that calls more than
ten minutes long are a luxury—for which, perhaps, something
extra could be charged. Even then, they are not happy about
having their calls timed; but they can see that the more lines
tied up by long calls, the bigger the company's investment and
the higher the price for all service.

Here are the makings of an ugly political issue in the years
ahead. In a time of persisting inflationary pressure, people may
become increasingly reluctant to give up anything they regard
as free, while the costs of expanding a network to handle an exponential usage increase of 2 percent a year per line will press
hard on telephone company revenues. As we can see in the
bright windows of apartment houses where the utility costs are
included in the rent, people find it hard to restrain their consumption of what comes for nothing.

The worst abusers of the present telephone pricing structure
are not individual householders, but banks and credit bureaus.
The telephone companies know banks turn on the dataset in the
morning and simply leave it connected to the computer all day,
effectively gaining a dedicated private line for the price of single
business base-rate service. Credit bureaus have found that one
of the most valid checks on creditworthiness is an individual's
willingness to list his home telephone number on his application. Every fall and winter, as people appear in Florida asking
for credit, the credit bureaus overload the long-distance circuits
with calls to Directory Assistance in the cities of origin of the
credit applicants. These calls cost the credit bureaus nothing,
and the telephone company has not yet figured out what to do
about it.

There is a minor, approximate corollary of Gresham's Law
which states that in the exchange system of the mind, data tend
to drive out thought; but I don't want to live that way, and neither, I assume, do you. Therefore, I should now like to set aside
these data and try to form some part of a structure for thinking
about the telephone and the uses of time.

As a centennial necessarily reminds us, the telephone as an
extension of self did not spring full-grown from anyone's brow;
as the attitude of those over fifty-five indicates, for at least its
first fifty years the telephone was foreign in everyone's life.
Until less than fifteen years ago, it was a minor factor in in

ternational contacts; and it is really only in this decade that the
self has been able to extend overseas, comfortably, through
telecommunications.

For the telephone to fulfill its extension functions requires intelligibility, ease of access, and privity. The first is, of course,
crucial. Very few people are so bilingual that they can originate
a call unconcernedly in a country that does not speak their language. For a good fifty years, as everyone now forgets, this was
an impediment to confidence in the instrument within our own
borders: in a nation of immigrants (which is not the same as a
nation of ethnics), you simply couldn't be sure people would
understand what was said. The great expression of this concern
survives in a recording of the famous vaudeville routine about
Cohen on the telephone, in which a man with a Jewish accent
tries and fails to get his landlord (a bank) to repair the consequences of a recent windstorm.

On a more subtle level, the elimination of information blocks
from telephonic communication has been accomplished by the
creation of a code language employed primarily on the telephone. Perhaps linguistics experts could do more work on this,
because it is an interesting and important problem in cultural
dissonance. Those of us who talk with Englishmen on the telephone know how much more likely we are to get something
wrong from that call than from a face-to-face meeting. For now,
I will lay down an a priori law: The more confident one is of
being understood on the telephone, the more likely one is to be
enthusiastic about using it.

Ease of access is equally crucial, and requires a positive answer to two questions—does the person you wish to call have a
telephone, and can the system put you through quickly? Barring
that certainty, the telephone cannot replace face-to-face contact
in the conduct of business (or of love affairs, for that matter);
and some of the most delightful habits of foreign countries rest
on the reality or on the culturally lagged belief that telephone
service is an unreliable substitute for direct contacts. While in
England last year preparing a British edition of my book *The
Bankers*, I was the beneficiary of this lovely cultural lag. In
America if I can charm someone who can introduce me to other
sources of possible information, I can hope to have him pick up
his telephone and call a friend about me. In London, my hosts

at the Bank of England clapped on their top hats and walked me around the city to meet the people they thought I should know. The qualitative superiority of this procedure is great, just as an interview in person is orders of magnitude better than an interview on the telephone. I never use the telephone for interview purposes if I can possibly do otherwise. But it does not economize on time. (See Chapter 18 for a fuller discussion of this.)

The substitution of telephonic for face-to-face contact is never total even in a city like New York. We still have a garment district and delightful streets where all the diamond merchants or feather sellers or used typewriter traders congregate. (We have lost my favorite, however; it was a street in the Bronx not far from Yankee Stadium, where men walked around with papers in their pockets representing the ownership of specialty bakery shops and brokered these properties back and forth on the curbstone every afternoon after the work of baking was done.) I also seem to remember a place called Madison Avenue; and it does not seem an accident that the home offices of the three television networks are within four blocks of each other, all on what has not historically been a particularly fashionable avenue.

Still, as other chapters in this book describe, neighborhoods are diffusing and one can survive today in almost any field without the opportunity for regular face-to-face contact. This is a fundamental factor in the reorganization of our cities, now proceeding all over the country. Because of the telephone, a great deal of the shooting-the-breeze element of professional and guild affiliation has moved upstairs (as they say on Wall Street)—an interaction much more important to the organization of economic effort than most sociologists or economists recognize.

The third necessity is the most fascinating and the most frequently ignored: the privity factor. Ramond speaks of "the intimacy of telephone communications" and insists that "one feels safer talking through an instrument he can cut off at will than he does conversing face-to-face."[4] He cites the story of a reporter who secured a telephone interview with a sniper who was shooting people on the streets, despite the fellow's occasional protestations that he was really too busy to talk on the phone. Most market research that involves asking people to answer questions has now moved to the telephone—partly because so many people won't allow interviewers in the house any more, partly because an interviewer working at a telephone can key answers directly into a computer as they are produced, but partly also because researchers have discovered that they can ask more penetrating and possibly embarrassing direct questions when the respondent utters the answer into a telephone.

A force advancing the quality of privity has been the dial system; one of the reasons older people have never come to regard the telephone as an extension of self is that when they began using the phone they constantly required mediation by an operator. There were, however, pleasures in having an operator available. The island where we spend the summer was the last place in New York to get dial phones. When a call produced no answer, it was a great comfort to have the operator say, "I think Lenis told me she was going to her mother's today; let's try there." On another occasion, I picked up the phone to ask the operator whether anyone had reported a missing beagle, because one had moved in with us; she replied, "Oh, that's my dog; when you get tired of him, throw a pail of water in his face." We miss that.

Increasing privity has generated other losses, too, the saddest being the departure of the party line which gave so much free entertainment to the aged poor and so much education to adolescents. More serious has been the arrival of the obscene telephone call, the deliberate violation of privity made safe by the elimination of the operator, the use of the telephone-extended self for purposes of panicking and disgusting someone else.

Privity has created a psychological reliance on the telephone as a source of emergency help (one of the standard terrifying incidents in television *policiers* is the knife cutting the telephone line). It is interesting to note, however, that few people actually use the well-advertised 911 number in an emergency. Someone might investigate that, too. Among the theories offered is that in suburban and rural areas emergency services are so deconcentrated that the 911 operation has never been made really effective, while in urban slums a frightened householder would rather reach an individual at a desk in the local precinct than trust an impersonal central exchange. It may be that people in an emergency wish to call someone they know, but that's a guess; this is another area where one would like to see more research.

The importance of privity in expanding the telephone's value can be seen vividly in the quantum leap that came to international usage with the shift to direct dialing across national boundaries. One must not put too much weight on any one factor: there were price reductions, too, and a great improvement in ease of access through the multiplication of circuits. But direct dialing itself seems to have been the prime mover, freeing a large class of users from their reliance on an operator who spoke a language not their own. As an extension of self, the telephone is now international; this, too, should have consequences.

Among these consequences, if I may speculate, may be widespread exportation of that American invention, the executive secretary. Today's American secretary is by no means just a stenographer, as her male predecessors were at the turn of the century: she is a gatekeeper, in much the same sense that the managing editor of the newspaper or the producer of the television news show is a gatekeeper. By her mastery over the passage of telephone calls, she influences her employer's use of time as no worker comparably placed ever could before. Admittedly, she works under instruction, but the instructions are often imprecise and the actual executant power is in her hands. As many a boss has had to learn, the executive needs her more than she needs him. The source of her authority is less the knowledge of the files than her grip on telephony.

A proposed study at AT&T, which probably won't be done, asks at what level in various companies and industries an executive no longer makes or receives his own telephone calls. I commend it to sociologists as perhaps the most significant single mark of status in the daily-work component of this society. One of the elements that maintains my self-image as a working stiff is that I do not have a secretary. I make my own calls, and I answer the phone when it rings. Thus, I remain among the victims on the dark side of the telephone experience: like other forms of telecommunications (television is the most notorious), the telephone permits others to impose their time scale upon you. One can hang up, but it's a gesture of supreme rudeness; escaping a bore is normally easier when a spatial dimension can be exploited.

Changing technology has recently created yet another status symbol in the office—the speakerphone—not because it is expensive, but because the microphone is directional. Everyone in the room hears the conversation, but normally only the boss (at whom the device is aimed) can successfully participate; everyone else is much less intelligible. Thus, the lower status members of the meeting take their time as witnesses to a dialogue of others, and their own contribution must be filtered through an intermediary or echoed through the room resonance.

On the other hand, the office workers can get even. Nobody can estimate the national total of personnel hours spent conducting personal affairs on an office telephone. Psychologists are unsure whether the loss of time and attention represented by personal calls is greater than the resentments felt during a campaign to eliminate such calls. "I've got to hang up now, because . . . " must be one of the most frequently spoken lines on the telephone network during office hours. In the few systems where usage-sensitive pricing is in effect, the local companies have been introducing special services. One of the most popular of these services is Dial-a-Joke; most of the callers seeking amusement turn out to be calling from office phones. Among the central differences between white-collar and blue-collar work in our society is that the white-collar worker has access to a telephone on the job.

What lies ahead for the telephone as for the rest of us is adjustment to the possibilities offered and the procedures imposed by the computer. The telephone, at least, is getting ready: those currently unusable eleventh and twelfth buttons on the touch-tone pad are there for a reason. In a world of bits, a twelve-button system is redundant as a source of alphanumeric information. A dataset in every home is probably unrealistic, but some symbiosis between television and the telephone touch-tone pad looks increasingly likely. It will have a significant effect on the uses of time.

The most interesting innovation on the horizon is the outdoors telephone—the instrument that can be taken into the backyard, out on the picnic, or maybe into the street. As the user moves about, his "number" is shifted from cell to cell, according to the retransmitting station that handles his current location. The telephone would become not only an extension of one's self, but an extension of one's shadow, too. Technically this thing is about ready; politically, it requires more help from those at the FCC who allocate spectrum space than it is yet able to command. The market for this service begins, of course, with the doctors and others who now carry a beeping device. It may be expandable far beyond that. After all, a telephone is a much more useful device than a CB radio, and many people want one of those. I must say, though, that before I would carry that sort of instrument I'd insist on a way to turn it off. The telephone may not have reached saturation yet, but some of us users certainly have.

## NOTES

1. John Brooks, *Telephone* (New York: Harper & Row, 1976), p. 345.
2. Charles Ramond, *The Art of Using Science in Marketing* (New York: Harper & Row, 1974), p. 249.
3. For supplying information for this paper, and for explaining what they supplied, I am grateful to Larry Garfinkle, Chris Whitsley, Robert Harkay, Belinda Brandon, Dean Beman, and Laurence Beck in various divisions of AT&T.
4. Ramond, *The Art of Using Science in Marketing*, p. 250.

# 11

### Latent Functions of the Telephone: What Missing the Extension Means

Alan H. Wurtzel
and
Colin Turner

On February 27, 1975, a fire in a major switching center of the New York Telephone Company left a 300-block area of Manhattan without phone service for twenty-three days. The fifteen-hour blaze silenced twelve exchanges, disconnected 144,755 phones, and disrupted the communication for 90,399 Ma Bell customers.

Although the local press and national news media covered the event, their projection of "what missing the telephone means" was anecdotal and preconceived.[1] Such partial scrutiny has long characterized the study of the telephone in academic and journalistic circles. With the rapid development of radio in the 1930s, systematic inquiry into the functions and effects of the telephone were displaced by a mounting preoccupation with more dynamic channels of mass communication. Even through the 1960s, though the steady growth of the "personal message system" outstripped the faltering expansion of the public media (Maisel, 1973), no concurrent increase in research occurred. Aronson (1971) cites "ninety-odd years of scholarly neglect, not to say disdain," in beginning his comprehensive treatment of the "social consequences of the telephone"—the only discussion of its kind since 1906.[2] If, as he suggests, we continually fail to examine the aspects of social life most taken for granted, then in the "invisible environments" where communication researchers are purported to live, the ring of the telephone will remain especially imperceptible.

The present study, in the spirit of Berelson's benchmark survey of reactions to missing the newspaper (1954), attempts to take advantage of the Manhattan phone blackout to examine a number of speculations regarding the telephone's significance.[3]

According to Aronson, in an urban setting the home phone functions to reduce loneliness and anxiety, increase feelings of security, and maintain cohesion within family and friendship groups. The central assumptions behind such a role concerns what Ball (1968) has called its "centrifugal effects." As family and friends are scattered geographically by mobility and change, ready access by telephone is made to compensate for the loss of shared environs even while facilitating the dispersion.[4]

Feelings of isolation and uneasiness are allayed by the potential to call or be called at any moment. Under these conditions, the telephone's impact on the use of other media and the frequency of face-to-face communications is an open question for Aronson. What both he and Ball take as axiomatic, however, is the telephone's decentralization of our urban lifespace into a matrix of "intimate social networks" dubbed "psychological neighborhoods." They assume that a normative, albeit impalpable, environment exists for everyone with a residential telephone—to which we add, but perhaps for only as long as it can sustain the potential for immediate interaction. Given a situation like the Manhattan blackout, will feelings of isolation and uneasiness surface in people abruptly deprived of phone service? Or will they value, instead, an unexpected freedom from intrusion? Will they turn to other channels of communication and increase their face-to-face encounters? Will they miss the telephone and both admit to and demonstrate a need for the latent functions ascribed to it? In formulating a questionnaire for residents of the blacked-out area we hypothesized the following:

1. *The social role of the telephone*: If the phone plays a particular role in our daily communications, then that role should be demonstrated by how those deprived of phones attempt to restore their personal networks by using outside phones (emergency street phones or business and pay phones outside the blackout area) rather than using other communicative modes like letters or telegrams. Further, if a primary role of the telephone is to maintain one's psychological neighborhood, then one should

not only miss the phone, but miss it most for calls to and from family and friends.

2. *Psychological functions of the telephone*: If, as Aronson suggests, the telephone is used to reduce loneliness and anxiety and increase feelings of security, we might expect loss of service to generate feelings of isolation and uneasiness and decrease feelings of control over daily life. Few people, correspondingly, should value their temporary freedom from telephone intrusions.

3. *Compensatory behaviors during the blackout period*: If personal use of the telephone displaces face-to-face communication, then people deprived of phone service should visit and be visited more frequently during the blackout period. Similarly, without their primary mode of mediated socialization, people should tend to compensate by increasing consumption of other media, particularly channels that most readily provide a parasocial substitute for personal interaction.[5]

## METHOD

Immediately following New York Telephone's announced restoration of phone service, a trained team of interviewers placed calls to 600 randomly generated telephone numbers.[6] The four-digit suffix was generated randomly for two reasons: (1) the telephone company declined to release a list of all affected numbers, and (2) an estimated 25 percent of all New York telephones are not listed in the city's directories. Recent research (Fletcher and Thompson, 1974) has suggested that random generation of phone numbers controls for unlisted numbers and yields approximately the same sampling distribution as alternate methodologies such as stratified sampling.

New York Telephone did indicate that all twelve exchanges affected by the fire contained approximately the same number of listings. Fifty numbers were randomly generated for each of the exchanges. In the course of placing the calls, 89 of the 600 numbers were discarded for being nonresidential, and 192 of the generated listings were found to be either disconnected or not in service; 97 respondents refused to cooperate; 32 numbers yielded no answer after at least five separate recalls. The completion rate based upon an eligible total of 319 was 190, or 60 percent.

To establish baseline levels of telephone use, respondents were asked to estimate the number of calls they made and received on their home phones during an average day and the total amount of time they spent daily on their home phone.

Attitudes toward the medium were sampled in two questions. The first asked subjects how they felt in general about using the telephone. Responses were measured on a four-point scale from "Enjoy using the telephone and use it at every opportunity," to "Avoid using the telephone as much as possible." The second question asked respondents to rate how necessary they thought the telephone was in their daily lives. Replies were chosen from a five-point scale spanning "Absolutely necessary" to "Not necessary at all."

To focus upon the telephone's social role, subjects were asked whether they missed the phone and, if so, what kinds of calls they had most missed being able to make and receive. A third question was asked about what other means the respondents used to communicate with others during the blackout.

The telephone's psychological functions were tapped by asking subjects whether they agreed with each of a series of six statements describing reactions to being without the phone. The statements were:

1. I felt uneasy without the telephone.
2. I enjoyed the feeling of knowing that no one could intrude on me by telephone.
3. Life felt less hectic without the telephone.
4. Life felt more frustrating without the telephone.
5. Having the phone back now, I feel more in control of things.
6. I felt isolated without the telephone.

To discover whether loss of phone service resulted in a compensatory increase of real and parasocial communication of an interpersonal nature, subjects were asked whether during the blackout period they had (1) done more visiting and/or had been visited more often than usual, and (2) used more mass media than normal. If they had consumed more media, they were asked which kinds.

Finally, subjects were asked basic demographic information on age, sex, and occupation.

**Table 1**
Telephone Use

| Amount of Use (No. of calls/time on phone) | Calls Made | | Calls Received | | Time on Phone | |
|---|---|---|---|---|---|---|
| | N | % | N | % | N | % |
| Low (0–1/less 15 min) | 20 | 10.5 | 22 | 11.6 | 69 | 36.3 |
| Medium/Low (2–3/15 min–½ hr) | 38 | 20.0 | 32 | 16.8 | 43 | 22.6 |
| Average (4–6/½ hr–45 min) | 92 | 48.4 | 90 | 47.4 | 16 | 8.4 |
| Medium/High (7–9/45 min–1 hr) | 17 | 8.9 | 15 | 7.9 | 30 | 15.8 |
| High (10 +/more 1 hr) | 23 | 12.1 | 31 | 16.3 | 32 | 16.8 |
| Total | 190 | 99.9 | 190 | 100.0 | 190 | 99.9 |

**Results**

*Telephone use*: Table 1 indicates the results of the first series of questions concerning normal telephone use. Slightly less than half the sample reported making and receiving from three to five calls per day. There was a significant relationship between the number of calls made and received; individuals who said they made a large number of calls also reported receiving a large number ($\chi^2 = 161.99$, $df = 16$, $p < 0.001$).

Almost two-thirds of those questioned said they spent less than thirty minutes per day on their home phone. There were no significant relationships between telephone use and the age, sex, or occupation of respondents.

*Attitudes toward the telephone medium*: Despite low figures for reported usage, the telephone was nevertheless perceived as a necessity of daily living. Of those questioned, 90 percent felt that the telephone was, in some measure, "necessary"; 48 percent characterized it as either "very necessary" or "absolutely necessary." As Table 2 indicates, the sample was also very positive in its attitudes toward using the medium. Thirty-three percent replied that they "enjoyed using the telephone and used it at every opportunity"; over half reported that they used it

**Table 2**
Attitudes Toward Using the Telephone

| Statement | Number | Percent |
|---|---|---|
| Enjoy using the telephone and use it at every opportunity. | 63 | 33.2 |
| Use the telephone whenever I have to. | 98 | 51.5 |
| Dislike using the phone but use it when necessary. | 22 | 11.6 |
| Avoid using the telephone as much as possible. | 7 | 3.7 |
| Total | 190 | 100.0 |

"whenever necessary." Only 15 percent of the sample indicated negative attitudes toward telephone use.

There was a significant relationship between perceived necessity and attitudes toward phone use. Not surprisingly, those who reported more positive feelings about using the medium also declared the phone to be far more necessary than those who expressed less positive attitudes toward telephone communication ($\chi^2 = 56.98$, $df = 12$, $p < 0.001$). Perceived necessity was also highly associated with the age of the respondent; as age increased, the need for the phone also increased ($\chi^2 = 31.07$, $df = 12$, $p < 0.005$).

*Social role of the telephone*: As expected, relatively few people increased their use of other modes of communication during the phone blackout. When asked how they managed to communicate with others during the blackout, 48 percent reported using emergency street phones provided by the phone company, and 33 percent (virtually everyone in the sample with a daily occupation) said they had made calls from work. Only 10 percent reported writing more letters during that period and less than 2 percent stated they had communicated via telegram. This supports the notion that the home phone has a distinctive role in communication behavior.

This role's nature appears to lie, as Aronson contends, in the maintenance of one's "psychological neighborhood." Four out of five respondents admitted missing the phone, and the calls they missed most were those to and from their primary social

**Table 3**
Types of Calls Missed by Respondents

| Type of Call | Missed Ability to Make Call | | Missed Ability to Receive Call | |
|---|---|---|---|---|
| | N | % | N | % |
| Friends | 89 | 36.3 | 108 | 44.8 |
| Family | 76 | 31.0 | 89 | 36.9 |
| Business | 39 | 15.9 | 27 | 11.2 |
| Medical | 24 | 9.7 | 9 | 3.7 |
| Shopping | 10 | 4.1 | 2 | 0.8 |
| Other | 7 | 2.9 | 6 | 2.5 |
| Total | 245 | 99.9 | 241 | 99.9 |

(Totals do not sum to 190 since respondents could reply to more than one type of call.)

relations. Table 3 presents the relevant data for all subjects. Of those in the sample who reported missing the phone, 47 percent said they most missed being able to call friends; 57 percent reported missing the ability to receive such calls. Family related calls were the next most frequently missed. Forty percent of those who missed the phone said they most missed calls to family members; 47 percent of them reported missing the ability to receive such calls.

*Psychological functions of the telephone*: As Aronson suspected and the results in Table 4 confirm, a working telephone reduces loneliness and anxiety, and increases feelings of security. Over two thirds of the sample agreed that living without the phone made them feel either "isolated" or "uneasy," 53 percent reported both reactions. Life was also more "frustrating" for over half the respondents. Most dramatically, 72 percent of all subjects agreed that they felt "more in control of things once their phone service was restored." Forty-five percent of the sample felt isolated, uneasy, and less in control during the blackout period.

Surprisingly, the supposition that few people would value their temporary freedom from intrusions proved false. Forty-seven percent of the sample agreed that "life felt less hectic" without the telephone, and 42 percent said they "enjoyed the feeling of knowing that no one could intrude on me by phone."

**Table 4**
Psychological Functions of the Telephone

| Statements | Agree | | Disagree | |
|---|---|---|---|---|
| | N | % | N | % |
| 1. I felt uneasy without the telephone. | 129 | 67.9 | 61 | 32.1 |
| 2. Life felt more frustrating without the telephone. | 113 | 59.5 | 77 | 40.5 |
| 3. I felt isolated without the telephone. | 128 | 67.4 | 62 | 32.6 |
| 4. Life felt less hectic without the telephone. | 90 | 47.5 | 100 | 52.6 |
| 5. I enjoyed the feeling that no one could intrude on me by telephone. | 80 | 42.1 | 110 | 57.9 |
| 6. Having the phone back now, I feel more in control of things. | 136 | 71.6 | 54 | 28.4 |

Those who preferred the lack of intrusion were also less likely to feel frustrated by the deprivation ($\chi^2$ corrected $= 4.49$, $df = 1$, $p < 0.05$). An interesting conflict in reactions, however, is posed by the extent to which those who enjoyed the hiatus were nevertheless likely to report negative feelings toward it. For instance, 59 percent of those who enjoyed the blackout for its lack of intrusions still agreed that they had felt uneasy without the telephone ($\chi^2$ corrected $= 4.60$, $df = 1$, $p < 0.05$).

In general, responses to the six questions on psychological functions were unrelated to age, sex, or occupation. There were a number of noteworthy relationships, however, between some of the statements and other variables pertaining to telephone use and attitudes toward the medium. For example, those who use the phone more felt more isolated during the deprivation period. Fifty-two percent of those who spend less than fifteen minutes a day on the phone felt isolated, compared to 81 percent of those who spend over an hour. Similarly, the more necessary a person considered the phone to be, the more likely he was to feel cut off during the blackout. Though 47 percent of those who said the phone was "not very necessary" experienced isolation, 86 percent of those who perceived the medium as "absolutely necessary" reported the reaction ($\chi^2 = 15.00$, $df = 4$, $p < 0.005$).

The perceived necessity of the phone was also found to be

associated with feelings of lost control over daily events. The more necessary a person declared the phone to be, the more likely he was to agree that he felt in control again once service had been restored ($\chi^2 = 37.72$, $df = 4$, $p < 0.001$). In addition, respondents who had access to phones at work tended to feel less out of control during the biackout period than those who could not call from a place of business ($\chi^2$ corrected $= 5.68$, $df = 1$, $p < 0.025$).[7] Finally, with regard to perceived necessity, individuals who enjoyed not being intruded upon also tended to rate the phone as being far less necessary than those who did not value the respite from intrusion ($\chi^2 = 11.02$, $df = 4$, $p < 0.05$). In summary, attitudes toward phone use and the responses indicating the telephone's covert functions were useful predictors of a respondent's degree of dependence upon it.

*Compensatory behaviors during the blackout period*: Aronson's suspicion that personal telephone use displaces face-to-face communication was supported, though not strikingly. Thirty-four percent of the sample responded positively when asked, "Did you find yourself visiting more people than usual during the blackout?" An accompanying question, "Did people visit you more than usual during the same period?" elicited an identical percentage of affirmative replies. Sixty-seven percent of those who increased their amount of visiting also reported being visited more themselves ($\chi^2$ corrected $= 44.45$, $df = 1$, $p < 0.001$).

The prediction that the sample, if deprived of its prime mode of mediated socialization (i.e., phone conversations), would tend to compensate for that loss of interpersonal contact by increasing consumption of other media was also weakly supported. When asked, "Did you find yourself using more media than usual during the blackout?" 31 percent of the subjects said that they had. But fully half the sample neither visited nor used mass media more often during the blackout.

The hypothesis that the forms of mass communications most readily substituted for personal interaction would be the ones that were most often turned to was supported. Television and radio were named most frequently by those who said they had used more media; people cited television three times as often and radio twice as often as they did newspapers, the third most frequently named medium.

## SUMMARY AND DISCUSSION

To most people on Manhattan's Lower East Side, there was no
satisfactory alternative to the telephone. Essential to an urban
life–style, the residential phone had been used without hesita-
tion when available and was missed when withdrawn. From
most in the sample, neither the exchange of letters nor the one-
way flow of mass communications could be made to substitute
for the *immediate interaction* provided by the telephone.[8] Residents verified this distinctive role by either turning to outside
opportunities for calling or tolerating the temporary dissolution
of their personal networks. Consequently, they experienced a
loss of the telephone's functions and felt variously isolated, uneasy, and less in control. Perhaps the significance of these latent
dependencies is best indicated by the fact that almost half of
those who rated the phone as less than necessary still reported
feeling isolated without it.

The modest boost in television and radio consumption among
deprivees suggests that these were used to some degree as surrogates for social contact. In the absence of data detailing specific program choices, it would not be justifiable to infer that
increased usage was solely related to the simulated intimacies of
electronic media. The need to hear the latest news on the blackout could just as easily account for the rise in consumption. Yet
the fact that one in three people who had lost an interactive
means of contact resorted to passive reception of other forms of
mediated communication remains a finding not entirely dismis-
sable.[9] The image of people turning from a disconnected source
of personal exchange to the unresponding faces, voices, and
printed words of our mass informers and entertainers suggests a
certain mutability among our communicative needs; that is an
area for future investigation.

The equally modest jump in visiting suffers less ambiguity of
interpretation. A third of the respondents either visited more
often, or took additional time to reenter "real space"; presumably those visits which were made involved social relations in
the immediate residential area. An answer to Aronson's query
on the telephone's effect upon the frequency of face-to-face
communication is thus suggested: the telephone seems to reduce the amount of unmediated socialization among friends and

family who still live near the caller. A person's "psychological neighborhood," in this case, would not be just a mental landscape beginning at the borders of his actual neighborhood, but one that superimposed itself upon his immediate environs, drawing him into a home-based telephonic web and out of the kind of street life that reduces isolation and makes a neighborhood a more supportive community.[10]

The telephone, in other words, both gives and takes away; though it may reduce loneliness and uneasiness, its likely contribution to the malaise of urban depersonalization should not be underestimated. Such ironies are now an old story: a technological device eventually is used in solving a problem it has helped create. [11]

Aside from this wider issue, the fact that people missed receiving calls more than making them clearly focuses upon what is central to the telephone's use as an instrument of urban adaptation. Though the difference in figures no doubt reflects the availability of the emergency street phones for outgoing calls, it also underscores the social role of the phone not only as a means of immediate interaction but of what might be called *imminent connectedness* as well. This means that for people who miss the phone, their lost accessibility to others is just as important to them as their suspended option to reach others without delay. The telephone, in short, can be seen as a two-way street through the psychological neighborhood. Additional research, we suspect, will show the frustration reaction related most strongly to the inability to make calls, responses of isolation and uneasiness to the suspended potential for receiving them, and the feeling of lost control to both aspects of the medium's bidirectional utility.

When taken in context with the kinds of calls missed most and the finding that people judge the phone more necessary as they grow older, that widely distributed control reaction suggests that the telephone's capacity to reduce loneliness and anxiety is better interpreted as the measurable effect of a still more fundamental function. One out of five Americans changes his address each year, and the dispersion is compounded by time (Toffler, 1971). The older a person becomes, the more frequently each of his friends and family members will have moved. Under these conditions the traditional extended family becomes the overextended family, strung out on telephone wires and occasional letters. The boundaries of one's social reality are no longer rooted in contiguous space but in a kind of *symbolic proximity* that short-circuits distances into dial time and replaces the supportive nature of daily interactions with the telephone's potential for instant contact. Though only an internalized "conceptual environment," as the ecologist Dubos (1972) would say, such an essentially subjective construct nevertheless conditions one's sense of socialization as surely as the tangible influences of the front stoop, the back fence, and the girl next door. To abruptly suspend these supports must be to some extent as threatening to the reliability of one's interpersonal linkages as waking one morning to find vacant lots where neighbors' houses had stood the night before.

In sum, ubiquitous feelings of lost control suggest that loss of telephone contact is an assault upon the way sample members conceived of and structured their social reality. Although Aronson and Ball suggest such reactions by their concept of the psychological neighborhood, they may not grant it the importance the present data suggest. Instead, they stress functions that depend on the telephone's ability to support *immediate* interaction and *imminent* connectedness. We propose that these are complementary to a still more primary psychosocial function of the medium, the maintenance of *symbolic proximity*.

The minority who enjoyed the blackout's lack of intrusions served to qualify (though not deny) this fundamental function of the telephone. Together with the response of those who found life less hectic without the phone, the sense of relief from intrusion reveals an ambivalence toward the medium that relates to the tradeoff of distance for time and privacy for accessibility. Though the telephone fosters mobility by promoting instant contact, it also annexes an individual to all those who have his phone number; though it dispels isolation by providing open channels, it also puts a person at the mercy of others' communicative needs.[12] The telephone might bolster the urban dweller's feelings of control, but it exacts a price in interruptions and the unchosen investment of time. Such compromises seemed less tenable to those who enjoyed the blackout's freedom from intrusions. Whether a similar ambivalence remained in those questioned (pointing to a latent *dysfunction* of the tele-

phone), or whether a distinctly different psychological orientation toward the value of mediated socialization versus personal
solitude is indicated could not be detrmined from the available
data. Along with the following, that question offers a focus for
further investigation.

Our population was preselected by a freakish and chance
event; the twelve affected exchanges were all concentrated in
one sector of a very diversified city. Furthermore, the concept of
the "psychological neighborhood" fits hand in glove with an
area like Manhattan's Lower East Side. Its successive waves of
migration and shifting socioeconomic stratification all but guar-
antee the degree of dispersion and mobility required to create
the need for "symbolic proximity." Therefore, a further sampling of both socially scattered and stabilized populations in urban, suburban, and rural settings is needed to determine
whether the telephone's latent functions are altered significantly
by such situational variables. One cannot recreate circumstances
in which each of the requisite subsamples are deprived of home
phones for an extended period. However, by using some of the
preliminary findings to guide formulation of a more sophisticated questionnaire, it should be possible to tap those thoughts
and valuations surrounding the telephone that remain initially
recessive in its undeprived users.

Such a broad-based survey might also attempt to settle the
questions raised here concerning other communicative behaviors presumed to be compensatory to telephone use. Specifically, further attention might be directed toward determining (1)
what media content is selected by people deprived of their telephone, and (2) the parameters of gratification associated with
consumption of such selections. The results could then be used
to further elucidate the notion that people deprived of one technological aspect of socialization might transfer their communicative needs to another medium, even when the substitute offers
only vicarious approximations of the original interaction.

Within the small body of telephone research, there are additional suggestions for investigation beyond those arising from
the present study. The invisibility of the telephone may have
passed its peak transparency with the wire-tapping revelations
of recent years; for too many social scientists, however, the tele-

phone is still a medium which is answered more often than it is questioned.

## NOTES

1. *Time* (March 24, 1975, p. 73) and *Newsweek* (March 31, 1975, pp. 61–62) stressed the blackout's impact upon businesses and service organizations. *Newsweek* devoted most of its space to New York Telephone's expensive repair activities. A feature article by P. Hagen in the *New York Times* (March 16, 1975, p. 9) quoted deprivees on the phone's necessity and its ability to create feelings of security. Both the *Time* and the *New York Times* stories carried statements by university professors lauding their regained privacy.

2. An article by Ball (1968) is also of note. Though it does not have the breadth of the Aronson study, its discussion of the telephone's impact upon interpersonal communication is salient and thorough.

3. Wright (1974), the functional analyst of mass communication situations, has noted that circumstances in which the normal operation of a medium has been disrupted are particularly beneficial in illuminating the social functions of the affected channel for its users.

4. McLuhan (1964, pp. 271–272) should also be read for his implications along this line of thought.

5. Glick and Levy (1962, pp. 141–170) describe a "para-social relationship" as a vicarious and unreciprocated state of pseudocommunication that television viewers typically generate and maintain between themselves and the "personalities" of the video medium.

6. Berelson (1954) used lengthy field interviews to measure the impact of the 1945 newspaper strike in New York. While that field technique provides an opportunity for extensive questioning, it was felt that for the present study early contact via the deprived medium itself would be especially valuable in maximizing respondent sensitivity and recall.

7. This working segment of the sample also tended to rate the phone as far less necessary ($\chi^2 = 10.62$, $df = 4$, $p < 0.05$) suggesting perhaps that having an alternate means of access had screened them from the sensitizing deprivation that would have made them more aware of both the telephone's necessity and its importance as an agent for reducing frustration and increasing one's sense of control.

8. Our conclusion that the telephone is a communicative mode without a satisfactory alternative among other media initially appears to dissent from a growing consensus among those social scientists who study the role and impact of the media from the "uses and gratifications" perspective. These researchers (Katz, Blumler, and Gurevitch, 1974) have recently tended to stress that different people can readily use the same media in different ways and that though there is a rough "division of labor" among the media, several different channels can typically be

used to satisfy the same need. Weight must be given, however, to other findings of some of these same researchers. When the need to be satisfied is for immediate personal contact, then face-to-face interaction, especially with friends, becomes preferable to substitute symbolic socialization via the media (Katz, Gurevitch, and Haas, 1973). Immediate personal interaction is, in fact, the very form and content of the telephonic medium; because within the realm of mediated experience no other mode of exchange combines the attributes of instantaneousness and real person-to-person contact, the principles of differential and substitute usage cannot be applied to the telephone in the same way that they can be applied among one-way mass media.

9. Rosengren and Windahl would be less likely to dismiss such findings. They would no doubt point to their conclusions (1972) that primary among the reasons that people turn to television and radio is the need for surrogate interactive experience. When combined with their view that any type of humanly enacted content, informative as well as entertaining, can be made to serve that fundamental need for interaction, then the attribute of immediate access—the instant accessibility to a human presence via the mass media—can be proposed as the likely factor responsible for turning the deprivees to television and radio for parasocial gratification.

10. Since conducting the survey, we have encountered much anecdotal data on the increase of interactive street life and community awareness during the blackout period. These sources also noted that once the telephone system was restored, the interactions attenuated and the community awareness abated.

11. Of many recent writings along this thematic line, see especially Slater (1974, pp. 1–34) and Illich (1973, pp. 10–48) for their precise and unrelenting analysis of the double-bind inherent in most of our technological tools and the institutionalized systems behind them.

12. The anthropologist Edmund Carpenter (1973) relates the incident also noted in Martin Mayer's chapter of the mad sniper and the enterprising reporter who discovered the phone number of the beseiged house. The killer put down his rifle and answered the phone. "What is it?" he asked. "I'm very busy now."

## REFERENCES

Aronson, S. (1971). "The Sociology of the Telephone," *International Journal of Comparative Sociology*, 12 (September): 153–167.

Ball, D. (1968). "Toward a Sociology of Telephones and Telephoners," in *Sociology and Everyday Life*, edited by Marcello Truzzi. (Englewood Cliffs, N.J.: Prentice-Hall).

Berelson, B. (1954). "What Missing the Newspaper Means," in *Public*

*Opinion and Propaganda: A Book of Readings*, edited by Daniel Katz et al. (New York: Henry Holt and Company).

Carpenter, Edmund (1973). *Oh, What A Blow That Phantom Gave Me*. (New York: Holt, Rinehart and Winston).

Dubos, R. (1972). *A God Within* (New York: Charles Scribner and Sons).

Fletcher, James E., and Harry B. Thompson. "Telephone Directory Samples and Random Telephone Number Generation," *Journal of Broadcasting* 18 (Spring) 187–192.

Glick, I., and Sidney J. Levy (1962). *Living With Television* (Chicago: Aldine Publishing Company).

Hagan, P. "How Telephoneless New Yorkers Got Along," *New York Times* (March 16, 1975) p. 9.

Illich, I. (1973). *Tools for Conviviality* (New York: Harper and Row).

Katz, E., J. Blumler, and M. Gurevitch (1974). "Uses of Mass Communication by the Individual," in *Mass Communication Research*, edited by W. P. Davison and F. T. C. Yu (New York: Praeger).

Katz, E., M. Gurevitch, and H. Haas (1973). "On the Use of Mass Media for Important Things," *American Sociological Review* 38 (April), pp. 164–181.

Maisel, Richard (1973). "The Decline of Mass Media," *Public Opinion Quarterly* 37 (Summer), pp. 159–170.

McLuhan, Marshall (1962). *Understanding Media: The Extensions of Man* (New York: McGraw-Hill).

"Rewiring the Big Apple." *Newsweek* 85 (March 31, 1975), pp. 61–62.

Rosengren, K. E., and S. Windahl (1972). "Mass Media Consumption as a Functional Alternative," in *Sociology of Mass Communication*, edited by Denis McQuail (England: Penguin).

Slater, P. (1974). *Earthwalk* (Garden City: Anchor Press).

"The Great Phone-Out." *Time* 105 (March 24, 1975), p. 73.

Toffler, A. (1971). *Future Shock* (New York: Bantam Books).

Wright, C. (1974). "Functional Analysis of Mass Communication Revisited," in *The Uses of Mass Communications*, edited by J. Blumler and E. Katz (Beverly Hills: Sage Publications).

# 12

## Women and the Switchboard

### Brenda Maddox

"Well, if I called the wrong number, why did you answer the phone?" Thurber's cartoon shows Woman at her worst, an arrogant primitive.[1] In suggesting that women are stupid about telephones, it is wrong. Since the telephone's invention, women have been especially good at making it work. The world's telephone system has run for a century on female labor, and much of its equipment has been assembled by dextrous female hands.

Thurber was right, however, in showing a woman considering the telephone as an extension of her private thoughts. Many even embody it with a personal presence of its own. The late Marilyn Monroe challenged a companion: "Do you know who I've always depended on? Not strangers, not friends. The telephone! That's my best friend."[2] For other women the phone is more than a friend; it is a mother, or at least the line through which they keep in touch with mother. (Is it a coincidence that the nickname for the world's largest telephone company is Ma Bell?)

Before examining in detail the employment opportunities created for women by the telephone, it would be good to explore the broader subject of women's relationship to the telephone. Films and popular songs tell us that for most of the twentieth century, the telephone was the avenue of approach from male to female. She could be a coy sweetheart ("Give me a kiss by wire"), a call-girl ("Butterfield 8"), or an outright victim ("Sorry, Wrong Number," "Dial M for Murder"). Whatever the circumstances, the woman tended to be passive. If she wanted a man, she was supposed to wait by the telephone but not initiate the call herself. A woman writer recalled being taught by her mother that calling a boy was a mistake from which she might not recover. "I think I thought my hand would fall off," she said.

While the sexual revolution of the sixties has done away with some of these old rigidities, the telephone is still an instrument of male sexual advances. Women who live alone try to hide from the obscene telephone caller by listing only a neuter first initial in the telephone directory.

While some obvious truths about women and the telephone are known to all of us, the fact remains that very little substantive research has been done on this subject. In contrast, *Psychological Abstracts*, which compiles an international index of publications in the behavioral sciences, shows column after column, from country after country, of papers on various aspects of television viewing. But its listings under Telephone are scarce and tend to be confined to two problems: the psychological strains of the telephone operator's job and the new counseling services, called "hotlines," which give advice over the telephone to would-be suicides, drug addicts, and others in distress.

A tantalizing glimpse into what might be done lies in a small study said to be the only one of its kind made within the Bell System. Called "Social Factors Relating to Increased Telephone Usage," it was conducted for the Management Sciences Division of the American Telephone and Telegraph Company by the Response Analysis Corporation of Princeton, New Jersey, in 1970. As it drew upon only thirty-one of Bell's millions of customers, all of whom lived in the Bronx, the study is hardly definitive. But most of them were married women, and what they reported sheds an interesting light on why women pick up the phone.

Three major social factors were responsible for their heavy telephone use. They were afraid of crime in the streets, they were confined to their homes by small children, and their relatives lived farther away than they had a generation before. As a result, these housebound women relied on the telephone for services of many kinds, from doctors to television repairmen, and for deliveries of goods from medicine to pizza. They telephoned for babysitters and sometimes used the phone itself as the sitter (leaving the phone suspended over the baby's crib while they went next door for coffee, with the other end of the open line on the neighbor's kitchen table). But above all, they called their mothers, sisters, and mothers-in-law. (Ironically, these calls to relatives, probably essential to their emotional health, were regarded by the women as frivolous. They felt guilty when the bill came in.)

Surprisingly, another major reason for using the phone was to call their husbands at work, not only to relieve loneliness but to share information and speak privately, which was not always possible in front of the children when their husbands came home. The telephone appeared to keep family life going while the husband was absent. One woman said, "When something happens with my son that I think is just adorable, I'll call my husband. He's coming home at 10:30 at night and I could tell him then, but I'd lose the excitement by then . . . . At 10:30, after four hours went by, I don't even begin to say it."

The telephone served a similar function when mothers went out to work. A switchboard operator reported a rush of late afternoon calls as children rang to ask: "Mommy, what should we have for supper tonight?"

These are just pinpoints of light, however. The telephone's social and emotional uses may come to be studied with the attention they merit as the telephone comes to be regarded as an essential appliance, subsidizable by welfare agencies.

The very invention of the telephone can probably be traced to two women. It does not require a psychoanalytic imagination to suspect that Alexander Graham Bell was inspired first by the deafness of his mother and then that of his wife. His father, Alexander Melville Bell, may have been moved in the same way. He wrote to his grandchildren how he had been attracted to his wife by her disability: "I found the lady very pretty, slim and delicate looking and with the sweetest expression I think I ever saw. But she was deaf, and could only hear with the help of an ear-tube. My sympathy was deeply excited."[3] Alexander Graham Bell, the couple's second son, became adept at communicating with his mother in sign language and in a kind of speech spoken low and close to her forehead. Later Bell married one of his pupils, pretty Mabel Hubbard, who had been left deaf by scarlet fever when she was five.

## WOMEN AND THE SWITCHBOARD

One of the major social effects of Bell's invention was to open a
vast new field of employment for women, not only in the United States, but in Canada and northern Europe, where the telephone put down early roots. The invention of the telephone
came at a time when women were beginning to seek white-
collar work outside the home. The sewing machine, the match,
and the cookstove had lightened their domestic chores, and
there was less call than ever on their labor in agriculture.

In the United States in the first half of the nineteenth century,
factory work was something respectable girls might do. There
were supervised boarding houses in mill towns such as Lowell,
Massachusetts, for girls who could not live at home.[4] In time,
however, with the flood of new immigrants from abroad, factory work acquired a lower social status. In any event, the hours
were long (60 to 84 hours a week) and conditions unattractive.[5]

The rise of free elementary schools created new jobs for wom-
en in teaching, a process accelerated by the Civil War, which
took men out of the classrooms. But by 1870, with nearly two-
thirds of elementary teachers female,[6] the field was full, and
teaching jobs were scarce. One other major alternative was serving customers in dry goods stores, but again the hours were
long, the pay poor, and the exposure to the public more than
was thought correct for a girl of good family.[7] And, because of
the loss of life in the Civil War, many of them would never
find husbands and wanted to find work, for spinsters were an
economic liability to their parents.

The swift transition of the telephone from a curiosity to a
commercial service came as a godsend to these young women. It
offered easy entry to white-collar work, with no special training.
The switchboard provided jobs not only in the public telephone
service but in business offices as well.

When the first commercial switchboard went into operation,
in New Haven, Connecticut, in January, 1878, boys were employed as operators. This was logical enough for males had been
used in the forty-year-old telegraph industry. Two to four boys
had to work together to complete a call, as they dashed from
board to board to make connections.[8] They also swept the floor,
heaped coal on the fire, and collected bills from subscribers.

But boys did not last long. The first woman operator, Miss Emma Nutt, was hired by the Telephone Despatch Company in Boston in September of that year. Women operators were on the job in New York by 1881 and the female's superiority was so immediately apparent that within ten years 3,000 women were working as operators.[9] Boys had been virtually eliminated, even at night.[10]

The boys, it seems, were rude. They talked back to subscribers, played tricks with the wires, took St. Patrick's Day off, and in the words of one of the early female operators, were "complete and consistent failures."[11]

Histories of the telephone are full of explanations about why women were faster, politer, and more capable than men. In the perspective of the 1970s, however, the praise given to women sounds very faint indeed, and the complaints against the boys genuinely admiring. For example, we have this eloquent assessment from a historian of the British telephone service which, like the American, very soon came to prefer females at the switchboard: "No doubt boys in their teens found the work not a little irksome, and it is also highly probable that under the early conditions of employment the adventurous and inquisitive spirits of which the average healthy boy of that age is possessed, were not always conducive to the best attention being given to the wants of the telephone subscribers."[12] In sum, it was soon appreciated that "the work of successful telephone operating demanded just that particular dexterity, patience and forebearance possessed by the average woman in a degree superior to that of the opposite sex."[13]

Dexterity, patience, forebearance. One other attribute of the female worker was rarely mentioned: cheapness. She worked at anything from a half to a quarter of what was paid to men. In the thirty-year period beginning in 1880, the average American male laborer earned between $10 and $20 a week, the female between $4 and $7.[14] Early women operators began at $10 a month.[15] What is more, women were so eager to find work that they were occasionally used as strikebreakers and were generally indifferent to unionization. When American industry was alarmed at the outbreaks of violence that accompanied the labor unrest of the late nineteenth century, the prospect of a new industry largely dependent on womanpower was not unattractive.

Thus, for many reasons telephone operating became woman's work. In Europe as well as North America, moreover, it was work for spinsters; virgins may not be too strong a word. Only very young girls were recruited: the international range of age limits ran between 17 and 20 years. And almost everywhere girls were expected to resign upon marriage.[16]

While the Bell System never had a formal policy against the employment of married women, and company records mention the payment of death benefits to husbands as early as 1913, girls were expected to leave when they married. For many years and particularly during the Depression, American society resented working wives, on the assumption that they were taking jobs away from men. In areas such as New England, for example, there was additional pressure from the Catholic Church which disapproved strongly of married women leaving the home to work. There the single-girls-only policy tended to be observed very rigorously. The New England Telephone Company did not employ married women until 1942—not knowingly, that is. In many places, it was quite common for girls to keep their marriages secret; married women do not seem to have been acknowledged openly on the rolls of even the more liberal Bell companies until 1929.[17]

A survey taken among European telephone authorities in 1910 showed that Sweden, Norway, Switzerland, and Britain did not employ married operators.[18] Neither did Germany—with the exception of childless widows under 30. France permitted them and so did Belgium, albeit reluctantly, "as they are so often incapacitated." Belgium's statistics substantiated its regret, for the average annual number of days of sick leave taken by married women operators was two to three times higher than that taken by single women. Belgium encouraged retirement upon marriage by paying a special bonus.

In Britain, the Post Office also paid what it called a marriage gratuity. But conditions were so strict that many girls were ineligible. The Post Office's records contain the sad story of Miss E. Nelson, a telephonist who left to marry in the early days of the first World War. She was denied the gratuity because she was four days short of the required six full years of service.[19] In vain did Miss Nelson, through her union, plead that "The short notice is regretted, but is due to the unexpected arrival of my intended husband and short leave granted for our marriage." In fact, she had worked two months longer than six years as an operator, but the rules did not allow either her first month, as an unpaid learner, nor her second, as a trainee at five shillings a week, to be counted. As full-time telephonists then earned about two pounds six shillings a week, the marriage gratuities, which were upward of £30, constituted a considerable dowry.

From the start, telephone work was distinguished by its respectability. The main reason was that the girls were protected from exposure to the public. Also, they had to be educated and well-spoken, capable of answering the customers' unending requests not only for telephone connections but for information on the time, weather, and transportation schedules.

The early operators saw themselves as quasi-schoolmistresses and dressed accordingly. One recalled how she wore, for the first day on the job in New York in 1881, "a garnet dress of cashmere, the waist was very well-boned . . . and a turnover white linen collar fastened with a bow."[20] Another found that on her first day mice had eaten her lunch. "Thereafter I brought my lunch in a tin box. It was not considered dignified to carry your lunch, so I tucked my tin box in a black satin bag."[21] Because it certainly was not dignified to go out for lunch and brave the male street crowd, the telephone companies soon began to provide lunch rooms, establishing a tradition of mothering female employees that has remained with telephone service almost everywhere.

That these were daughters of the middle classes impressed a British team of observers sent early in the century to study the recruitment methods of the much-admired American telephone system. The American practice, according to the team's report, was to inquire carefully "as to the position and manner of life of the parents and other relations of candidates. The object is not only to secure good operators, but also to increase the reputation of telephone work as a reputable employment for girls of a good class."[22] In passing, the British committee noticed that in all the cities visited, "girls of Irish parentage educated in America were the most satisfactory operators. Their voices and manners are pleasing and they show tact and an intelligent alertness in dealing with subscribers."

What these stringent requirements added up to was a very

high turnover. Recruitment has been a perennial problem for telephone authorities from the beginning. The capable girl interested in telephone work has not, in the past, been overly ambitious or she might have gone into teaching or nursing. The very qualities that made them good operators made them good wives. Turnover rates have always been high and the search constant for more efficient and effective ways of selecting and training qualified girls.

Such was the demand of the telephone companies for female labor that by 1902 women equaled men in the Bell System.[23] Because of the typewriter as well as the telephone, the percentage of American women working in jobs other than agriculture or domestic service rose from 20 percent (when it was first surveyed by the United States Census) to 42 percent. The Census Bureau noted in 1902 that telephone work was particularly attractive to the educated girl.[24]

As a consequence of the numbers involved, increasing attention was paid to the health of operators. In 1915, a field investigator from the U.S. Commission on Industrial Relations remarked upon the incidence of nervous breakdown among operators and attributed it to the rigid discipline and strain at the telephone exchanges.[25] As a remedy, Miss Nelle Curry recommended that operators be restricted to a working day of six hours, that none younger than 18 be employed, and that they be paid a minimum of $55 a month. (Miss Curry's major conclusion—that female telephone workers, because of their youth, docility, and short stay in the job, were unlikely to look to organized labor to protect their interests—was proved wrong in a few years' time, as operators flocked to join the International Brotherhood of Electrical Workers and the National Federation of Telephone Workers.)[26]

The unhealthy tensions of the operator's day were seen in Britain, in 1911, as a cause of the high absenteeism among telephonists. A medical committee identified several irritants. One was the perpetual jumping up and down and stretching, as the girls plugged and unplugged cords. Others were the constant wearing of head and chest equipment, the occasional electric shocks delivered by the apparatus, and—above all—the dealings with an impatient public, ignorant of the workings of the telephone. An operator was "constantly smoothing out difficulties and is often the subject of abuse or reproach."[27] Its recommendations were for shorter hours, abolition of split shifts, and rejection of any girls who were anemic.

It was at about this time that telephone authorities began to put into effect generous programs of sick benefits and vacation pay. The New York Telephone Company in 1883 had given its women operators a week's vacation with pay and in 1891 added a Christmas bonus of a $10 gold piece.[28] Other Bell companies introduced sick pay. By 1911, the British General Post Office provided free medical attendance for its operators. It gave full pay for six months' illness. Switzerland, Germany, Norway, Sweden, and Belgium held comparable policies.[29] In 1913, the Bell System established an overall employee benefit plan, replacing the various arrangements made by the individual companies.[30]

## A JOB DONE BY ROTE

The chief satisfaction of the work to the operators was the contact with the public, irritating as that could be. For the rural telephone operator, this satisfaction endured well into the twentieth century. She had social status as well as genuine power in her community, knowing as she did the whereabouts of the doctor, the fire chief, and the contents of most people's telephone calls. However, the urban operator's work quickly lost the personal touch. There were too many subscribers to be known by name, and in the interests of efficiency, supervisors were placed to monitor the operator's work.

The first supervisor in the Bell System, Miss Katherine Schmitt, began in 1881 and did not retire until 1930. She designed the training methods, with their emphasis on good diction, in which generations of Bell operators were schooled. At the end of her career, Miss Schmitt defined the ideal: "The operator must now be made as nearly as possible a paragon of perfection, a kind of human machine." [31]

Anyone who has ever listened to telephone operators talk among themselves will know that "the supervisor" was once an awesome figure. She sat behind the girls where she could see them but they could not see her. She, or a team of higher-ranking operators, monitored the performance of a girl and listened in on her telephone calls. A woman operator at a London exchange several decades ago recalled what her first supervisor was like: "If she didn't choose to say good morning, she didn't. She just sat there like the old Queen Mary when you went in. And we would never call her by her Christian name."[32]

The constant search to improve efficiency soon led to what came out of the operator's mouth. Lists of set phrases were drawn up. The operator was not allowed to use her own words. She had to recite, as taught, "I'm sorry that li-on is busy." Should a situation arise not covered by the phrase book, she had to refer the call to the supervisor.

Again the view from across the Atlantic was admiring. The British Post Office adopted the American standard expressions in 1908. The object was, an official explanation says, to get around the delays caused by operators giving diffuse explanations. But the press, which greeted the news with great cheer, saw another advantage. Perhaps telephone service could be made courteous at last. So much bad language was being exchanged over the wires that, the *Birmingham Evening Despatch* declared, "the telephone in England has largely destroyed what was once a nation of good-tempered people."[33] The *Irish Times* blamed the telephone itself: "Its function encourages, and even requires, a graceless brevity of expression . . . and a man is apt to carry these faults into the less mechanized relations of life." Welcome then, said the *Irish Times*, to the standard expressions "with their liberal requirements of 'Please' and 'Thank you.'"[34]

As the decades went by and the demand for operators increased, their training became more and more routinized. In the 1930s, requirements were still strict: the Bell System, in large cities, rejected two out of three applicants because they did not meet the qualifications for health, intelligence, eyesight, and temperament.[35] When they were hired, they began right away on a practice switchboard and underwent intensive drills in locating lines on the board and understanding speech in the faint and garbled forms that it often reaches operators. And particular care was taken to ensure that each girl trained to the specific needs and traffic conditions of her office.[36]

In spite of, or perhaps because of, the regimentation of the job, women seemed to enjoy the work. By 1932 the Bell System provided work for nearly 3 percent of the female labor force, ex-

cluding farm workers and servants.[37] They liked the security
and the sense of being looked after by a public utility. They enjoyed the companionship of other women who were not rough.
And somehow they retained a sense of performing a public service. The famous advertisements of the Voice with a Smile,
showing a neat, proper young woman, blandly pretty, appealed
to operators as well as to the public.

## AUTOMATION

By 1946, virtually a quarter million women were working as
operators for the Bell System. Since then, because of the introduction of automation in various forms, their numbers have declined—to be balanced, however, by the rise in numbers of jobs
in telephone companies' clerical staffs. In truth, the telephone
operator's job was threatened by automation right from the beginning. The automatic switch, which connects two parties
without the aid of an operator's hand, was invented by Almon
B. Strowger in 1889 because, legend has it, he wanted a "girlless
cussless telephone." Strowger, an undertaker in Kansas City,
suspected that local telephone operators were shunting calls for
his funeral establishment to those of his rivals. While the Bell
System began developing the dial (which is what the automatic
switch meant to the subscribers) as early as 1900, it did not
make its first large installation until 1914, in Newark, New Jersey.[38] Bell's first all-automatic exchange opened in Omaha,
Nebraska, in 1921. When the fiftieth anniversary of the telephone came around, only 20 percent of the system had been
converted to dial, but the pace quickened after that. Wherever
the dial replaced the manual system, the gain in efficiency was
about 50 percent per operator.

Unfortunately, the changeover happened to coincide with the
onset of the Depression. Yet the much-feared loss of jobs among
telephone operators never materialized. While AT&T, parent of
the various Bell operating companies, had been criticized for
continuing to pay dividends to stockholders while reducing its
payroll by nearly 185,000 during the lean years,[39] it drew praise
for its management of the conversion from manual to dial exchanges. Mrs. Frances Perkins, Secretary of Labor, called it an
"almost perfect example of technological change made with a

minimum of disaster."[40] Although the dials ostensibly made some operators obsolete, the constant growth in the volume of telephone traffic and the rapid turnover among operators made it possible for Bell to avoid mass layoffs. Yet it is obvious that the dial went a long way toward removing the operator's personal contact with the public; electrical impulses replaced the caller's instructions to the operator.

Modern operators have lived through a technical revolution comparable to the introduction of the dial, with the appearance of the cordless switchboard. Instead of cords and plugs, today's operator faces lights and keys. She does not choose which call to answer. Automatic call distributors feed them to her automatically. The increasing reliance on computer technology in central exchanges allows customers to dial most of their own long-distance calls, and the operator comes in only briefly on person-to-person and reverse charge calls. Depersonalization has increased. Some operators who have worked with both kinds of switchboard unhesitatingly prefer the old model, with its physical activity, almost like weaving. "You don't get the same sense of accomplishment," one woman admitted.[41] On the other hand, automatic dialing, now extending to international calls, has freed operators for work on information services where they have lengthy contact with the public and must rely a great deal on ingenuity and personal knowledge. By 1980, the Bell System estimates that 84,000 of its operators will work on information services.[42]

But the work, at best, is monotonous. This plain truth caught up with the Bell System, and the British Post Office as well, in the prosperous 1960s. Turnover rates escalated as there was a shortage of labor and operators could find better-paying jobs in business and industry. With the exodus of the middle classes to the suburbs, the pool of female labor that existed in very large cities tended to be less well-educated and submissive than before, and many were immigrants from rural areas or abroad and lacked the basic knowledge of the city's place names and accents that the job requires. The turnover, at some exchanges, exceeded 100 percent per year.

Yet it was undeniable that the jobs paid well and offered excellent fringe benefits. Might there be something wrong with the nature of the job itself? Led by Dr. Robert Ford, director of

**Table 1**
Traffic Operators in the Bell System

|      | Men   | Women   |
| ---- | ----- | ------- |
| 1935 | —     | —       |
| 1940 | 994   | 107,381 |
| 1945 | 56    | 171,383 |
| 1946 | 60    | 223,764 |
| 1950 | 72    | 208,067 |
| 1955 | 16    | 205,160 |
| 1960 | 8     | 159,946 |
| 1965 | 5     | 148,041 |
| 1967 | 0     | 153,133 |
| 1970 | 167   | 165,461 |
| 1974 | 7,725 | 134,198 |

Note: Selected years from Bell System's personnel figures on Nonmanagement Employees.

work organization at AT&T, the whole company became a testing ground for the motivation theories of Dr. Frederick Herzberg, the industrial analyst, who held that the strongest sense of job satisfaction derived not from pay or extra benefits, but from the work itself.[43] People like to work well, he said. By Dr. Herzberg's definition, the telephone operator's job was a prime example of job impoverishment. Girls were hired for their intelligence, then prevented from using it. The rigid routines and deference to superiors forced on them were counterproductive and outdated.

After a decade, the final verdict on AT&T's campaign for job enrichment is not in. Some parts of its vast system were more responsive than others. Yet there is no doubt that it has cheered up telephone operators. They can talk like human beings; regional and ethnic accents are recognizable. They can handle their own emergency calls without referring to the supervisor. They can leave their posts without asking for permission, and trips away from the board are no longer counted. Bell's employee publications now stress that operators require more tact and judgment than formerly.[44] An operator from Mississippi said, "In the past, when a new practice would come in, they used to sit us down and read it to us like a bunch of kindergarten chil-

dren. Now they pass the practice around and we read it ourselves . . . . All in all, there is a lot more trust shown and we're
finally being treated as the adults that we really are."[45]

## SEXUAL DISCRIMINATION

It was not until the 1970s that the telephone operator came to
appear a classic example of discrimination by sex. Why were
only women operators? Why were so few women in the better-
paid slots of the telephone system, especially in the upper levels
of management?

Legislation against sexual discrimination in employment hit
the Bell System hard. The habits of nearly a century were challenged by the Equal Employment Opportunity Commission. In
January 1973, AT&T signed a consent decree with the EEOC,
promising to aim at special goals for the hiring and promotion
of women and minority groups. It also made some conspicuous
gestures, such as appointment of the first woman to its board,[46]
to demonstrate that women were welcome at the top as well as
the bottom.

For all the well-publicized pictures of women climbing telephone poles, the task of shifting the sexes out of their traditional niches in telephone service has not proved easy. In 1975,
AT&T acknowledged that it was having trouble attracting wom-
en to outside craft jobs. It said, almost plaintively, that "most of
the women the telephone companies have traditionally attracted
have tended to be women who are interested in inside work."[47]
It would have to aim at the woman who before would never
have thought of working for a telephone company.

By 1975, AT&T was able to report that one third of its employees at the first level of management and 4 percent of those
at the top were women. But the EEOC's judgment was that the
record overall was not good enough, even though the fault was
not always AT&T's. New and tougher goals were set for hiring
and promoting women and members of minority groups.

In Britain, where telephone growth has been slower and employment policies more conservative, the process of undoing
sexual segregation at the switchboard was not required by law
until December 29, 1975. On that day, both the Equal Pay Act
and the Sex Discrimination Act came into force, with major im-

**Table 2**
Plant Craft Forces in the Bell System

|       | Men     | Women  |
|-------|---------|--------|
| 1940  | 58,984  | 16     |
| 1945  | 54,906  | 457    |
| 1950  | 108,259 | 69     |
| 1955  | 132,796 | 439    |
| 1960  | 141,511 | 353    |
| 1965  | 152,059 | 362    |
| 1970  | 198,267 | 1,939  |
| 1974  | 205,826 | 11,978 |

Note: Selected years from Bell System's personnel figures on Nonmanagement Employees.

plications for the telephone service. Unlike the United Sates, which switched to an all-female force before the turn of the century, Britain has kept night work exclusive to men; the only break in this pattern came during the two World Wars. Even then, the unions objected very strongly either from prejudice against women or from a fear that men would lose jobs to women because men received higher pay.[48] The Post Office's archives contain a bundle of agonized correspondence showing the Postmaster General wondering, at the end of World War I, how he could keep his promise to turn the night girls out when the men came home from war. It was the old story: the women, he said, were quicker, more accurate, deft, skillful, and attentive than the men. The ideal, therefore, was an "all-woman night staff." But the union would not hear of it, and the women were ousted. The same process happened during World War II. In recent years, as British policies advocated eliminating the differences between men's and women's pay, the male night operator's higher wage was explained as a bonus for working at unsocial hours. Under the new laws, women will be able to apply for night work and men for day work, but there is still old protectionist legislation restricting women's night work to be overcome.

## WOMEN IN MANUFACTURING

For their same long-suffering virtues and more delicate finger action, women have been a mainstay of telecommunications equipment manufacturing. In the United States, women workers may have fared better at Western Electric, the supply subsidiary of AT&T, than they did in the operating companies. Western Electric boasts that it has given equal pay for equal work throughout its corporate history, which goes back to 1856.[49] Women worked at braiding cords and winding coils, in the main, but also, as the twentieth century progressed, were used as chemists, systems analysts, managers, engineers, and economists. By 1900 there were 100 women at Western Electric; by 1920, there were 8,900. By mid 1975, women comprised 35 percent of the total of all employees. Yet of this work force in 1975 fewer than 400 women were supervisors, and Western Electric promised the EEOC that it would pay compensation to 2,000 women for promotions or raises they would have had had they been men.[50]

In the mid 1970s in Britain, an estimated two-thirds of the workforce engaged in making telephone equipment are women; they are just beginning to receive pay equal to that of men. "Your damned equality," said a representative of the industry, "comes at a time when the industry, the Post Office, and the country can ill afford it."

Such a remark evokes the early days of the telephone service, a time less ashamed of its sexual stereotyping. But it is necessary to keep a historical perspective. Women often did not do work equal to that done by men because such work in the days of heavier machinery was often beyond their capacity. Women did not seek advancement because they expected to be kept at home by child-raising for most of their lives. Men got the jobs that led somewhere. A British Postmaster put it most succinctly when, in 1924, he decided to give girls the work of breaking up old telephone instruments: "The job offers no opportunity for training boys either in constructive mechanical work or for storekeeping duties. I therefore propose to substitute girls on it." Moreover, during most of the lifetime of the telephone jobs were scarce. There was a general belief that married women should not deprive a breadwinner or a spinster from a job. It also should be clear that, drudgery or not, women enjoyed the work of telephone operating. It gave them good jobs during a century which believed that woman's place was in the home and laid down very strict conditions for their working outside it. It is still doing so in the developing world, where the telephone is belatedly making its way. And if there are comforts in the telephone operator's job, it is a sign of progress that men are now allowed to enjoy them, too. Perhaps the most interesting statistics yet to emerge as AT&T struggles to right the historic imbalance of the sexes in its job assignments is that in 1974, it employed 7,400 male telephone operators. In 1957, before anybody had heard of women's liberation, there were none.

## NOTES

1. James Thurber, *The Thurber Carnival* (London: Penguin, 1945).
2. W. J. Weatherby, *Conversations with Marilyn* (London: Robson, 1976).
3. Robert V. Bruce, *Alexander Graham Bell and the Conquest of Solitude* (London: Gollancz, 1973), p. 14.
4. Elizabeth F. Baker, *Technology and Women's Work* (Columbia University Press: New York, 1964), pp. 10–16.
5. Ibid., p. 13.
6. Samuel Eliot Morison and Henry Steele Commager, *The Growth of the American Republic* (New York: Oxford University Press, 1930), Vol. II, pp. 115–116.
7. Baker, *Technology and Women's Work*, p. 65.
8. Frank B. Jewett, "The Telephone Switchboard—Fifty Years of History," *Bell Telephone Quarterly*, VII (1928), p. 9.
9. Frank B. Jewett, "Opportunities for Women in the Bell System," *Bell Telephone Quarterly* (1932), p. 32.
10. Thomas W. Lockwood, Memorandum on Night Operators to the Bell Telephone Company, October 19, 1891.
11. Katherine M. Schmitt, "I was Your Old 'Hello Girl,'" *Saturday Evening Post*, July 12, 1930, p. 3.
12. F. C. C. Baldwin, *The History of the Telephone in the United Kingdom* (London: Chapman and Hall, 1938), pp. 269–270.
13. Ibid.
14. Morison and Commager, *American Republic*, p. 241.
15. Schmitt, "Hello Girl," p. 3.
16. Report of the Committee of Medical Officers, "Conditions of Working of Telephonists" (London: His Majesty's Stationery Office, 1911).
17. I am grateful to the information department of the American Tele-

phone and Telegraph Company for searching out this information on what was custom, rather than policy, and which varied widely throughout the Bell System.

18. Report of the Committee of Medical Officers, 1911, p. 22.

19. Correspondence between the General Post Office and the Postal and Telegraph Clerks' Association, London, August–September, 1915.

20. R. Barrett, "The Changing Years as Seen from the Switchboard," *Bell Telephone Quarterly* 14 (1935), p. 105.

21. R. Barrett, "First Lady of the Switchboards," *Long Lines*, April 1951.

22. Extracts from official reports on American telephones, 1904–1906, British Post Office archives, Doc. E28782/1909F6.

23. Laura M. Smith, "Opportunities for Women in the Bell System," *Bell Telephone Quarterly* 11 (1932), p. 34.

24. Baker, *Technology and Women's Work*, p. 69.

25. Nelle B. Curry, *Investigation of the Wages and Conditions of Telephone Operators*, U.S. Commission on Industrial Relations Report, 1915, pp. 6–7.

26. Baker, *Technology and Women's Work*, pp. 371–383.

27. Report of the Committee of Medical Officers, 1911, p. 5.

28. New York Telephone Company, Information Department 3-10.

29. Report of the Committee of Medical Officers, 1911, p. 22.

30. Barrett, "First Lady of the Switchboards," *Long Lines*, April 1951, p. 214.

31. Schmitt, "Hello Girl," p. 120.

32. Brenda Maddox, "Good Jobs for Girls," *Telecommunication Journal*, December 1975, p. 711.

33. Editorial, *Birmingham Evening Despatch*, December 9, 1908.

34. Editorial, *Irish Times*, December 10, 1908.

35. H. C. LaChance, "The Training of Telephone Operators," *Bell Telephone Quarterly* 10 (1931), pp. 12–16.

36. Ibid.

37. Smith, "Opportunities for Women," p. 35.

38. Jewett, "The Telephone Switchboard—Fifty Years of History," *Bell Telephone Quarterly*, VII, 1928, p. 149.

39. Joseph C. Goulden, *Monopoly* (New York: Revised Pocket Book, 1970), p. 240.

40. Baker, *Technology and Women's Work*, p. 317.

41. Maddox, "Good Jobs for Girls," p. 712.

42. "The Changing Role of the Telephone Operator," *Bell Telephone Magazine*, Vol. 46, November–December 1967, p. 6.

43. "Let's Talk about Job Enrichment," *Bell Telephone Magazine*, Vol. 52, May–June 1973, pp. 16–21.

44. "Changing Role," p. 4.

45. "Job Enrichment," p. 19.

46. Information sheet on compliance with EEOC's consent decree, American Telephone and Telegraph Company, 1975, p. 2.

47. Eileen Shanahan, "AT&T is Penalized Anew for Job Bias," *New York Times*, May 14, 1975.

48. When in 1915 it was first proposed to put women on night duty, at 22 shillings versus the male 30 shillings a week, the Postal and Telegraphs' Association wrote to the Postmaster General saying that, even in spite of the war, "It is thought by my committee that men would be available for this work if the Department would offer a wage sufficiently high to attract them." (Letter from General Secretary, March 17, 1915.)

49. "First There was Sara," Western Electric film script, October 6, 1971, p. 2.

50. Western Electric news release, May 5, 1975.

# 13

## The Telephone in New (and Old) Communities

Suzanne Keller

Someone invented the telephone
And interrupted a nation's slumbers
Ringing wrong, but similar numbers.

Ogden Nash

The German Social philosopher, Georg Simmel, once speculated what a calamity it would be if all the clocks of the nineteenth century metropolis suddenly stopped telling the time. We know that it would be even worse if the telephone system ceased to function. When the women of Iceland went on a one-day strike for International Women's Year on October 24, 1975, the biggest problem was communications—with telephone service at a virtual standstill.[1] And in March 1975 a switching center fire knocked out twelve exchanges in a 300-block area of New York City; some 100,000 residences and 8,500 business and professional offices were left without service, with consequences examined in Chapter 11 by Wurtzel and Turner. Innumerable problems resulted not only for commerce, but also for the elderly, disabled, and those caught in emergencies. Without doubt, the telephone has become indispensable and modern life inconceivable without it.

Given the telephone's ubiquity and centrality, it is striking how ignorant we are about its impact on our lives; anecdotes abound, but there is little systematic study. Berelson and Steiner, in their massive compendium on human behavior, are silent regarding the telephone.[2] In fact, few of the machines that have transformed modern life—the elevator, the motor car— have been adequately studied for future record.

The telephone is a curious cultural artifact. On the one hand it is seen as a link to distant people and places, but on the other it is a symbol of loneliness. One is reminded of the famous 1920s song "All Alone by the Telephone" and of the last image of Marilyn Monroe, hand on the phone, reaching for life while withdrawing from it. "Why," asks Marshall McLuhan, "should we feel compelled to answer a ringing public phone when we know the call cannot concern us? Why does a phone ringing on the stage create instant tension?" His answer is that the telephone is a participant form of communication that demands a partner with all the "intensity of electric polarity."[3]

From these preliminary aperçus, it is clear that the telephone can break our contacts with the world just as it can create and sustain such contacts.

In this paper I will concentrate on the ways telecommunications have helped create links among people, and what these links imply about the nature of modern communities.

## TECHNICAL AND HUMAN SYSTEMS OF COMMUNICATIONS

It seems hard to realize that once human communication depended largely on face-to-face contact. This had a number of important consequences; it reduced the radius of connections within and between settlements and fostered a dependency on physical continuity and proximity that survive in our fantasies as the earmarks of "true" communities.

In contrast to that era, we now have advanced techniques that extend us in space and time and enable us to transcend these communities of place. Despite our dazzling innovations, however, we are not yet free from our ancient dependencies on the local and physically near. One of the most interesting questions is the meaning of "near"—once human yardsticks are displaced by electronic ones.

## COMMUNICATION CHANNELS IN HUMAN COMMUNITIES

Communication is not only necessary for the formation of human communities, it is also indispensable for sustaining them.

We know that a growth in the size and scale of communities is always, and perhaps necessarily, associated with a change in the number and kinds of communications channels. One difference between villages and cities concerns the typical configurations of networks in each. In villages, networks tend to be overlapping, comprehensive, and in that sense closed, whereas in cities they are nonoverlapping and open-ended. Such open-ended networks complement other tendencies of urban life toward variety, diversity of interests, and pluralistic standards and styles. Elizabeth Bott, in her study of British urbanites, found a clear-cut difference among her respondents regarding the type of friendship network they belonged to. More urbanized residents had friends throughout a large and diversified area; these friends of the respondents did not necessarily know one another or have any common interests beyond knowing the respondent. The less urbanized city dwellers, however, belonged to networks of kin and friends—all of whom were acquainted and had connections and shared interests beyond their knowledge of the respondent. Not surprisingly, the closed networks are also spatially confined and localized, whereas the more urban networks spread far afield and could surely not be sustained without the telephone or some equivalent medium.[4]

This supports the notion that we really must supplement the study of territorial communities with studies of telephone communities, especially since these are expected to proliferate in the future. Such studies would extend our understanding of a whole range of phenomena involving the interplay between technical and human systems of communication.

Our ignorance about the ways these two systems connect may account for the large gap that continues to exist between our technical capacities to transport people and goods and our incapacity to coordinate them with specific personal schedules and programs. Despite the pervasiveness of telephone contacts it would be difficult to describe, in detail, a typical urbanite's daily telephone commute. We have an intuitive awareness of the growing significance of telecommunications, but we still do not know much about the hows and wherefores of its uses.

## ON POSSIBLE USES OF THE TELEPHONE

The various reasons for telephone use in daily life boil down to two: *instrumental* and *intrinsic* uses. Among the instrumental uses are the resort to telephones in times of household crises or emergencies such as illness, accidents, or unwanted intruders. Closely allied to this is the telephone as an aid to safety both in and outside the home. Thus the New York City Welfare Department now permits welfare recipients to have telephones. Mothers on welfare tend to live in dangerous neighborhoods, which makes telephone protection not a luxury but a necessity.

Another growing telephone use is for the purchase of goods ranging from daily groceries to Wall Street stocks. Wall Street business is conducted over the telephone and the modern Stock Exchange is inconceivable without it.

Finally, the telephone is the daily messenger; we use it to make appointments for dates and dinners, reservations for trips and outings, to transmit information and advice ranging from inquiries about the weather to counsel on intimate personal matters.

Requests for information are made by only a minority of telephone users—90 percent of the calls for information are made by 30 percent of the people, and the most popular requests are for time, weather, and timetables. In New York, the ten most frequently requested numbers are for Penn Central, Eastern Airlines, American Airlines, the Department of Motor Vehicles, Pan Am, Macy's, the Port Authority, TWA, United Airlines, and Western Union.[5] Here the telephone is clearly a means to geographic mobility.

Unlike television or radio, the presence of a telephone is a link to needed services and people; thus it helps to reassure shut-ins, newcomers, and individuals living alone. Americans can today call for a sermon, a pep talk, medical advice, astrological forecasts, and advice to the lovelorn.[6] There are "telephone reassurance programs" consisting of daily calls to the elderly living alone to see whether they need something. If there is no answer, someone is sent to the house immediately. The greatest fear of these isolated older persons is of dying alone without anyone knowing about it. Unfortunately, many poor, elderly persons do not have telephones and lack the money needed for the initial deposit and upkeep.[7]

Medical advice is also frequently sought by phone. In Hungary, for example, "Dr. Telephone" gets some 20,000 calls a week for information and therapy.[8] Hotlines (discussed more fully in Chapter 20 by Lester) have become a staple in many communities in this country, dispensing advice on marriage, drug, and drink problems to individuals who have no one else to turn to, who cannot relate well to people they know, or who wish to hide their problems from intimates. Requests for legal advice constitute another large number of calls. Confidentiality and anonymity appear for many to be the attractions of such telephone relationships.[9]

Among the *intrinsic* uses of the telephone are the social contacts it facilitates between friends, relatives, neighbors, and clients. While mobility and distances have greatly increased, speed of connections has been greatly improved. In this sense it is true that the telephone is "one of the basic instruments holding people together."[10] The volume of such contacts has increased enormously over the last few decades. In 1969, Americans participated in 350 million telephone calls each day. Long-distance and overseas calls have greatly increased in the last twenty years and the number of telephones—150 million in 1970—has increased fourfold since 1950. These linkages connect not only individuals, but also entire communities; they thus help forge a common national and international culture. When the telephone first came into widespread use around the turn of the century, its role in erasing traditional barriers between town and country and in linking widely scattered communities was often remarked upon.[11]

Such a proliferation of wanted contacts, however, also implies a proliferation of unwanted contacts. People protect themselves by having unlisted numbers, leaving phones off the hook, or interposing secretaries as shields and go-betweens. It would be interesting to know how large and diversified a community becomes before such devices are adopted. One wonders whether there now is a new breed of telephone hermits and how these might differ from the more traditional variety.

Such general descriptions of telephone usage obscure significant variations by region, season, and social class. There are homes where virtually every member has a telephone in his or her own room; this is the case in the planned community I am studying where 80 percent of the teenagers live in households with three or more phones. There are others where a single phone may be shared in a hallway by thirty tenants. Still others, even today, have no private phone available at all. Although more than 95 percent of all households have telephones now, one in twenty is still without one, and this group is disproportionately represented among older and poorer citizens.

Telephone usage also varies by season. In one study of Canadian housewives, telephone usage decreased between winter and summer as personal visits replaced telephone visits.[12] In other cases, telephone contacts proliferate along with personal contacts.

## TYPES OF TELECOMMUNITIES

Since the primary interest here is in entire communities, let us try to specify how the telephone helps create communities.

1. *The telephone as a creator of spontaneous communities.* A study by Gaertner and Bickman dispelled the myth of the callous urbanite by charting the willingness of strangers to extend help requested by telephone. Callers would dial a random number and say they were stranded on a highway with a broken-down car and had used up their last dime to call the present number in the hope of reaching a garage for help. They then asked the listener if he would be kind enough to dial the correct number for them. Quite a few received help in this way.[13] Another study compared people's readiness to permit strangers to enter their homes to make a telephone call. Compared to those from cities, small-town residents are inclined to be more trusting and helpful. Of interest is the extent to which such spontaneous communities can be created instantaneously by use of telecommunications. We may see a rise in such spontaneous and short-lived "telecommunities" in the future.

2. *The telephone as a creator of therapeutic or altruistic communities.* In a summary and codification of community responses to disasters, Allen Barton concluded that one of the crucial ingredients for the emergence of a communal helping hand was the ex-

istence of central communications nets without which, despite
good intentions and available resources, rescue efforts and information monitoring were almost impossible. Poorly coordinated communications systems can greatly, perhaps irrevocably, set
back the rallying and recovery of a community stricken by a disastrous fire or flood. Accessibility to existing equipment proves
as crucial as the availability of such equipment. "For most organizations, the loss of telephone service is literally disorganiz-
ing."[14] But such loss of accessibility is precisely what follows
the panic that clogs highways or telephone connections in times
of crisis.

Barton makes the "ability of modern societies to create an
emergency social system" for responding to calamities like fires,
earthquakes, or air raids, "uniquely dependent on long-range
instant communications."[15] In outlining a "model of the therapeutic community response," communications play the expected
crucial role in activating "mutual help in situations of collective
stress." Formal and informal channels of communication carry
the stories of victims to those spared the catastrophic events,
thereby helping to form a network of need and concern among
them.

Despite Barton's general acknowledgment of the importance
of communications in collective emergencies, he does not discuss the role of specific media or modes, presumably because
the original studies on which his codification was based did not
discuss them. But a common tendency is to overemphasize the
significance of direct personal contact, and of physical proximity
for communication. Barton observes, for example, that "the closer an individual is to the location of the impact (of the disaster),
the more likely he is to have direct contact with the victims."[16]
With the availability of telecommunications, however, anyone
would seem to be close to the impact, being virtually only
moments away from reports, pictures, and eyewitness accounts
of the disaster. Barton implies that the relevant proximity is
territorial rather than electronic. In another example, he suggests that people "are more interested in nearby events than in
remote ones." But what is remote and near in our global village? In an electronic era, nearness and distance connote quite
different meanings.

Further proof of the importance of electronic accessibility versus physical proximity to events comes from studies of collective panic and crisis situations. When calamity strikes on planes, ships, or in nightclubs, access to communications media can be as crucial as access to exits. For one thing, an anxious crowd needs to be supplied with information and reassurance; without them they readily panic. But there is more to it than that. The number of channels is also crucial. When crisis strikes in an enclosed space, it is apparently better to have no exits at all than to have too few exits. One overloaded or inaccessible channel of contact or escape—be it a fire exit or a telephone line—may set off the very panic it was designed to prevent. Where there are no exits at all, then no hope for survival is generated. Where there are insufficient or inadequate channels, however, hope and desperation exist side by side; the simultaneous chance for escape and the terror of being trapped may make the single road to life paradoxically and tragically the path to death.

In view of this it would seem desirable to plan adequate communications systems in new communities to prepare them for eventual emergencies. It is not at all certain that we are building such safety factors into the communities now under way. We are still not truly convinced that we are entering an age of telecommunications where a clogged telephone channel can be as serious as a jammed highway or a blocked exit in a burning building.

In this connection, the Robert Wood Johnson Foundation has recently supported the setting up of a telephone system in thirty-two states and Puerto Rico as a means of summoning emergency care for the 700,000 victims of heart attacks—one-half of whom die before reaching a hospital—and for the 115,000 victims of accidents each year. Such an emergency program would include phoning in the necessary information to the hospital from the scene of the emergency, thus preparing the setting for the patient and saving valuable time. [17]

3. *The telephone as the means to suburban sociability*. In Park Forest, according to William H. Whyte, the telephone was clearly important in getting the community organized and mobilized for all kinds of projects. It was also instrumental in speeding up "suburbia's phenomenal grapevine" and useful in spreading

scandal and bad news which travels more readily along telephone channels than from person to person.[18]

It has long been known that the siting and design of dwellings (where they face, where stairs are placed, and where common footpaths and exits intersect) play an important, if contin-
gent, role in the formation of social relations and groups. The
telephone's role has not been singled out for its possible contribution, but Whyte noted that morning visits among housewives—forerunners of later friendship cliques in the various
courts—depended upon the telephone in important ways.
"When wives go visiting," he observed, "they gravitate toward
the houses within sight of their children and within hearing of
the telephone, and these lines . . . crystallize into the court
'checkerboard movement.'"[19]

In time even here there was need for escape, as the hectic
pace of the initial sociability began to take its toll and people
did the unforgivable—"they don't answer the phone."[20]

## THE TELEPHONE IN NEW COMMUNITIES

Norbert Wiener once remarked that "society can only be understood through a study of the messages and the communications
facilities which belong to it."[21] If this is true, then in our preoccupation with transportation systems we may be going about
things the wrong way. Telecommunications is not only ignored
in the design of new communities, it is largely neglected in the
study of old ones. And even in readers and compendia devoted
to the topic of communications, one often looks in vain for
some mention of it. Radio, TV, and newspapers have been
extensively studied and discussed (largely because of the pioneering efforts of Paul F. Lazarsfeld and his students), but the
telephone is conspicuous by its omission.

As for what might be done in the future, let us keep in mind
that between now and the year 2000 hundreds of new communities (ranging in size from 10,000 to 500,000 people) are being
planned for the U.S. alone. Whatever else, their potential contribution as laboratories for social and technical innovations is inestimable. Telecommunications might be included in a number
of ways in our speculations and designs for the future.

## Communications in the Home

In his study of proxemics and invisible personal bubbles and buffers, Edward Hall has suggested that cultures, groups, and individuals vary considerably as to the distances considered comfortable or appropriate for interpersonal, face-to-face communication.[22] To my knowledge, this work has not been extended to the bubbles of privacy required by different types of telephone users. But if our personal distance bubbles are not standardized, then why should we expect the telephone to be standardized—either as an artifact or in terms of various settings and purposes for use? In particular, since the telephone is a two-way channel of communication and an active medium permitting some exercise of individual will and control, personal and subjective factors should be at least as important here as they are in the cars people drive.

The problem of standardization plagues many of our ideas about suburbia and new towns. We assume, for example, that everyone's family conforms to the American ideal of a residentially independent married couple and their young children. But many studies have shown that only about half the families are standard in this sense. In addition, even standard families vary greatly in their uses of internal domestic space, their tastes in furnishings and appliances, their habits of entertaining and leisure, the amount of illness in their households per week or per month, and in their attitudes toward noise and privacy. How many of the accidents that occur in the home, for example, result from racing for a telephone that is inconveniently placed or too far out of reach?

As Daniel Bell has pointed out, one of the earmarks of modern society is "the loss of insulating space." This has pointed relevance for the location and accessibility of telecommunications within the home. While sheer availability of telecommunications is important, it is equally important to consider their location; privacy is indeed in short supply in the modern American home. Soundproofing, in particular, seems greatly neglected.[23]

Furthermore, families also differ in how much time they spend at home, what they expect from family life, and how much togetherness they desire. An early study by a Cornell team identified nine different value orientations among house

dwellers, a number of which are directly associated with communications behavior and demands. Those devoted to leisure-
time reading, record listening, and conversation (including, one
would suppose, telephone conversation) stress privacy and
noise control and the desire neither to overhear nor to be overheard. Those devoted to physical health or to social climbing,
on the other hand, would stress quite different design features
and communications devices.[24]

### Communications Outside the Home

Moving from the private dwelling to more public spaces, it
seems that the number and location of public telephones is significant. In Resurrection City, for example, which was virtually
constructed overnight, and with minimal resources, public telephones were part of the indispensable infrastructure and were
located in the most public place on Main Street.[25]

One new community studied by the author had no public
telephones (except one in the Management Office for visiting
contractors) even though the community already had 500 families living there. The reason given was that the telephone
company's policy was not to install public telephones unless
guaranteed a minimum income. If public use were to fall below
this amount, the difference would have to be paid by the town
management. In practice such a policy goes back a long way. In
1895, when the village of Wanseon, Ohio, petitioned the Central
Union Telephone Company for service, they discovered to their
dismay that the main office in Chicago refused their application, because "Wanseon, with less than 3,000 [is] entirely too
small to warrant the installation of a plant."[26] One could argue
that because telephones are now necessities, public telephones
should be provided as a courtesy service at locations of the
greatest public convenience.

### Other Communications Needs in New Communities

In the process of developing coherence and continuity, communities go through several phases; communications planning
might do much to help in the transition from one phase to the
next. Four phases of adjustment to a new community have been
identified: (1) separation from the old familiar community and
feelings of isolation and strangeness in the new settlement; (2) a

frantic period of neighboring and socializing to counteract the loneliness, the inadequacies of the unfamiliar setting, and the personal anxieties engendered by the move; (3) a settling-down process followed by a retreat to the home and selective attention to neighbors and to community organizations; the latter eventually leads to (4) a more urban texture of life in middle-class suburbs versus a more provincial reaction in working-class suburbs.[27] None of these phases of adjustment has yet been related to the accessibility to various communications media and channels.

In the first phase one finds frequent references to the isolation of new residents. They have not yet been caught up by life in the new setting; the isolation is discussed almost solely in terms of inadequate transportation to the places and kin left behind or of inadequate human relations in the new setting. One would think, however, that the telephone would alter both of these reactions. While it may not be able to replace the daily visits between British Mum and her working-class daughter, it can bring them into daily emotional contact. Where there is access to a telephone within the new home, phase 1 should see an upsurge of telephone visiting and greatly speed adjustment.

Phase 2, which involves most face-to-face contact, should also benefit from any available technical aids to communication. Phase 3 may not occur at all without judicious use of the telephone. Phase 4, in which territorial proximity gives way to interest-based contacts, is also bound up with use of the telephone.

Other adjustment problems in new communities, albeit without any temporal phasing, involve an especially important age group—the adolescents. A major portion of Levittown's adolescents, for example, considered the community Endsville because there was not enough for them to do. Their chief complaints centered on a lack of sports, play, and meeting facilities, a lack of transportation to get to facilities that were available, and lack of privacy at home. All of this convinced them that Levittown was not planned for them, as indeed it was not. At home their rooms were too small to entertain and they lacked privacy and soundproofing to talk with ease. Presumably this includes talking over the telephone, a favorite pastime for young people.[28]

Similar complaints have been voiced in most other new towns both in the United States and abroad.

In addition to the young, elderly citizens must not be neglected in the planning of new communities. Such aids as instant signaling devices to convey sudden illness or other emergencies, "programmed automatic telephone dialing of critical numbers" for the housebound, and other conveniences have been proposed.[29] Old people like to sit and watch others, which is their way of participating in the active life around them at a pace they can manage. Perhaps they might also be served by making it possible for them to sit and hear conversations—anonymous, street conversations piped into their homes.

## SOME CHANGES AHEAD

If we are to integrate telecommunications and life in new communities in some explicit way, we must consider some of the broader trends and developments for the years ahead which would affect both.

In a famous essay, Melvin Webber discusses one of the significant developments affecting future communities in terms of moving from communities based on territoriality and proximity to "communities without propinquity." These would replace the current place-dependent communities by breaking down the territorial confinement of past communities. Webber does not specifically analyze the role of telecommunications in the formation of such communities, but it is obviously closely bound up with developments in communications.[30] Surely the radius of interconnections is bound to increase the radius of interest and cohesion in the global village of tomorrow and affect human behavior in ways still largely unexplored.

In this connection the telephone has long been seen as a decentralizing influence, for it permits spatial dispersion while encouraging interpersonal cohesion. Thus McLuhan, with his customary penchant for the exotic, suggests that the call-girl is a creation of the telephone just as the red-light district is its victim.[31] As J. R. Pierce, who has written extensively on communications in the future, reminds us, we "use the telephone *because* we have interests that lie beyond the home, the family, and the neighborhood." [32]

Spatial dispersion, even with a telephone close at hand, may nonetheless spell loneliness for those who have no one to call, and this loneliness should not be casually dismissed in the current urban aggregates of transients and strangers. Some, unsuited by personality or talent to meet others, may well remain deeply isolated from human contacts even while surrounded by telecommunications and the bustle of urban life. Indeed, most accounts of new communities, even as they stress the newfound sociability of the incipient suburbanite, also refer in passing to some who remain isolated from the main hub of invitations and communality. But here again the telephone could be used far more effectively as a public service than has occurred to us in our individualistic culture. On *Sesame Street* not long ago, one of the characters who needed to complain simply "dialed a grouch" just as some are already dialing a prayer in real life or dialing a last SOS when contemplating suicide. Such dialings might not only be expanded to meet a variety of needs now ignored but might also be formally and explicitly organized from without. A new role of telephone confessor, counselor, and therapist might thus be installed, responsible for calling the community's elderly, ill, or isolated as a sign that somebody does care.

We might here turn our attention to other proposed innovations involving telecommunications in the future. In addition to picture phones and Magnafax, there is the projected art of "telefaction" which would permit us to feel and manipulate objects at a distance via telefactor gloves. [33] In view of such novelties as bank-a-phone, education by television, shopping via closed circuit TV displays, and computer-telephone diagnosis of medical ailments, it may also be possible to conduct a major portion of certain types of work from one's home and perhaps dispense with the office, the department store, and the central city core altogether. [34]

Even more dramatic projected developments in science and society will transform life as we know it today. They also will help usher in the "second industrial revolution," the electronic society, or the automated world. This society of tomorrow, according to many serious observers, will be a society of communications, moving information and images as we now move people and goods. "We are," writes Robert Theobald, "moving from an order based on transportation and production to one based on communication, in which decisions and their results become simultaneous."[35]

There are really only two basic modes of creating contacts between human beings and their world. One brings human beings to the experience and the other brings the experience to human beings. [36] We have lived by the first, and we are moving toward the second—toward the electronic encounter with the world. This should permit us to experience the world in a far less fragmented manner than heretofore and may restore to us a wholeness and richness which the first industrial revolution destroyed. Of course it may also make us all more stationary, but with the difference that our senses would be in constant and instant touch with any part of the world and its happenings.[37]

Unfortunately, few of these projected possibilities have yet found their way into the minds or plans of the planners. All too many still think largely in terms of territories divided into stable residences from which people move to and from work, to and from shopping, and to and from amusements. Only a minority are thinking along the lines just mentioned.

I do not wish, in this brief foray into futurism, to ignore or minimize unforeseen problems that the electronic society may usher in. We already suffer from communications overload, both in terms of what existing channels can carry and what our nervous systems can process.

But the problems go even deeper. They touch on basic ways of perceiving and responding to the world and involve a basic reorientation in our designs for living. They challenge all those props of our existence we had thought immutable. Among these are the values of "settling down" in one place and one routine, of owning things, of having one occupation, one house, and one spouse for life—in short, all the idols of domestication: permanence, security, familiarity, and continuity. The world ahead, as conceived early in this century by H. G. Wells and in more recent years by Arthur Clarke and Buckminster Fuller, is just not like that.

Instead, as John McHale has suggested, we will increasingly

be renting rather than owning things as we develop a culture geared to mobility and improvisation. A whole ethos of personal possessiveness—of land, house, car, and kin—may thus recede and with it a whole chunk of human history.[38]

In line with this McHale sees a shift to new work patterns in round-the-clock cities, thereby breaking the artificial and by now superfluous dawn-to-dusk rhythm of agricultural and early industrial societies. He sees the coexistence of many alternative types of communities and settlements, each presumably with its own characteristic mix of communications media. Among these he singles out the mobile and flexible instant city, the university city which is inconceivable without advanced telecommunications, the festival city, recreation and museum cities, and experimental cities where new life styles may be explored by young and old.

As a final thought, perhaps we will not inhabit cities or communities in the future at all, no matter what their shapes or attractions. Having discarded the idea of fixed roots and definite boundaries, the idols of the foyer and the comforts of the hearth, perhaps we will turn into a species of exotic insects, stationary nomads, going everywhere without moving from one spot, in instant contact with any and everyone, armed solely with ourselves, our personal computers, and our portable telephones.

## NOTES

1. "Iceland: Women Strike," *New York Times*, October 25, 1975, p. 34.
2. Bernard Berelson and Gary A. Steiner, *Human Behavior* (New York: Harcourt, 1964).
3. Marshall McLuhan, *Understanding Media* (New York: McGraw-Hill, 1964).
4. Elizabeth Bott, *Family and Social Network* (London: Tavistock, 1957).
5. Andrew Tobias, "Sorry, Right Number," *New York Times*, October 4, 1972, pp. 77–87.
6. Robert A. Wright, "Urge to Dial Answered by Recorded Message," *New York Times*, November 20, 1971, p. 33.
7. "Phone is Lifeline to Homebound Aged," *New York Times*, October 23, 1973, p. 29.
8. Raymond H. Anderson, "In Need of Advice, Hungarians Dial a Psychologist," *New York Times*, February 17, 1973, p. 20.

9. Henry Jordan, "Social, legal problems flood hotline headquarters," *Daily Princetonian*, October 11, 1974, p. 4.

10. Ben J. Wattenberg and R. M. Scammon, *This U.S.A.* (New York: Doubleday, 1965), pp. 251–253.

11. Ray Brosseau and Ralph K. Andrist, *Looking Forward* (New York: American Heritage, 1970), p. 81.

12. William Michaelson, "Space as a Variable in Sociological Inquiry: Serendipitous Findings on Macro Environment," paper delivered to Annual ASA Conference, 1969.

13. Stanley Milgram, "On the Experience of Living in Cities," *Science*, March 13, 1970, pp. 1461–1468. See also, Suzanne Keller, *Twin Rivers, Study of a Planned Community* (School of Architecture, Princeton University, Sept. 1976).

14. Allen H. Barton, *Communities in Disaster* (New York: Doubleday Anchor, 1970), p. 170.

15. Ibid., p. 171.

16. Ibid., p. 218.

17. Lawrence K. Altman, "Phones to Speed Emergency Care," *New York Times*, May 23, 1974, p. 19.

18. William H. Whyte, *The Organization Man* (New York: Simon and Schuster, 1956), p. 394.

19. Ibid., p. 379.

20. Ibid., p. 389.

21. Robert W. Prehoda, *Designing the Future* (New York: Chilton, 1967).

22. Edward Hall, *The Silent Language* (Garden City, N. Y.: Doubleday, 1959).

23. J. E. Montgomery, "Impact of Housing Patterns on Marital Interaction," *The Family Coordinator*, 19, No. 3, July 1970, pp. 267–274.

24. G. H. Beyer (ed.), *Housing: A Factual Analysis*, Chapter 7 (New York: Macmillan, 1958).

25. John Wiebenson, "Planning and Using Resurrection City," *Journal of the American Institute of Planners*, November 1969, pp. 5–411.

26. Brosseau and Andrist, *Looking Forward*, p. 66.

27. Suzanne Keller, *The Urban Neighborhood* (Magnolia, Mass.: Peter Smith, 1968), p. 72.

28. Herbert Gans, *The Levittowners* (New York: Random House, 1967), p. 206.

29. M. Powell Lawton, "Planner's Notebook: Planning Environment for Older People," *Journal of the American Institute of Planners*, March 1970, pp. 124–139.

30. M. M. Webber, "Order in Diversity: Community Without Propinquity," in L. Wingo (ed.), *Cities and Space* (1963), pp. 23–54.

31. McLuhan, *Understanding Media*, pp. 233 ff.

32. J. R. Pierce, "Communications," in Daniel Bell (ed.), *Toward the Year 2000* (Daedalus, 1967), pp. 909–1021.

33. Edward N. Hall, "The Anomalies of Urban Requirements," United Aircraft Corp. (1969).

34. Olaf Helmer, "Simulating the Values of the Future," in Kurt Baier and N. Rescher (eds.), *Values and the Future* (1969), pp. 193–214.

35. Robert Theobald, *An Alternative Future for America* (Swallow Press, 1968), p. 140.

36. Don Fabun, *The Dynamics of Change* (Englewood Cliffs, N. J.: Prentice-Hall, 1967), pp. 1–30.

37. C. F. Pierce, in "Communications," p. 921, comments: "Hopefully, in the future we will be able to live where we like, travel chiefly for pleasure, and communicate to work."

38. John McHale, "Future Cities: Notes on a Typology," *The Futurist*, Vol. III, No. 5, October 1969, pp. 126–130.

# III

## The Telephone and The City

# Editor's Comment

Of all the telephone's effects, none is more dramatic than its impact on the ecology of the city and countryside. For thirty years beginning in the mid 1890s, every few months someone wrote about how the telephone was rescuing farmers from rural loneliness. They could learn when the price was right for bringing crops to town, call the doctor when a child was sick, or call for help when there was a fire. Life on the farm, the writers said, would become tolerable. In 1905 someone wrote "with a telephone in the house, a buggy in the barn, and a rural mail box at the gate, the problem of how to keep the boys and girls on the farm is solved."

At the time that he wrote, 34 percent of the American work force were farmers. Little did he realize that seventy years later only 4 percent of the work force would remain on farms. Nonetheless, the phone did appeal to farmers. Primitive systems were widely established, sometimes using barbed wire fences as signal carriers. At one time early in the century, the rural state of Iowa had the country's highest telephone penetration rate. Yet, the phone did not reverse (even though it slowed) the massive migration from the country to the cities.

The telephone's direct effects on urban life are equally important, and equally ambiguous. In the introduction we noted the central question: has the telephone fostered dispersion and the growth of suburbia, or has it fostered overcrowding in the core city? The common sense answer might be the former. If the telephone has made it possible to form communities without contiguity, move to the suburbs without losing touch with those left behind, and operate businesses from outlying low-rent locations, then how can one doubt the phone's key role in urban sprawl?

Yet all the authors in this book who consider the question (Jean Gottmann, Ronald Abler, Alan Moyer, Ithiel Pool, and Bertil Thorngren) conclude that the phone's relationship to the city is far more complex and not what it seems. As noted in Chapter 6, the telephone contributed as much to the practicality of tall buildings and to core city density as to dispersion.

Jean Gottmann, the Oxford geographer to whom we owe the concept of megalopolis, clarifies the complexity of the processes that are changing the structure of modern cities. Megalopolis is not the decomposition of the city into an undifferentiated urban sprawl. Gottmann describes that pattern of human ecology as "antipolis" and argues that what is emerging (partly with the telephone's help) is a complex and genuinely urban megalopolis—not antipolis.

Both Gottmann and Abler examine the changing structure of the work force—specifically the growth of quaternary occupations—and its effect on patterns of settlement and the dependence of those occupations on adequate means of communication. They also examine the changing spatial and organizational relationships among business offices, manufacturing plants, and marketplaces. These processes of ecological change have shaped the modern city.

Abler traces these processes in U.S. data on the occupational distribution of the population, the phone system's geographic spread, the declining communication distance between places, and the location of activities within urban complexes.

J. Alan Moyer covers some of the same ground in an intensive case study of the history of Boston. Its spread began before the telephone, largely along streetcar lines. The advent of the phone greatly helped processes that had already begun, making suburban living and business activity more feasible. The precise form of the telephone billing system made a large difference in which businesses would function well and where. It was not only the telephone but the system's organization that influenced patterns of urban growth.

In the following section, Bertil Thorngren discusses some of these issues as he examines who communicates to whom on the phone (we shall return to that topic later). He also supports the notion that the phone is not antipolitan; it facilitated urban growth.

"Facilitate" is perhaps the key word. The phone could make life better on the farm or make it easier to move to the city. It made possible suburban operation of urban businesses; it also made possible downtown operation of businesses away from plants and customers. It is rare that one can identify a unique effect of the telephone as such.

In this respect the telephone is not like a new railroad, a new oil field, or the invention of the cotton gin; each of those has a specific use for a specific industry in a specific location. As such, its social impact can be described more definitely; each pushed society in a certain direction. The telephone is quite different. It is a facilitating device with a myriad of uses for a myriad of people and has thus magnified whatever processes were taking place in society at a given time. Since societies are neither unified nor consistent, the telephone often contributed simultaneously to quite opposite developments.

# 14

## Megalopolis and Antipolis: The Telephone and the Structure of the City

### Jean Gottmann

The telephone has for some time been at the heart of a debate about the design of the modern city. The basic question under discussion is whether in the present evolution toward an information society, the gathering of people in cities is still necessary. As telecommunications improve, could the massive and dense concentration in large urban centers, and the nuisances this concentration entails, be done away with and replaced by a scattered habitat held together by networks of wires and waves? The telephone experience seems to be crucial in the matter because of all such networks it is the most ancient and the most widespread; moreover, of all the means of person-to-person communication it provides the greatest feeling of closeness and intellectual intimacy. This is a momentous debate because it examines the play of forces affecting the geographical pattern of settlement and because it probes deep into the stucture of urban society.

The role of the telephone in the evolution of the urban way of life has been considerable and may still be increasing. Nevertheless, a search of the libraries reveals amazingly little scholarly analysis of these issues. Mostly one finds brief statements interspersed in the literature concerned with the impact of communications on location or on the traditional role of distance or on spatial differentiation. Many of these statements are projections into the future, based on fragile assumptions of what the writers believe people want.

Some opinions hold that the improvement of personal communications brought about by generalized telephone use fosters the growth of selected transactional centers and the sprawl of

vast urban systems. If these opinions are right, the progress of
the telephone has helped and perhaps caused megalopolitan formations. A more commonly encountered view is that the telephone encourages geographical scatteration of the places where
telephone users live and work and that this trend will develop
until it brings about a complete dispersal of settlement and the
dissolution of compact cities. If the former opinion may be
called promegalopolitan, the latter could be termed "antipolitan." This antipolitan view begins to deny the need for traditional urban agglomeration, for the old *polis* of the Greeks.

These two views seem to be in clear conflict. Which one is
correct? On the one hand, in North America (understood as the
United States and Canada), by far the best telephone-equipped
section of the world, urban sprawl and metropolitanization have
progressed faster and more widely than anywhere else. On the
other hand, on this continent some 44 percent of the population
live in big cities—that is, in urban agglomerations of 500,000 inhabitants or more. This percentage, calculated for 1975 by the
United Nations Statistical Office, has risen since 1960 when
it was 36 percent. Most of that big-city growth developed together with the excellent facilities of the telephone and other
telecommunication networks. In other parts of the world, urban
concentration can generally be shown to be proportional to the
intensity of telephone use.

Let us attempt a cold-blooded look at the role of the telephone
in the present evolution of the physical and social structure of
the city. This is an involved matter in which many forces besides telecommunications are at work. It may well be that the
role of the telephone is considerable, but the case is not so simple or one-sided as it has too often been assumed since the days
of Alexander Graham Bell.

## THE TELEPHONE AND THE FUNGIBILITY OF SPACE

Direct verbal communication between persons is an essential
element in the foundation of human society; it is at the basis of
the family system and of the negotiation of transactions. Until
the nineteenth century, all communications were greatly imped-
ed by distance. In the second half of that century, distance began to be conquered by the spread of diverse new technologies

and organizational advances. As the technology of steamers and trains improved, the Suez Canal was opened in 1869 and the first transcontinental railway completed in the United States. The improvement of continuous transportation networks around the globe, sewing together the mushrooming centers of population, created the need for more transactions and for more and better communications. The birth of the telephone in 1876 is historically sandwiched between the establishment of the Universal Postal Union in 1875 and the invention of the Gramophone by Edison in 1877. In three successive years, the means of communicating between people scattered around the globe and of recording communications were revolutionized by new prospects of three different kinds. The telephone, which made possible a quasi-instant connection between people located at a distance from one another, seemed destined to modify the relationships built into society by distance and the partitioning of geographical space.

Alexander Graham Bell and many other Americans saw deep social and political changes resulting from the spread of telephone service. On the occasion of the telephone's fiftieth anniversary in 1926, Arthur Pound expressed this faith strongly in a book entitled *The Telephone Idea: Fifty Years Later*. Pound was right in stressing that "words are most precious freight" and that the telephone was "getting the message through" faster and better than could have been done previously, but he went much further in assessing the impact of the telephone on widely scattered populations. He saw it as a "cohesive force for the Nation," as an "antidote for sectionalism," and as an "invigorator for trade." The first and third qualifications have certainly been supported by the experience of the last century. The second statement about sectionalism is rather debatable.

These apparently were Bell's beliefs from the start. He is credited with saying in 1876 that "some day all the people of the United States will sing the Star Spangled Banner in unison by means of the telephone." When inaugurating the Long Lines transcontinental connection in 1915, Bell spoke from New York with T. A. Watson in San Francisco. Pound reports it was said that "had the telephone system reached its present perfection previous to 1861, the Civil War would not have occurred. The wires would not have let the North and South drift so far apart." This was said in January 1915. Later that same year telephone lines began to be used in the trenches dug on both sides of the front dividing Europe in World War I. Sixty years later in 1975, I received a circular in my Oxford office to be kept near my telephone, instructing me what to do in case a call came saying that a bomb had been planted in the building.

Like so many other Promethean dreamers, Bell and his associates had an exaggerated confidence in man's motives, in the workings of society, and in the ease with which the differences recorded on political maps could be overcome. Mankind has always had need of prophets announcing a Golden Age, but to expect the telephone to bring this about was and remains an overestimation of the power of technology; and, what is especially significant for my purpose here, of the impact of distance.

Indeed, excellent modern telephone systems get the message through, but what does the message carry? And to what extent can it modify intentions at both ends of the line? Overcoming distance does not necessarily bring closer together the points of view and patterns of interest established at the places separated by distance. The arrival of the message may exacerbate or precipitate the conflict, rather than pacify it, if conflict there is. However, pursuing this sort of exercise in logic would not be fair to what the prophets of the telephone wanted to convey in their romantic pronouncements. The telephone enabled the voices of individuals to penetrate space or other physical obstacles to its propagation and to request an immediate response. It did not cancel out distance or organization of space but modified the use or effect of both.

The development of technology has never been aimed solely at saving human labor and reducing physical exertion. It has also been aimed at making geographical space, the space inhabited by mankind, *fungible*. If this condition were achieved, an infinity of problems that have always plagued individuals and society would be resolved. The fungibility of space would mean that every point in that space would for all practical purposes be equivalent to any other point. Geometrical space is fungible unless otherwise qualified. Geographical space is not, both because of physically diversified conditions and because of the differentiation and partitioning added by man-made economic, social, and political organization. But in the aspiration of men and women, some day, when paradise would extend on earth, every place in it would enjoy all the virtues and advantages that a place in the sun can provide. No one would then have reason to covet anyone else's location. Peace would reign, assuming, of course, that inequality in the distribution of material goods and benefits is the only source of conflict.

Exploration and technology were largely conceived as tools for bringing closer such a unity and uniformity of space. The objective of a large sector of science has always been to elucidate the general laws of human behavior and to standardize that behavior. This latter stage, it has been thought, would be easier to achieve if geographical space, with all its resources, could be made fungible. It is debatable whether modern improved means of transport have actually advanced the fungibility of space. There can be no doubt, however, that modern telephone systems, with their use of wires and waves, switchboards and computers, cables and satellites, have made the space they serve more fungible for communication purposes. It has become possible, in principle, for individuals located anywhere in that space to converse with one another.

To the extent that the pattern of settlement is determined by man's need to communicate with others and to obtain information quickly, it must be affected by the telephone.

## THE PATTERN OF SETTLEMENT AND THE USE OF THE TELEPHONE

The telephone has immensely improved the faculty of isolated people to communicate with others outside their household. By 1926, as noted by Arthur Pound in his book on the first fifty years, the telephone was already "a fundamental fact of American farm life; . . . the farmer's first aid." Indeed, to people settled in a scattered pattern, in relatively isolated buildings, the telephone gave a heightened sense of security against hazards and provided an escape from loneliness. What was true of those dispersed on farms was also true of categories of individuals isolated by physical or social circumstances restricting their movements outside their home. There are several such categories: invalids, the elderly, young children, residents of places where streets or roads are reputed to be unsafe. The last situa-

tion arises, alas, with increasing frequency in a variety of dense
settlements. The rapid increase in the use of telephone lines that
led to the crisis in the New York City network in 1969–1970 has
partially been explained by the fear of the rising criminality in
that city. It caused many inhabitants to stay at home and call
instead of visiting friends and relatives.

Such help to the lonely must be recognized as a great social
virtue of the telephone. It relieves scattered settlements which
experience certain drawbacks due to relative isolation. This
could not, however, be enough to foster scatteration. One must
recognize that the use of the telephone also relieves some aspects of isolation which in recent times have been characteristic
of certain large, dense urban centers. In fact, the telephone can
be described as being first aid as much for individuals located
in high-rise towers and in massive central concentrations of
people as it is for those on isolated farms. The number, frequency, and urgency of contacts to exchange news or opinions
is usually greater in the business conducted in central districts
than in farming routine. Moreover, as there is usually some cost
involved in every call and the cost is smaller within a dense
community, one may expect much heavier use of the lines per
telephone installed, and even per capita, within large agglomerations than in the rural countryside and between distant
locations in general. Statistical counts usually support the hypothesis that the intensity of telephone traffic decreases with
distance and with the dispersal of settlement.

There are, however, significant anomalies in this general pattern. The anomalies affect the structure of the rate system, but
they also provide interesting clues to relationships between patterns of settlement and the telephone. The flow of telephone
calls reaches higher intensity, for instance, between cities concentrating white-collar workers and transactional activities than
between cities of similar or even larger size but concerned
mainly with blue-collar work. In the megalopolis on the northeastern seaboard, this relationship between concentration of
offices and higher intensity of telephone use was clearly documented in my book (*Megalopolis*, 1961, pp. 582–597). The flow
of calls was particularly great between cities having a common
specialization, for instance, Hartford, Connecticut, and Newark,

New Jersey, because both have headquarters of large insurance companies.

Certain categories of work use the telephone more than others in the pursuit of their routine. Communications or information-oriented work seems to need the telephone most, and this can easily be illustrated by the density of telephones and of the flow of calls in centers of administration, business, and research. The main growth sector of the labor force in the last thirty years has been in white-collar occupations, particularly in the occupations called quaternary which process, analyze, and distribute information. The telephone's role in the actual operation of these categories of work, largely conducted in offices and laboratories, has been considerable. To what extent has it determined the location and the geographical distribution of offices and laboratories?

Most students of office location recognize the telephone as the main factor which allowed geographical separation between office work and the other stages of business it administered, such as production, warehousing, and shipping of goods (Daniels, 1975). The office could be moved from the premises adjoining the plant to a section of town or even to another town where it could find an especially favorable hosting environment. As office work underwent rapid fractioning due to specialization and division of labor, its various parts could easily be located in clearly separate spaces owing to the excellent communication provided by the telephone network. As the material processed by many offices was being increasingly stored in the memories of electronic robots, communication with these stores was provided by the telephone network. The stored material was thus equally available at every place equipped with a terminal connected with the central computing equipment.

The telephone appears to have made office work footloose, liberating it from old locational shackles. The result has been the concentration of offices in selected districts of cities and towns and also huge congregations in certain selected cities. This trend produced the specific architectural form of the skyscraper and the skyline. It should be recognized that lofty, dense skylines exist as much owing to the telephone as to the elevator. Such a dense and massive piling up of office and servicing activities could not function without the widespread, diversified, and incessant flow of communications within the conglomeration and outside it that is obtained through the excellence of the telephone network.

The telephone's impact on office location has thus been dual: first, it has freed the office from the previous necessity of locating next to the operations it directed; second, it has helped to gather offices in large concentrations in special areas. The concentration of office work and of the related but diverse services catering to the needs and fancies of office workers has been, particularly in countries of advanced economy, a major factor of urban growth and urban redevelopment. The modern city is increasingly becoming a white-collar, transactional center as manufacturing, production, and warehousing move out to suburban or small town locations. Office concentration is recognized as one of the reasons for centralization and congestion, with their attendant high costs, nuisances, resentments, and other problems. Therefore, office decentralization has come to be considered a desirable policy, especially in areas of huge agglomeration, such as megalopolis in the United States, or the London, Paris, and Amsterdam metropolitan regions in Western Europe, Tokyo in Japan, etc.

One of the main arguments constantly put forward around the world for office dispersal is the excellence of present telecommunications, and the anticipated technological improvements. It is said that in an information-based society, large cities essentially founded on their transactional and communication functions will gradually dissolve. The telephone network supplemented by television, telex, cassettes, computers, and other means of distributing audiovisual materials would eliminate the need for physical presence at meetings, conferences, and other collective events. As a study of offices in England suggests: "Office work may become a cottage industry" (Cowan, 1969). If offices follow the outward trend of manufactures, little would remain of the centrality of large modern cities; all recent trends of urbanization would then be reversed, and a dispersed pattern of settlement would prevail.

The telephone can thus be presented as a factor of concentration in the location of quaternary work as well, apparently, as a factor of its dispersal; so great is its impact that it may be oper-

ating both ways at the same time. If we admit that such a con-
flicting duality of influence is correct, the conclusion may be
that the patterns of settlement are determined by other factors
which direct people to use the telephone in one way or another,
depending on the way of life and city structure they choose.

To base a forecast of the dissolution of the "white-collar
cities" on the impact of improved telecommunications requires
acceptance of several assumptions: first, that access to the
material transmitted by these means of information will fully
satisfy most people for their work and leisure; second, that isolated living with good communications will satisfy most people;
third, that the quality of personnel and the availability of
adequate labor resources could be maintained by remote control; fourth, that the vast expenditure of energy and materials
necessary to operate and maintain the networks of supply and
the movements of a dispersed population would not be too costly; and last but not least, that most individuals have no reason
for frequent recurrent presence in urban centers other than
efficiency of work.

None of these assumptions appears realistic. Rather the experience of the last 100 years indicates that the telephone has been
used in the evolution of settlement in diverse ways but mainly
as a help in the development of larger metropolitan systems
with a more diversified and complex structure. Excellence of
communications has made possible the more variegated, multiple, partitioned structure of modern cities. The telephone has
not made space fungible; it has not modified human nature
much; however, it has permitted a spatial and political restruc-
turing of cities of considerable portent.

## THE TELEPHONE AND THE EVOLUTION OF COMMUNITIES

The rapid and enormous growth of metropolitan systems has
consumed large tracts of space. An agglomeration that grows
from 200,000 inhabitants to a million or from one to five million
must devour space. This consumption of acreage will depend
on many considerations, including cost of land, provision of
services, taste for certain forms and densities of residence, pre-
dominant economic activities, and ease of transport and communications. Within the framework of the last factor listed, the

generalization of the individual motorcar and of the telephone
have actively aided suburban sprawl.

Growth at the rate and scale of the modern metropolis implies
the coexistence within the system of a variety of people and
economic activities. As has generally been the rule in cities for
some 4,000 years, this variety causes a division into compart-
ments specialized in housing certain groups or activities. In the
dynamic circumstances of recent times, especially in countries of
advanced economy, the telephone has increased this spatial division of labor and society: the home could be more distant
from the place of work, the office of a firm from its plant, the
consumer from his supplier. The telephone provides, when
needed, quasi-immediate verbal communication between all
these interdependent units at minimal cost. However, it is only
one of many indispensable networks of electricity, water supply, sewerage, road traffic, retail trade, schools, hospitals, and
so forth. Nevertheless, it would have been very difficult for
these complex and integrated networks to work in unison without the telephone, which made possible the constant and efficient coordination of all the systems of the large modern city.

It has also greatly improved the rhythm at which city life develops. Here again, the telephone alone cannot create the city,
its rhythm, its excitement. Large cities have functioned long before the telephone: Babylon, Jerusalem, Rome, Constantinople,
Canton, Paris, London, and even New York and Boston. The
telephone has helped make the city better, bigger, more efficient, more exciting. It alleviates loneliness, it helps arrange encounters. It saves its users time, trips, and frustration. A young
couple living in Zagreb, Yugoslavia, who are waiting for a
home telephone to be installed, told me that a telephone line is
more desired and more difficult to obtain in Yugoslav cities
than a car. Asked what use they would make of it once it was
installed, they said mainly to make appointments. The major urban message the telephone carries is still the same as the first
call of Bell to Watson: "I want to see you."

However, the theory that dispersal of settlement is forthcoming because of telecommunications networks is founded on the
belief that this type of distant connection will be able to replace
face-to-face meetings to the general satisfaction. It is said that
just as the elderly replace visits by telephone chats and just as

office workers discuss matters and exchange information by telephone, more and more will do likewise. They will obtain materials by telephone or in the form of recordings; they will hold conferences first by telephone (as many executives do in government or large corporations) and later on closed-circuit television. Hence the office will become "a cottage industry."

In fact, the office has always been to some extent a cottage industry. Authors, editors, and a variety of professionals have preferred doing at least part of their work at home or even in their country houses; some prefer to work on premises where they cannot be reached by telephone, in order not to be disturbed. Such cases are more frequent in Europe than in America, for the use of the telephone and the need for privacy are cultural attitudes, differing from one country to another. Even in America, however, individuals are learning to protect themselves against intrusion by telephone, and communications are being delayed as the caller waits to be called back.

In countries as different as Sweden and Ireland, two experts on office work (Dr. Thorngren and Dr. Bannon) agree that "the use of the telephone by all types of office groups would seem to be confined to contacts . . . which are characteristically fast, frequent, well-structured and, while being important, tend to be related to coordination and routine functions" (Bannon, 1973).

Transacting quaternary business by wire does not at present appear to give results as satisfactory as does physical presence. As Reid documents in Chapter 18, communicating over an electronic network may satisfy individuals who know one another intimately and trust one another fully. Between individuals who do not have a close relationship, communication over networks usually leads to more personal contacts. This need of physical presence, a major factor of urban centrality, seems to be imbedded in human psychology and in the highly competitive character of a dynamic society (Gottmann, 1970).

As a larger sector of employment goes into transactional work, it becomes important to assess the environment hosting such activities. Transactional performance is more difficult to evaluate than productivity in the case of goods or simpler tertiary services. The higher level of transactions are performed in an environment providing amenities and entertainment. Transactional work that involves responsibility in research, analysis, decision-making, or education is done by relatively well-paid and discriminating personnel. A center of quaternary transactions is normally a large consumer of entertainment, and in competitive situations this may be an important part of the transaction. It is no simple coincidence that Broadway is in Manhattan and that the performing arts flourish in certain centers of wealth and power.

It is to be expected that routine transactions will be increasingly mechanized and automated owing to telecommunication networks. This trend will hardly cause a quantitative decrease in the demand for face-to-face contacts, but these may come to be reserved for an elite or for specially chosen occasions. As the numbers of transactional operations and personnel increase, the demand for communications of all sorts snowballs. Barriers and partitions must be built between the callers and the person or place called. A movement in that direction has started: the gadgetry and personnel related to the telephone are partly aimed at filtering communications demand.

## THE TELEPHONE AND THE TRANSHUMANT SOCIETY

Our examination of the telephone's influence on patterns of settlement and location of activities does not lead us to indulge in the hypothesis of a final scatteration of settlement and dissolution of nucleated cities. The telephone, rather, helped to bring people together; it has also helped to maintain or establish linkages despite geographical dispersal when the latter was caused by other influences shaping the way of life. In one way the telephone must be considered as fostering "megalopolitan" formations rather than "antipolitan" scatteration. There is, however, another modern trend in developed Western countries which could be termed "antipolitan" in a sense and which could hardly have developed so much without good telephone service.

This is the trend of a large and increasing proportion of North Americans and Western Europeans to move frequently between two or more places of abode. I do not allude here to the crowds who commute to work, but to individuals who have more than one residence and perhaps more than one place where they perform work useful to their occupation.

These situations proliferate around us. It is becoming increas-

ingly difficult for competitive quaternary personnel to get all
their work done in one spot if they can obtain additional information, advice, or relaxation outside the normal location of employment. Very few will feel secure enough not to avail themselves of such additional opportunity. There are also benefits for
individuals in our society to have and perhaps own more than
one residence. One residence may be close to the main location
of work, another away in the countryside or in a small town for
vacation periods, a third for intermediate purposes may be located either in a central city (where useful complementary contacts can be made recurrently) or in a peripheral weekend place
for brief periods of relaxation and entertaining.

Census figures do not account for this new nomadism or rather "transhumance." These people move within the week or the
year between several places in more or less regular fashion. The
legal domicile may be at the spot most advantageous in terms of
voting and taxation. The census will usually count individuals
as living at their legal domicile, even if they spend most of their
days and nights at another residence, classified as "secondary"
in legal terms.

In France, for instance, there are many advantages, even for
notables of large cities, to claim as their official residence a
country home which may be in a small and perhaps distant
town. The smaller place derives a twofold advantage from such
absentee citizens counted there by census; it adds to the subsi-
dies received from the central budget and to its official political
weight. In the last twenty years the central city of Paris has officially lost about 600,000 residents according to census count, yet
in the same area and period the total number of housing units
has increased by 210,000! This expansion was largely due to
new construction, and there are few vacancies.

How is this discrepancy to be explained? Partly the expansion
of housing may be caused by transients coming at frequent intervals to transact business in Paris. Partly it is a different phenomenon: many persons who spend most of their time in Paris,
who in fact live and work there, are officially domiciled elsewhere. The distance between the two places of abode may be a
few dozen or hundreds of miles. How many people have already adopted such a transhumant, geographically pluralistic
way of life is not known, but the indications are that their num-

ber is large and growing. The trend affects almost all large cities in France, but not in France alone; it may be true even in Massachusetts.

The town of Antibes, on the French Riviera, was originally called Antipolis in ancient times. The word did not mean "against the city" but rather "the city opposite" or "the advanced city." Antipolis was meant to designate a new stage of a growing network centered on a metropolis. The present transhumant society, within megalopolis or between a megalopolis and an antipolis, could not exist in today's tightly organized professional communities without good linkages by telecommunications and particularly by telephone. To this category of urbanites the telephone has given, as it did for isolated farmers, invalids, and bureaucrats, a heightened sense of security. Owing to the telephone network, the transhumant remains within easy call of his main center of living and activity. Frequent and even irregular moving from one place to another does not endanger the system of connections on which the individual's life is founded. This mobility may, however, seriously undermine the political and social structure of the large city's community when its leadership becomes transhumant. With its increasingly pluralistic habitat, our society needs to rethink the structure of the city more seriously than ever (Gottmann, 1976).

These problems are due to the concurring effects of several factors. They might not have developed, however, without the opportunity provided by the telephone. In many ways the telephone has preserved and enhanced large-city growth and the concentration of transactional business. It has also opened up other possibilities which weaken or reshape the structure of communities. It may be that the social impact of the telephone is so difficult to assess because it is such an adaptable and unobtrusive tool, the use of which is molded by individuals and society to the pursuit of diverse aims.

## REFERENCES

Bannon, Michael J. (1973). *Office Location in Ireland: The Role of Central Dublin*, (Dublin: An Foras Forbartha.)

Boorstin, Daniel (1974). *The Americans: The Democratic Experience* (New York: Random House).

Cowan, Peter, et. al. (1969). *The Office: A Facet of Urban Growth* (London: Heinemann Educational Books).

Daniels, P. W. (1975). *Office Location* (London: Bell & Sons).

Gottmann, Jean (1961). *Megalopolis: The Urbanized Northeastern Seaboard of the United States* (New York: Twentieth Century Fund).

_____ (1970). "Urban centrality and the interweaving of quaternary activities," *Ekistics* (Athens), May 1970, pp. 322–331.

_____ (1976). "Paris transformed," *Geographical Journal* (London) 142, Part I, pp. 132–135.

Pound, Arthur (1926). *The Telephone Idea: Fifty Years Later* (New York: Greenberg).

Thorngren, B. (1970). "How do contact systems affect regional development?" *Environment and Planning* 2, pp. 409–427.

# 15     The Telephone and the Evolution of the American Metropolitan System

## Ronald Abler

Stepping over to the telephone, he called up the hotel of the town, and ordered a dinner to be cooked and sent to his home forthwith. "We have a telephone in every house, for use in all the everyday affairs of life," said he. This seemed an improvement even on New York. . . .

Titus K. Smith, *Altruria*, 1895

The dreams of late nineteenth century utopians have been realized. In Titus Smith's day only the wealthy could afford telephones. Local service existed in most cities and towns, but long-distance circuits were still rare. Since then, telephone service has spread to all the nation's cities and throughout most of its rural areas. Eighty-seven percent of the nation's households and 90 percent of metropolitan households had telephone service in 1970.[1] We use the telephone in all daily affairs of life, averaging 540 million local calls and 37 million long-distance calls each day. The integral part of our lives the telephone has now become, has been the work of a century; the changes in national and individual life that demanded and depended upon advances in telephony have reached every part of the nation. The telephone has conquered distance as no other technology has.

## GEOGRAPHICAL EFFECTS OF TELECOMMUNICATIONS

The telephone is a *space-adjusting* technique.[2] Telecommunications (like transportation) can change the proximity of places by

Parts of this research were supported by the Central Fund for Research of the Pennsylvania State University.

improving connections between them. Other things being equal, an individual will have more contacts with people close and fewer with people distant from him. Thus, any technology that makes it easier to contact people at a distance makes it possible to communicate as though that distance had been shortened. The space-adjusting technology of telephony has now been applied on such a massive scale that for some purposes the nation has become a single, highly interdependent communications network. Much the same can be said of the entire globe. In its ads the Communications Satellite Corporation emphasizes its ability to "take you practically anywhere in milliseconds" by showing pictures of the major cities being pulled together by its circuits (Figure 1).

In the course of a century, the telephone has almost succeeded in providing complete *time-space convergence*. Time-space convergence is the average rate at which places approach each other in the relative space created by a space-adjusting technology. In 1776, for example, a journey between Edinburgh and London took 5,760 minutes. In 1966, the trip could be made in 180 minutes. Thus London and Edinburgh have converged upon one another at an average rate of 29.3 minutes per year (5,760−180/ 190 = 29.3).[3] Because telephonic communication is effectively instantaneous, it can produce much greater space convergence than transportation technology.

In the early days, for example, placing a long-distance call was time-consuming; it took fourteen minutes to effect a transcontinental call in 1920,[4] and was expensive since it also took eight operators. By 1930, improvements in network structure had cut average service speed to 2.1 minutes; further advances brought an average service time of 1.4 minutes by 1940.[5] Operator toll dialing reduced average connection time to about 1.0 minutes in the 1950s, and customer direct-distance dialing now makes it possible to have a call switched through the national network in less than thirty seconds. The introduction of electronic switching into the toll network will make nationwide telephony virtually instantaneous, leaving only the negligible difference between the times needed to enter seven digits for a local call and ten digits for a long-distance call separating local and long-distance connections in time-space. Assuming an average service speed of thirty seconds for a transcontinental call in

**Communications
Geography:
The world is being
pulled together**

# Via Satellite

An artist's imagination? Not at all. This is how it really is.

Comsat is helping pull the world together...putting far-away places on the main street of business, industry and commerce...giving people a front row seat to history, Live via Satellite...pioneering new potentials for U. S. domestic as well as international telephone, tele-vision, telegraph, data and fac-simile communications.

Comsat, a shareholder-owned communications company, oper-ates the satellites in the global system...U.S. earth stations for satellite communications . . . the COMSAT Laboratories and a wide range of related technical activi-ties that are creating new com-munications advances.

These are part of a world-wide satellite system to give you better communications.

More than 60 countries al-ready communicate daily with each other via satellite.

In the United States, if it's via satellite, it's via Comsat.

Write to Comsat's Informa-tion Office for the booklet, "Via Satellite, The Comsat Story".

COMSAT
CAN TAKE YOU
PRACTICALLY ANYWHERE
IN MILLISECONDS

## C✦MSAT

**Communications Satellite Corporation**
950 L'Enfant Plaza, S.W., Washington, D.C. 20024

**Figure 1**
A Communications Satellite Corporation advertisement. Courtesy of
Communications Satellite Corporation.

1970, the time-space convergence between New York and San Francisco averages about sixteen seconds per year since 1920 (Figure 2). Automatic dialers and a completely electronic toll switching network should produce complete time-space convergence and therefore complete independence of contact time from distance (Figure 3) sometime in the remainder of the century.

The telephone network has not yet been able to achieve complete *cost-space convergence*, yet massive progress has been made. Between 1920 and 1970, New York and San Francisco converged at an average rate of 28.5 cents per year as the cost of a three-minute, station-to-station call between the two places dropped from $16.50 to $1.35 (Figure 4). If convergence were calculated in constant dollars, the rate would be even higher because of inflation during the period. Although the telephone system has so far been unable to make costs completely independent of distance, it has come remarkably close to doing so, especially in comparison with early rate schedules and especially over distances in excess of 500 miles (Figure 5).[6]

Consumer response to the telephone network's ability to provide nearly instantaneous communication at ever lower costs has been staggering. Americans now make as many local calls in a month as they did throughout all of 1920, and it takes them just over two week's current toll call volumes to equal the nation's total 1920 long-distance traffic.[7]

The impact of increased capacity for information exchange over short and long distances on the nation's metropolitan re-

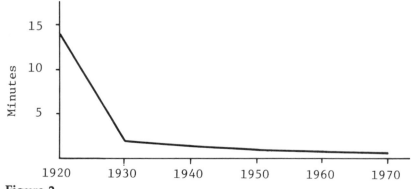

**Figure 2**
Telephone time-space convergence between New York and San Francisco, 1920–1970. Calculated from sources cited in text.

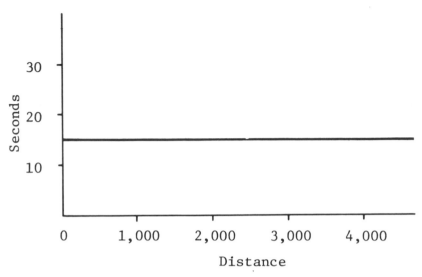

**Figure 3**
Idealized telephone time-distance relationship.

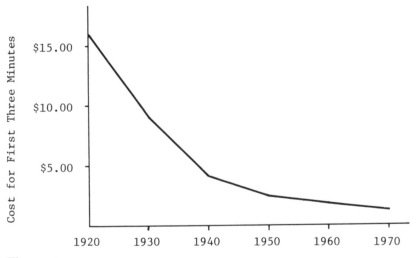

**Figure 4**
Telephone cost-space convergence between New York and San Francisco, 1920–1970. The curve is a linear approximation of many step-like changes. Source: *Historical Statistics of the United States*, p. 481.

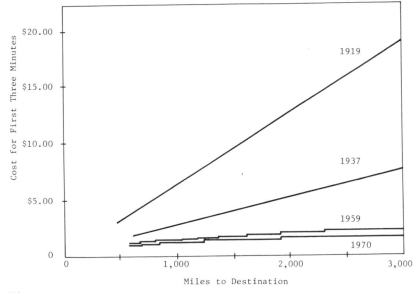

**Figure 5**
Telephone cost-distance relationship, 1919–1970. Sources: Leland L.
Johnson, *Communication Satellites and Telephone Rates: Problems of Gov-
ernment Regulation* (Santa Monica, Calif.: RAND Corporation, 1961).
Data supplied by AT&T.

gions has been profound. Its effects are especially evident in the
structure of the national economy and in the locations of business activities among and within metropolitan regions.

The nation now has an information economy. To illustrate the
importance of information activities in the United States today,
the nation's industries can conveniently be divided into five
classes:

1. A *primary* group consisting of establishments engaged in
extractive pursuits such as agriculture, forestry, and fishing.
This group is comparable to the traditional notion of primary
industries.

2. A *secondary* group of firms that manufacture tangible
products.

3. A *tertiary* industrial sector of firms providing tangible business or personal services such as transportation, maintenance,
repair, and custom production of goods (e.g., custom tailoring).

4. A *quaternary* category of industries including those engaged
in continuous, large-scale production of information; these are

routine information activities that do not call for major decision-making inputs, such as insurance companies and banks.

5. A *quinary* class of establishments engaging in control activities, including craft (as opposed to routine) information production and nonroutine decision-making. Government is the major quinary industry although private examples (e.g., management companies) exist.

When national production is tabulated according to this breakdown, the growing importance of the information (quaternary and quinary) industries is emphasized (Table 1). Whereas information industries generated a fourth of the nation's income in 1950, in 1970 they accounted for 40 percent of the national output.

The five-category industrial classification can fruitfully be accompanied by an occupational classification that divides the labor force into four categories:

1. *Routine tangible* occupations that involve repetitive production operations in any industrial sector.

2. *Nonroutine tangible* jobs in which tangible goods and services are produced but only on a custom and individualized basis (e.g., a barber or cabinetmaker).

3. *Routine intangible* workers who handle information in a repetitive manner—secretaries, clerks, and most lower- and middle-level management.

4. A *nonroutine intangible* occupational group consisting of those engaged in individualized, nonroutine decision-making; nonroutine intangible workers deal largely with information, but each decision, request for information, etc., is essentially unique.

A cross-classification of these industrial and occupational groupings applied to the nation's labor force produces a twenty-cell matrix providing useful ideas about the evolution of the information economy and the locational behavior of its components.[8]

In 1975, the information industries in the quaternary and quinary categories employed about a third of the nation's labor force (Table 2). Forty-six percent of the labor force now works at information processing (intangible) occupations. Employment by information industries and in the information processing

**Table 1**
National Income by Industry Sector, 1950–1970

| Industries | 1950 | | 1960 | | 1970 | |
|---|---|---|---|---|---|---|
| | Million Dollars | % | Million Dollars | % | Million Dollars | % |
| Primary | 17,378 | 7.3 | 17,161 | 4.1 | 24,511 | 3.1 |
| Secondary | 75,869 | 31.7 | 120,497 | 28.9 | 213,151 | 26.8 |
| Tertiary | 85,387 | 35.7 | 146,990 | 35.2 | 243,231 | 30.6 |
| Quaternary | 22,153 | 9.3 | 48,255 | 11.6 | 111,408 | 14.0 |
| Quinary | 37,329 | 15.6 | 81,871 | 19.6 | 198,938 | 25.6 |
| Total | 239,170 | 100 | 417,054 | 100 | 795,887 | 100 |

Source: *Survey of Current Business*. Totals include the following unclassified amounts: 1950, $1.054 billion; 1960, $2.280 billion; 1970, $4.648 billion.

**Table 2**

Industrial-Occupational Structure of the American Labor Force, 1975 (Percent of Total Labor Force in Each Industrial-Occupational Category)

| | Occupation | | | | |
| | Tangible | | Intangible | | |
| Industries | Routine | Nonroutine | Routine | Nonroutine | Total |
| --- | --- | --- | --- | --- | --- |
| Primary | 4.01 | 0.18 | 0.06 | 0.12 | 4.37 |
| Secondary | 9.87 | 5.29 | 3.64 | 3.64 | 22.44 |
| Tertiary | 4.95 | 18.41 | 10.52 | 6.86 | 40.74 |
| Quaternary | 0.76 | 7.69 | 7.31 | 8.39 | 24.15 |
| Quinary | 0.36 | 2.67 | 2.82 | 2.43 | 8.28 |
| Total | 19.95 | 34.24 | 24.35 | 21.44 | 100 |

Source: Compiled from U.S. Department of Labor, *Tomorrow's Manpower Needs* (Washington: U.S. Government Printing Office, 1969), Bulletin Number 1606. Column and row totals do not sum to 100 because of rounding.

occupations has expanded considerably since 1950 (Figures 6a and 6b).

The nation's increasing reliance on information industries and increased employment in information processing occupations would be impossible without the cheap, efficient telephone services that have evolved during the last few decades. A labor force in which close to half the workers earn their livings by reading, writing, talking, calculating, and deciding would be unthinkable without the telephone's information-carrying capacity.

The ongoing evolution of the nation's information economy is abetting new work patterns between and within cities. Increasingly, each of the twenty cells in the cross-tabulation identifies a distinct work unit that can now respond to its own locational needs and constraints. Once firms were single entities with production, distribution, clerical, and management tasks located in the same place, but modularity is increasingly evident. A large corporation today may have a production plant in one city, its national warehouse in another, its sales office somewhere else, and its corporate headquarters in yet another place. Telecommunications and air transport have made dispersed operations possible for business and government. Plants and offices scat-

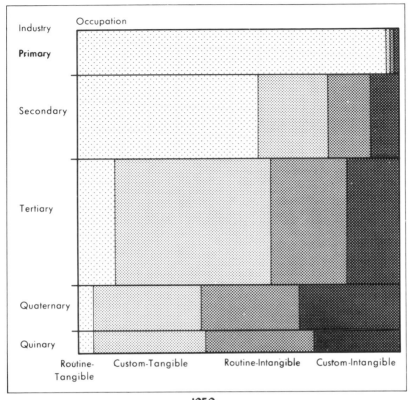

**1950**

**Figure 6a**
Industrial-occupational structure of the American labor force, 1950 and
1975. Sources: United States Census, 1950; U.S. Department of Labor,
*Tomorrow's Manpower Needs*, Bulletin No. 1606 (Washington, D. C.:
Government Printing Office, 1969).

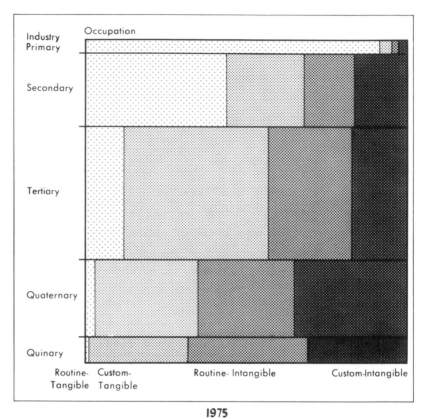

Figure 6b

tered over the nation—each at the best location for carrying out its particular task—can function as single productive units.

A corollary is that many advantages formerly possessed by information-rich locations, such as New York City, are disappearing. The external economies available to information industries in New York City have traditionally been one of its overwhelming locational attractions,[9] but New York City is no longer attracting information activities at the rate it once did. A study of trends in office employment revealed that the proportion of the nation's central administrative jobs (corporate headquarters employment) in the New York metropolitan region declined from 20.8 to 18.7 percent between 1954 and 1967; the share of the nation's central administrative jobs in the twenty-one largest metropolitan areas declined from 69 to 65 percent.[10] The losses are products of more rapid headquarters growth in small metropolitan areas rather than absolute losses in the larger centers, but corporate headquarters and other office jobs are in general dispersing among metropolitan areas. [11]

Metropolitan areas now scramble for corporate headquarters and their thousands of office workers the same way cities once courted manufacturing plants. Atlanta extolls its appeal as a headquarters city in full-page advertisements in national magazines (Figure 7). The New York State Department of Commerce fights back with full-page ads in the *New York Times* proclaiming New York City "Corporate Headquarters, U.S.A." and asking "How come New York has more corporate headquarters than the next eight cities combined?"[12]

The same competition prevails locally, with central cities struggling to keep information activities downtown and suburbs trying to lure them away and attract new growth. Despite increases in office employment in the metropolitan area, New York City lost an estimated 40,000 white-collar jobs between 1969 and 1974; of these, 20,000 were jobs in headquarters offices.[13] Similar trends are evident in metropolitan areas throughout the nation, and in most places office employment in suburbs is growing more rapidly than it is downtown.

The telephone's role in the dispersal of information activities at national and metropolitan levels remains murky. In the literature on the location and relocation of information activity, this factor is never mentioned by movers or those deciding where to

# Atlanta's appeal as a Headquarters City.

ATLANTA IS
HEADQUARTERS FOR
THE SIXTH FEDERAL
RESERVE DISTRICT

Few cities today offer quite the combination of advantages Atlanta does as a corporate headquarters location. Atlanta provides both the amenities of living and the necessities to support and serve large businesses. To live here is to enjoy life with far fewer of the problems of urbanization.

To work here is to have all the economic resources of a diversified, progressive city: transportation and communications accessibility, financial strength and expertise, abundant and reasonably-priced office space, technical and research skills, and above all the talents of a hard-working, forward-

looking people. 430 of the "Fortune 500" operate facilities here. Find out why Atlanta has such appeal for them. Contact: Paul Miller, Atlanta Chamber of Commerce, 1382 Commerce Building, Atlanta, Georgia 30303. 404-521-0845.

HARTSFIELD INTERNATIONAL
AIRPORT AFFORDS EASY
ACCESSIBILITY TO AND FROM
MOST MAJOR U.S. CITIES

**Figure 7**
Advertisement to attract corporate headquarters.

locate new facilities, which suggests that adequate telephone service is taken as a given. This is especially true at the metropolitan level, where the general quality-of-life criteria seem to be more important than hard-headed business considerations.[14]

The telephone's major impact on locational behavior has probably been at the national level. National telephone service has been a technological reality since 1915, but the costs of toll calls limited their widespread use of the network, even for business, until well after World War II. Rate reductions for long-distance calls in the 1950s and 1960s complemented the ongoing evolution of the nation's economy by making it feasible to manage and coordinate widely dispersed facilities. The net effect has not negated the advantages of locating in metropolitan areas, but it has put all these areas on a more equal basis in competing for information activities.

At both the metropolitan and national levels, telephonic communication and its offshoots have acted as necessary rather than sufficient conditions of dispersal. The dispersal with integration that has occurred to date is unthinkable without telephonic communication, but that does not imply that the telephone *caused* dispersal in the normal sense of the word. The telephone network made it *possible* for dispersal, which had other causes, to proceed. On balance, the effects of the availability of telephone technology have been more revolutionary for the nation's metropolitan system than at the level of the individual city.

## TECHNOLOGICAL EVOLUTION

Achieving the efficient, inexpensive telephone service that is prerequisite to intermetropolitan locational freedom has been long and difficult. Until the 1890s, telephone service was largely confined to cities and their immediate vicinities. Signal attenuation problems had to be overcome before long-distance calling could expand significantly.[15] The toll network began with open wire strung on poles.[16] The predecessor of AT&T, the Interstate Telephone Company, was organized to build such a long-distance line from New York City to Boston in 1880; large-scale construction of an integrated toll network came later. In 1890, the AT&T network was still largely confined to New England, with one salient reaching north and then west in the Hudson–

Mohawk corridor and another extending southeast in the
Washington–New York corridor (Figure 8).

The network expanded rapidly after 1890, reaching Chicago
by 1892. The years 1894 and 1895 were devoted to filling in gaps
in the New York–Chicago corridor; very little expansion occurred in 1895. Two years later, the network had spread west
to the Missouri and into the South along several axes, and for
the first time, large enclaves with AT&T service appeared in
advance of the main network. On the original network maps,
they are shown connected to the main body of the network by
projected lines. Thus AT&T's organizational efforts spread in
advance of the integrated network in some places, for the connections among towns in the outliers are shown as completed.
Little extension occurred in 1898, but by 1900 the system had
filled in more of the East and was beginning to push west again
in Nebraska and North Dakota. At the turn of the century, the
nation's West remained outside the AT&T network, presumably
with only local or short-range toll service.

Four more years of network construction brought AT&T interconnection to large parts of the West. The coast and northern
interior were unified, but much of Nevada, Arizona, New Mexico, and the western Great Plains were not connected to the
AT&T system; nor were the eastern and western halves of the
network connected. Transcontinental service awaited the development of electronic signal repeaters, which became available in
1914. The first transcontinental line, consisting of four strands of
open wire, was completed in 1915.[17]

A map published in 1906 shows virtually no change in net-

**Figure 8**
Growth of AT&T toll network is shown by shading each county for the
year in which one or more towns in each county were connected to the
AT&T toll network. Thus the areas in black represent the approximate
extent of the AT&T network at the dates indicated. The maps presented
here are based on route maps published by AT&T; the last extent map
for each year was used to compile this series.

Because only AT&T maps were used, this series may understate
somewhat the extent of toll service in the nation, for it does not show
toll circuits of independent telephone companies. Interconnections
among independents were usually local, however, and of limited effectiveness because of AT&T's refusal to provide toll interconnection to
independent telephone companies before 1913.

work extent from the 1904 map, especially in the West. A 1916 map indicates the inclusion of virtually all counties in the East and some additions in the West but surprisingly little extension of the network into the plains and intermontane areas that were unserviced on the 1904 map. The 1916 map does shows the route of the transcontinental circuit, and a separate trunk circuit extending from New York to Miami. Network maps for 1923 and 1929 document a gradual filling in of the unserviced areas and the building of a trunk route network throughout the nation. With the exception of the transcontinental and New York to Miami trunks, the spread of the trunk network seems roughly to parallel the earlier network of short-haul connections.

Completion of the national toll network in the 1920s brought AT&T face to face with a host of technological problems. Signal attenuation with distance and in the multiple switches required to complete even short- and medium-distance toll calls was severe. The first nationwide plan for providing good random access service was put forth in 1925.[18] Implementation of the original plan and modifications, coupled with the progressive development of new transmission and switching techniques, culminated in the direct-distance dialing programs introduced in the 1960s. The application of electronic switching in the toll network begun in 1976 fulfills a persistent goal of network design since 1925; it virtually eliminates the differences between the handling of exchange and toll messages.[19] Only differences in price remain.

The fascinating history of telephone network evolution has only been touched upon here, but three important conclusions can be drawn from what has been reviewed.

First, it has taken a long time to develop efficient, nationwide telephone service. Within a year of Morse's 1844 demonstration of the practicality of his telegraph, some 30,000 miles of telegraph wire had been put into service. A transcontinental line was completed to San Francisco in 1861.[20] Telephony is a higher order of communication than telegraphy in that it conveys much more information; progress toward a national network was accordingly slower. Despite the rapid spread of local telephone service throughout the nation, long-distance service scarcely existed two decades after the telephone's invention and it was forty years before service on a continental scale became possible.

More importantly, only after World War II—almost seventy-five years after the invention of the telephone—did toll calls become available at prices that permitted widespread toll calling.

Second, the telephone network was and still is an expensive machine; the value of the nation's telephone plant in 1970 was about $60 billion, 6.1 percent of the 1970 gross national product.[21] Since 1900, investment in the nation's telephone service has consistently outpaced increases in gross national product except during the depression and World War II years (Figure 9). In 1970 alone, $10.1 billion was invested in telephone plant and equipment, an amount that was 12.7 percent of the $79.7 billion total business expenditures for new plant and equipment in that year.[22] Until the 1950s, large capital requirements meant high toll rates that discouraged consumer use. The ratio of local to toll calls stayed almost constant at 40:1 from 1900 through World War I; thereafter the ratio fluctuated from 31:1 in 1920, to 22:1 in 1945, to 27:1 in 1950. Since 1950 the ratio has declined steadily to 23:1 in 1960, 17:1 in 1970, and 15:1 in 1974, the latest year for which data are available.[23] Before toll service could achieve widespread use, rate reductions following automation of the nation's toll network had to reduce prices to levels competitive with other means of communication.

Third, the major technological obstacles to efficient and inexpensive nationwide network operation have been switching problems. There have been many important advances in transmission technology over the decades, and our national network could not exist without them. But increases in transmission capacity have generally preceded development of the switching devices needed to handle the traffic they generate because of the combinatorial nature of random access networks. The number of possible two-way interconnections in a telephone network is $N(N-1)/2$, where $N$ is the number of telephones in the system. Each telephone added to the network imposes astronomical numbers of new possible connections, for which switching capacity must be provided. In a network with 120,000,000 telephones, there are 7,199,999,940,000,000 possible pairs. Increasing the number of telephones to 120,000,001 increases the number of possible pairs to 7,200,000,060,000,000, which is 120,000,000 more than the number of possible pairs in the network before the new phone was added, for the new telephone

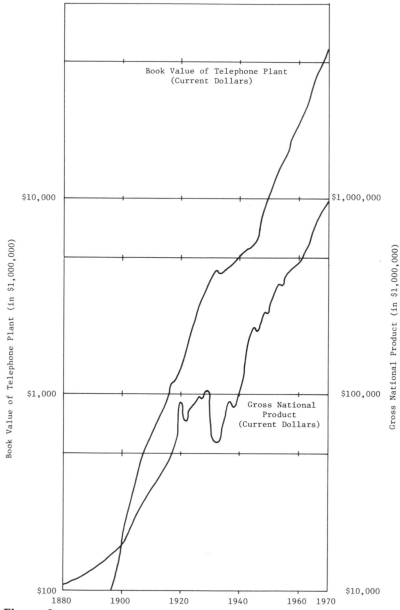

**Figure 9**
Book value of telephone carriers reporting to the F.C.C. and Gross National Product, 1880–1970. Sources: *Historical Statistics of the United States*, pp. 139, 481–482; *Statistical Abstract of the United States, 1974*, p. 374; *Statistics of Communication Common Carriers, 1972*, p. 19.

must be provided with a possible route to each of the 120,000,000 telephones already in the network. Careful design of the nation's hierarchy of toll trunks and switching centers makes it possible for a new telephone to reach any other telephone in the nation with a minimum of new routes, but the increasingly complex switching arrangements needed to provide nationwide, high-volume long-distance capacity are a direct consequence of the number of possible connections in the national network and the way that number increases with each new telephone.

## PROSPECTS

It is helpful to reflect upon the telephone's history and its effects on cities, for utopians are once again abroad in the land.[24] Whereas the descriptions of telecommunications and its impacts authored by Titus Smith and others (such as Edward Bellamy) were clearly utopian, modern utopians who are entranced with electronic communications write forecasts rather than fiction.[25]

If current commentators are to be believed, an electronic communications complex revolving around two-way cable television (CATV) will soon free the nation's cities from existing geographical constraints. The high proportion of the national and metropolitan labor force working at information-processing jobs and advances in telecommunications will, by 1990, allow managerial, professional, and clerical personnel to work at home, substituting electrical communications for face-to-face contacts which generally relate to routine matters not requiring physical presence.[26] Nicholas Johnson, former FCC Commissioner, described the ultimate vision as a "home communication center where a person works, learns, and is entertained, and contributes to his society by way of communications techniques we have not yet imagined—incidentally solving commuter traffic jams and much of their air pollution problems in the process."[27] John R. Pierce suggests: "We can live where we like, travel for pleasure, and communicate to work."[28]

In the context of the evolution of the telephone network, such forecasts and the general literature they represent must be classified as utopian. There is little doubt that over the very long run we shall enjoy the benefits electronic utopians describe, but we are at least as distant from them today as Titus Smith was from direct-distance dialing, for the fundamental constraints that governed the evolution of telephone service will also control the development of new telecommunications technologies.

Video communication is a much higher order of communication than voice communication. We can, therefore, reasonably expect that building a nationwide, random access video communications network will take longer than did the construction of the voice telephone network, which in turn required more time and money to build than the telegraph network. Current CATV technology delivers one-way, unswitched video signals of rather poor resolution. Despite the rapid advances being made in digital transmission via waveguides and optical fibers, I remain unconvinced that the nation will have anything even approaching two-way, random access video communications at the turn of the century. Random access will be particularly difficult to achieve because it will require switching technology that barely exists, on a scale, in terms of information carrying capacity, that is almost unimaginable.

The cost constraints are more serious than technological problems. An investment of $15,000 per video terminal has been estimated as the capital required for a nationwide video network.[29] E. Bryan Carne has estimated the aggregate cost of a nationwide video network at a trillion dollars.[30] The nation's 1971 gross national product was about a trillion dollars,[31] and in the same year, the telephone network was worth about $60 billion.[32] Additional evidence that building a nationwide video network will be a long, hard job can be adduced by historical analogy. In 1958 dollars, the 1971 gross national product was $750 billion, and the book value of the national telephone complex was $45 billion. In those terms, the nation first had a $45 billion gross national product about 1890, when the construction of the long-distance network began.[33] It seems to me that it will take a similar period of time to build a video communications complex with a value equivalent to the 1971 GNP.

Should a random access telephone system eventually provide the comprehensive services currently envisioned, its evolution may be accompanied by dispersal of information activities and people with information processing occupations. I doubt, however, that the causal role telecommunications will play in such dispersal will be anything other than what it has been in the past—namely, a permissive technology that is a necessary rather than a sufficient condition. Residential dispersal in and around metropolitan areas is as old as Roman villas and Versailles; the only thing revolutionary about urban dispersal in the nineteenth and twentieth centuries is the number of people who could afford it. The trend toward dispersing information activities, which might formerly have remained concentrated in the nation's largest metropolitan centers, is more recent and more dependent on telecommunications technology. At that scale also, dispersal is most directly tied to economic development and shifting settlement patterns. The existence of dependable telephone service does not *cause* such dispersal.

The nation's telephone services evolved with the nation's metropolitan regions in a highly complex and circularly casual relationship in which cause and effect are impossible to discern. Much is often made of the dispersal telephone services seem to make possible, but careful examination of dispersal reveals other equally important causes such as ongoing economic development, federal housing policies, and better roads and automobiles. Moreover, emphasis on space-convergence and dispersal overlooks the *space-intensification* that telephone permits; a modern high-rise office building relies on cheap, efficient telephone service just as much as a sprawling suburban office complex. The ambiguous role of telephone technology in the evolution of the American metropolitan system suggests that new telephone services and new communications media will be more evolutionary than revolutionary in terms of their effects on the nation's settlement patterns.

## NOTES

1. U.S. Bureau of the Census, Census of Housing, 1970. *Detailed Housing Characteristics. Final Report HC(1)-B1. United States Summary* (Washington: USGPO, 1972), p. 280.
2. Edward A. Ackermann, *Geography as a Fundamental Research Discipline* Department of Geography Research Paper No. 53 (Chicago: Department of Geography, University of Chicago, 1958), p. 26.
3. Donald G. Janelle, "Central Place Development in a Time-Space Framework," *Professional Geographer* 20 (1968), pp. 5–10.

4. James J. Pilliod and Harold L. Ryan, "Operator Toll Dialing—A New Long Distance Method," *Bell Telephone Magazine* 24 (1945), p. 102.

5. Ibid., pp. 102–103.

6. Letter mail complements telephone service by providing complete cost-space convergence.

7. Calculated from statistics in U.S. Bureau of the Census, *Statistical Abstract of the United States, 1974* (Washington: USGPO, 1974), p. 500; U.S. Bureau of the Census, *Historical Statistics of the United States, Colonial Times to 1957*. In 1920, 50,207,000 local calls and 1,607,000 toll calls were placed daily.

8. This cross-tabulation was developed with John S. Adams, Associate Professor of Geography and Public Affairs at the University of Minnesota. The cross-tabulation was compiled from Census data and from U.S. Department of Labor, *Tomorrow's Manpower Needs*, Bulletin No. 1606 (Washington: USGPO, 1969).

9. Raymond Vernon, *Metropolis 1985* (Garden City, N.Y.: Anchor Books, 1963), pp. 99 ff.

10. Regina Belz Armstrong, *The Office Industry: Patterns of Growth and Location* (New York: Regional Plan Association, 1972), pp. 28–30.

11. R. Keith Semple, "Recent Trends in the Spatial Concentration of Corporate Headquarters," *Economic Geography* 49, 4 (October 1973), p. 318.

12. *New York Times*, May 21, 1972.

13. Eleanore Carruth, "The Skyscraping Losses in Manhattan Office Buildings," *Fortune* (February 1975), p. 162.

14. Elizabeth P. Deutermann, "Headquarters Have Human Problems," *Federal Reserve Bank of Philadelphia Business Review* (February 1970), p. 6.

15. William C. Langdon, "The Beginnings of Long Distance," *Bell Telephone Quarterly* 10, 4 (October 1931), pp. 247–250.

16. Thomas Shaw, "The Conquest of Distance by Wire Telephony," *Bell System Technical Journal* 33, 4 (October 1944), p. 343.

17. Walter B. Emery, "A Talking World," *Journal of Communication* 4, 2 (Summer 1958), pp. 59–62.

18. H. S. Osborne, "A General Plan for Toll Telephone Service," *Bell System Technical Journal* 9, 3 (July 1930), pp. 433–437.

19. Ibid., p. 477.

20. Emery, "A Talking World," p. 57.

21. Federal Communications Commission, *Statistics of Communication Common Carriers, 1970* (Washington: USGPO, 1972), p. 19; *Statistical Abstract*, 1974, p. 373.

22. *Survey of Current Business, 1972*.

23. Calculated from *Historical Statistics of the United States*, p. 480; *Statistical Abstract*, 1975.

24. James W. Carey and John J. Quirk, "The Mythos of the Electronic

Revolution," *American Scholar* 39, 2 (Spring 1970), pp. 219–241; 3 (Summer 1970), pp. 395–424.

25. Titus Smith, *Altruria* (New York: Altruria Publishing Co., 1895); Edward Bellamy, *Looking Backward: 2000–1887* (New York: Signet Books, 1960).

26. Paul Baran, "30 Services That Two-Way Television Can Provide," *The Futurist* 8, 5 (October 1973), p. 204.

27. *How to Talk Back to Your Television Set* (New York: Bantam Books, 1970), pp. 110–111.

28. "Communications," *Science Journal*, October 1967.

29. Edward M. Dickson and Raymond Bowers, *The Video Telephone: A New Era in Telecommunications* (Ithaca, N.Y.: Cornell University Program on Science, Technology, and Society, 1973), p. 143.

30. "Telecommunication: Its Impact on Business," *Harvard Business Review* 50, 4 (July–August 1972), pp. 125–132.

31. *Statistical Abstract, 1974*, p. 373.

32. *Statistics of Communication Common Carriers, 1970*, p. 19.

33. *Historical Statistics of the United States*, p. 139.

# 16

## Urban Growth and the Development of the Telephone: Some Relationships at the Turn of the Century

J. Alan Moyer

Much of the spacial structure of a metropolis is the result of more or less independent location decisions by individuals and firms. Many factors influence those decisions (e.g., availability of jobs, amount and cost of space, proximity to family and friends, availability of amenities, accessibility and cost of transportation, availability of services, location of markets, material, and suitable labor, etc.), but attention in the existing literature has focused mostly on transportation. The impact of ships, rivers, canals, railroads, trolleys, automobiles, and trucks has been seen as a key to the development and growth of urban areas. By comparison, communications—specifically telecommunications—has received only passing interest from researchers and has been largely seen as a decentralizing influence. F. J. Kingsbury's early comments about the telephone's impact typify a prevailing view. He states:

Three new factors have been suddenly developed which promise to exert a powerful influence on the problems of city and country life. These are the trolley, the bicycle, and the telephone. It is impossible to foresee at present just what their influence is to be on the...distribution of population; but this much is certain, that it adds from five to fifteen miles to the radius of any large town.[1]

Communications as a factor has been treated mainly in discussions of physical proximity of firms and individuals. Communication costs are normally considered part of the "agglomerative economies" associated with physical proximity. Such proximity is necessary to the specialization of occupations that is at the core of urbanism. Specialization makes individuals depend upon one another and upon interaction in a variety of forms: direct tactile or visual contact, face-to-face conversation, written or electrical transmission of information, exchange of money, or exchange of goods and services.[2] Many activities are centralized largely to minimize communication costs. Population clusterings express the drive to reduce the costs of interaction between people who depend upon and therefore communicate with each other.[3] Urban structure and form can be viewed as a result of these interactions. "The city is man's greatest monument to the importance of communications. . . . The need for contact between man and man constituted the principal impetus for the creation of cities."[4]

A key question is how communication costs affect the decentralization process of the central city. Communication costs enter a firm's decision to move from the city core to a suburb. Costs are incurred in maintaining linkages and established ties with suppliers of raw materials, services, and labor and with customers in the core or to establish new ties.[5] To a degree, the same may be said about individuals. As individuals move in the metropolis, there are costs of communicating with those left behind or with those who have moved elsewhere.

In the late 1800s the telephone altered communication costs for both individuals and firms, leading to locational changes. What effect did widespread telephone service have on the decentralization of cities? Did it have an effect on the dispersal of residences from the core? What role did it play as manufacturing, retailing, wholesaling, and other economic activities moved to the periphery or left the metropolitan area? Did the telephone allow the residents of the metropolis to improve the quality of their lives in their new locations? The goal of this paper is to begin such an inquiry by investigating a single case, Boston in the late 1800s and early 1900s. Boston, where the telephone began, was an area of high telephone usage. Because of its geographical attributes, Boston grew from a small original core through a core-dominated metropolis to a multicentered metropolis on the fringe of the large "Bos-Wash" megalopolis; thus, the influence of the telephone on patterns of centralization and decentralization of the city can be observed.

In the late 1800s, many cities began a decentralization process with the growth of peripheral suburbs. Farms turned to street grids and houses grew in place of last year's crops. It was the beginning of the continuing decentralization process and also of extensive telephone use. The number of telephones in the United States grew approximately by a factor of ten in this period.

This was also a time of intense interest in the provision of telephone service. The Bell System faced competitors, regulators, political bosses, reformers, and progressives and militant cost-conscious subscribers simultaneously, which presented a challenge to Bell's management—and a nagging concern over the course of events by public officials. This led to sizable investigations of telephone service in many major metropolitan areas that were documented and are available for analysis. Boston's investigation was particularly rich, being based on extensive traffic measurements in the metropolitan area in the early 1900s. The traffic data provide insights into the phone's use both by subscriber location in the metropolis and by subscriber type (business, residence, etc.).

We shall first consider the process of growth of the city. Metropolitan telephone service growth in Boston can then be related to urban growth (and a few other "externalities") by examining the structure and change of telephone exchange areas, rates, the type and quality of subscribers, locations, and frequency of use.

## GROWTH OF BOSTON

In 1850, Boston was a small city with residences, businesses, and factories intermingled. It was a tightly packed seaport where people normally walked to their jobs, to stores, and to visit friends and relatives. Face-to-face communication was dominant; it was a walking city whose densely settled area was within two miles of City Hall; it included parts of the nearby towns and cities of Brookline, Cambridge, Charlestown, Chelsea, Dorchester, Roxbury, and Somerville.[6] Firms depending on each other for business or services clustered together and their employees lived nearby—a pattern of settlement designed to facilitate communication. The diversity of occupations within the city were mostly small-scale, skilled enterprises supplying retail establishments to meet local consumer demand; the city's character was commercial rather than industrial.[7]

With large scale immigration, a ready source of cheap labor, and with improved power sources, industrialization began to shape the city. Factories developed and immigrants ran the machines; by 1870, Boston became the factory-manufactured, ready-made clothing center.[8] Similar developments occurred in other industries. With a large influx of cheap immigrant labor, Boston grew and prospered.

The large numbers of immigrants and creation of big factories stimulated a land boom, with housing and factories competing for space. Factories requiring large amounts of space often built on the periphery of the walking city. Horse-trolleys and, later, electric streetcars reached out from the city to take passengers home to the surrounding countryside, partly relieving the demand for city housing. By 1900, radial streetcar lines extended six miles from City Hall, and a pattern of land development emerged. Many factories were built in the area peripheral to the walking city of 1850, which encouraged large numbers of low-paid workers to live in working-class and lower-middle-class housing nearby. The area made a ring around the old city (broken only by affluent housing in Back Bay), including South Boston, Roxbury, East Boston, Charlestown, and part of Cambridge. A transition zone or "zone of emergence" bridged the crowded tenements of the inner low-income portions of the city to the middle class and the wealthy suburbs farther out on the streetcar lines.[9] Beyond the streetcar lines to about ten miles, only relatively wealthy individuals could live and commute to downtown Boston on trains. (See Figures 1–3.)

Housing development was constrained to a convenient walking distance from the streetcar lines, and there was little transportation between streetcar lines beyond the close extremes of the walking city. Areas between radials remained undeveloped. As a result of this radial pattern, the metropolis covered a large area which would be the basis for growth as the land between radials filled in during subsequent decades. Except for a few locations in downtown Boston and a few scattered settlements on the close periphery, the wealthy lived on the outer reaches of the metropolis. The middle class lived in the area a medium distance from the center, leaving the poor in the center (see Figure 2.)

The location of classes remained relatively fixed during this period, but the movement of individuals, as a result of class mobility, was dynamic. Not only were there many immigrants

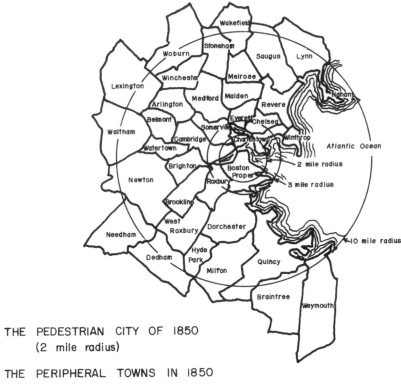

THE PEDESTRIAN CITY OF 1850
(2 mile radius)

THE PERIPHERAL TOWNS IN 1850
(3 mile radius)

THE NEW SUBURBS IN 1900
(10 mile radius)

**Figure 1**
Stages of Boston's metropolitan growth, 1850–1900 (from Warner, *Streetcar Suburbs*).

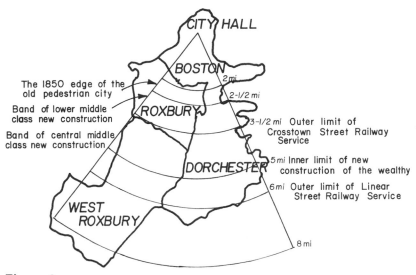

**Figure 2**
Suburban growth pattern/class patterns (from Warner, *Streetcar Suburbs*).

and migrants each year, but many were leaving for other towns in the metropolis and beyond. About one-third of Boston's population in 1890 had migrated in the previous decade.[10] Although the city's population was growing rapidly at the turn of the century (20–25 percent a decade), the metropolitan area was growing even faster (22–29 percent)[11]—a trend which has yet to reverse itself.

## EARLY YEARS OF TELEPHONE SERVICE

The telephone business developed slowly in the late 1870s and early 1880s through local franchises. The local franchisee obtained his franchise and leased his telephone instruments from the Bell Company. This method was chosen largely because of the company's lack of capital.

In more distant areas of the country, large areas or districts were assigned to agents who then employed subagents to develop local exchanges. Because local entrepreneurs lacked capital, the size of exchange territories was restricted, and smaller cities were largely developed before much was done in the larger cities.[12] This was a period of education for those constructing

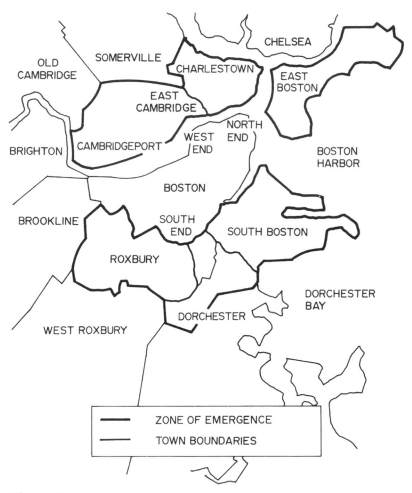

**Figure 3**
Zone of emergence (from Woods and Kennedy, *The Zone of Emergence*).

and operating the telephone exchanges. In this period practically all communication was local; there was little interconnection
between exchanges. Exchange owners did not appear to see the
business future of long-distance services in small cities and
towns, as people often lived within one or two score miles of
where they were born. According to Rhodes, the interweaving
of business and social interests over large areas had not yet begun,[13] but that was not true in cities; there, large population
movements occurred.

The telephone exchange owners were unlikely to interconnect
their exchanges, primarily for competitive and technological reasons. Each of the exchanges had a nonexclusive franchise in an
area, and exchange owners hoped to expand their areas to include territory in adjacent exchanges. Adjacent exchanges were
thus in competition; to interconnect would give the competitor
the use of the value of his plant. For the more established exchange, this was perceived to be bad practice—a policy that
later followed when Bell faced competition after the expiration
of its patents in 1894. Even with cooperative exchange owners,
interconnecting exchanges still would have been difficult because of the state of technology. In this period, a single line
cable with a ground return was utilized for transmission, cable
attenuation was high, and the circuits were too noisy to allow
practicable interconnection over the necessary distances.

At this period, with rates beyond the reach of the general
public, the telephone was chiefly an instrument of business and
commerce; the instruments were assembled by hand, and the
demand far exceeded the supply. Exchange owners sought to
maximize their profits with those instruments they could
obtain.

In 1878, the Boston banking community supplied money for
the expansion of the Bell company;[14] the control of the company
shifted to Boston financiers and remained with them for almost
thirty years.[15] Within two years, Bell and Hubbard had withdrawn completely from the business.

## FORMATION OF OPERATING COMPANIES—NET

The development of hard-drawn copper wire manufacturing
technology, as well as other advances, enhanced the ability to

talk greater distances. A retrospective assessment, made later by AT&T, said:

With the extension of the speaking limits of the telephone over connecting lines came also the necessity for the extension of the territorial limits of the exchange systems, the necessity of standardization, uniformity of apparatus and operating methods, and an effective common control over all. The necessity for system was the beginning of the Bell System.[16]

In the Boston area, the interconnection of exchanges began with the formation of the New England Telephone Company (NET) in 1883. NET was formed by consolidating companies that operated in Maine, New Hampshire, Vermont, and all but the southeastern part of Massachusetts. The advantages of consolidation were reduced expenses, uniform rates, prompter and easier connection, communication over longer distances, the elimination of competition, and, as NET elaborated, "an entire avoidance of the confusion and other inconveniences, both to the public and to the management resulting from the operation of various independent competing and not always friendly companies."[17] It was noted in the first annual report that Boston had a lower rate of telephone penetration than any other city (e.g., Portland, Lowell, Worcester, Springfield, and Concord); the parent company which was also the franchise holder had been concentrating on more pressing problems, probably the need for capital. Without large capital resources, it was unable to expand the market rapidly in Boston.

With the connection of various exchanges, the uncertain exchange boundaries of previous years came to be defined; tolls or extra charges were imposed for use of the interconnections. NET saw some major advantages: "The lack of a system which heretofore characterized the conduct of toll business throughout the entire territory has been superseded by the adoption of a tariff which it is believed will not only result in a greater net income, but will secure uniformity in charges and accuracy in rendering accounts."[18]

The capability of calling throughout the large metropolitan area enhanced the value of the telephone, but the unification of exchanges also caused an awareness that there were almost as many different rates as exchanges. In some exchanges, there were almost as many rates as subscribers. Classification of suburban[19] rentals showed 184 rates; in one city alone there were

53.[20] This situation caused considerable customer dissatisfaction and led to the adoption of uniform rates for the same classes of service. The rates were less in smaller cities and towns than in larger cities; NET justified this on the basis of the value of the service to the customer and the cost of providing the service. NET commented: "In the large exchanges established in cities of commercial importance, the average use of the telephone is greater, the service is more valuable and is furnished at greater cost to the company per station." [21] The company reported a decrease of nearly 3,000 subscribers (almost 20 percent), principally due to change of rates and the closing of unprofitable exchanges.[22]

Paralleling the interconnection of neighboring exchanges, long-distance or intercity lines linking large urban centers were developed. In 1879 the first intercity line was installed connecting Lowell with Boston, allowing the mill managers to communicate with offices in Boston.[23] Intercity line development proceeded slowly, technical problems were formidable, poor transmission curtailed usage. A Boston to New York line was completed in 1884, fell victim to the transmission difficulties of the era, and failed to provide the expected financial return. By the late 1880s, technological improvements and organizational developments[24] fostered the extension of long-distance circuits across the Northeast, and Boston had "instantaneous communications" to most of the hinterland. The development of long-distance circuits, the locational advantages they afforded communities in the network, and the interconnection of regional companies in the westward trek are topics treated in Ronald Abler's paper elsewhere in this book.

In 1886, the issue of telephone rates and services was discussed by the legislature, but no action to regulate the telephone company was taken.[25] The telephone company apparently lobbied against any regulation; they claimed that "hampered by restrictive rate laws, the service would inevitably deteriorate and the interests of the subscriber would suffer."[26]

The telephone users at this time were primarily business and industry. Many residential subscribers in the 1887 directory had a business listing with the same or similar name.[27] Within the Boston district (downtown Boston and suburban area), subscribers could make unlimited calls without extra charge, and suburban subscribers could restrict their unlimited calling area to the suburban zone; those suburban subscribers paid an extra 10 ¢ toll for each call to Boston.[28] The yearly rates are shown in Table 1. A special or private line was available only at the Pearl Street exchange, and subscribers over one mile from the exchange were charged extra.

The suburban district area was extensive;[29] it included almost the same area as the ten-mile radius indicated in Figure 1 as the 1900 metropolitan limit. (See Figure 4.)

NET's rationale for expansion was that the value of telephone services was relative to the number of other subscribers who could be called. In developing the suburban district, NET added additional exchanges—and therefore subscribers—to the system to increase the value of telephone service to its present subscribers.

The impact of rate schedules upon businesses' location decisions was insignificant: a firm could locate anywhere in metropolitan Boston without additional telephone costs, but firms probably did not include communication costs in location decisions at that time.

While the rarity of residential subscribers was partially due to telephone rates,[30] after the interconnection of exchanges individual telephone use greatly increased. The number of exchange subscribers rose 43 percent between 1888 and 1894, daily exchange connections increased 13 percent, toll connections increased 200 percent, and each subscriber's use increased 66 percent.[31] As evidenced by traffic growth far in excess of phone

**Table 1**

Yearly Rates for Boston District Subscribers, 1888

|  | Business | | Residence | |
|---|---|---|---|---|
|  | Full Exchange | Local Suburban | Full Exchange | Local Suburban |
| Special Line | $120 | 64 | $96 | 54 |
| Party Line 2 | 108 | 58 | 84 | 47 |
| Party Line 3 or more | 96 | 52 | 72 | 40 |

Source: NET, *List of Subscribers January 1, 1887, and Business Directory of NET&T—Boston Division* (Boston, 1887); NET, *List of Subscribers January 1888 and Business Directory of NET&T—Boston Division* (Boston, 1888).

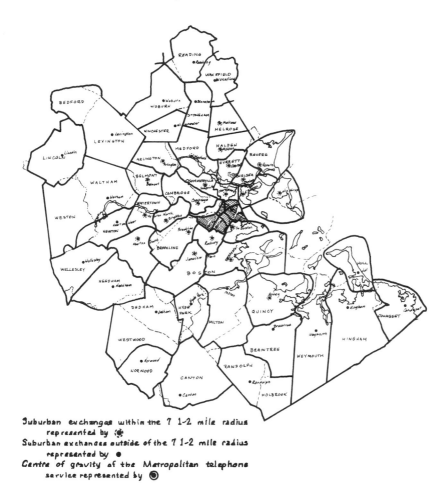

Suburban exchanges within the 7 1-2 mile radius
   represented by
Suburban exchanges outside of the 7 1-2 mile radius
   represented by ●
Centre of gravity of the Metropolitan telephone
   service represented by ◉

**Figure 4**
New England Telephone and Telegraph Company, Boston-Suburban
District (March 7, 1910).

growth, the subscribers had learned to use the phone to their advantage. In the eyes of NET, the phone was well on the way to becoming a necessity: "The man of business now uses his telephone not as an occasional convenience or as an alternative for other means of communication, but as the usual constant and effective agent for transactions requiring dispatch, accuracy, and clear interchange of thought.[32] With calling beyond the local exchange improved by organizational and technical improvements, subscribers moving from Boston to the suburbs could obtain the same communication capability at the same cost.

## COMPETITION

When Bell's patents expired in 1894, telephone service became fair game for anyone. To deal with the new competition, Bell offered an option of "measured service" at considerably less than the flat rate. This service was intended for small businesses and residential subscribers.[33] Within the NET area this policy speeded growth. The telephone began its gradual evolution from an exclusive instrument of commerce to one which served commercial and domestic uses. This transformation occurred slowly, but the change is reflected in the company reports. NET commented:

There has been a large increase in subscribers due to measured service.[34]

Telephone has shown to be of value in the conduct of business and as a convenience and safeguard in homes.[35]

Most noticeable feature of the development of our system during recent years has been the constantly increasing extent to which our service is coming into use in the small business establishments and for householders of very moderate means.[36]

This growing awareness that expansion would come through increases in residential subscribers stimulated competition in NET territory outside of Boston and its suburbs. The competitive policies developed by Bell were initially unsuccessful; by 1907, its share of the telephone market had fallen from the 100 percent in 1884 to 51 percent;[37] the growth in the industry during this period was explosive. The annual increase in the number of telephones was 19 percent in 1895 and 27 percent in 1897, 48 percent in 1899, 33 percent in 1901, 18 percent in 1903, and

23 percent in 1905 (21 percent annual rate of growth over 1895–1906),[38] in contrast to the 6.3 percent average growth in the period up to 1894.

The telephone's use in Massachusetts acquired a regional dimension. The Highway Commission commented:

From the greater length of line per subscriber, it is apparent that the telephone lines were more diffused in Massachusetts than the average. While the local messages were less than the average, the long-distance messages were much in excess, and it would seem that the business interests fully appreciate the importance of the speed and the certainty of this means of communication.[39]

By 1906 there was growing awareness of the impact of competition on the rates and services provided by the telephone company. There was also an anti-big business and anti-monopoly mood; this rapid growth led to legislation in 1906 (Chapter 433 of the Acts of 1906, Massachusetts) providing for general supervision by the Massachusetts Highway Commission of all companies transmitting intelligence by electricity. The Highway Commission was the natural body to involve, since all cables ran over or under the state's highways.

In the fall of 1906, an independent telephone company attempted to obtain a Boston franchise; capitalizing on public dissatisfaction, the Metro Home Telephone Company promised lower rates and better services. They proposed to charge not more than 5¢ between city and suburban stations,[40] and to install a dial system.[41] The franchise was granted by the aldermen in mid December and sent to the Mayor for approval,[42] but approval became a "hot" issue. The Mayor appointed an expert to investigate existing rates and services and the impact of two telephone systems in those cities that had them (about half the major cities).[43] The Boston Merchants Association, "one of the most powerful business organizations in the city," opposed the new franchise.[44] Bernard M. Wolf, local clothier and backer of the new telephone company, charged that Merchants Association members had a direct or indirect interest in telephone business and securities;[45] NET was also a member. Claims and counterclaims were made over the misuse and duplication of capital caused by a second company, the disruption caused by digging up streets, the duplication of facilities, and the need for competition. The Fruit and Produce Association and the Retail

Grocers Association endorsed two systems.[46] The Mayor vetoed the second phone franchise citing legal grounds. The aldermen overruled the veto, and the Mayor then refused to grant construction permits; anyone tampering with the streets would be arrested.[47]

The *Boston Post* did not take a position during this fight but did advocate a rate reduction and commented that Bell's rates were excessive.[48] The *Post*, joined by other "leading Boston merchants," petitioned the Highway Commission to revise rates and improve service in the Metropolitan District.[49] This petition joined one filed by suburban subscribers on December 18, 1906, asking for a reduction in toll charges from suburban districts to the metropolitan district (downtown Boston). The suburban petition, signed by thirty-three telephone subscribers, wanted tolls to downtown Boston reduced from 10¢ to 5¢ per call.[50] The grocers among the signers may have been concerned with toll calls to their supply houses, but they also might have been interested in cheaper local service. It was common for a grocer to call his customers daily to take their order for delivery. The grocers, perhaps enticed by the competitive telephone company's claims for rates and services, supported a dual telephone system.

The *Boston Post*'s two petitions addressing excessive rates and quality of service had (besides the *Boston Post*) twenty-seven different signers on each. The first petition (January 2, 1907) was signed by fifteen retail merchants,[51] six wholesale merchants of grain flour, hay, and butter and cheese, and two milling companies.[52] Eleven of the signers had private branch exchanges (PBX) in their businesses. The second metropolitan petition was signed by six physicians, one wool merchant, one manufacturing company, one bank, two wholesalers (books, dry goods), one architect, one derrick company, one hotel, one decorator, one importer, one retail merchant, and three real estate, insurance, and investment businesses.

On receipt of the petitions, the Highway Commission initiated investigations of NET rates.[53] When the investigations left many questions unanswered, the Commission decided to hire an expert, Professor D. C. Jackson, Chairman of the Electrical Engineering Department at MIT and President of the American

Institute of Electrical Engineers. He had just participated in a similar study in Chicago.[54]

After three years' study, Jackson proposed new rates and services on the basis of an inventory and appraisal of the property of NET and a traffic study. He advocated eliminating unlimited service in a wide area and substituting measured service to make phone service available to more people. Smaller zones were proposed where unlimited calling would be available. In the suburban district a zone would include the subscriber's town and contiguous towns except if a contiguous area was the metropolitan district. The same zone applied to measured service subscribers and additional charges were made for calls exceeding a maximum. Jackson found a wide variation in the rates per call for various types of subscribers. Business subscribers, especially those with flat rate service, often used the phone far more than residential ones. Thus costs ranged from 10¢ to less than 1¢ per call. Large users paid even less than ½ cent per call.[55] (Note that with a yearly rate of $162 [unlimited business rate], 105 calls daily would be necessary to approximate a 1½ cent per call cost on the basis of a six-day week and fifty-two weeks a year of business—about one call every five minutes.) The maximum calling rate was found to be more than 50,000 calls yearly, or about one call every three minutes.[56] Small users were paying part of the costs of the large users. Jackson's proposal, endorsed by the Highway Commission, was to eliminate flat rate (or unlimited service) in both the metropolitan and suburban districts.[57] The Massachusetts Highway Commission justified this action by the following reasoning:

The district to be covered by a given telephone rate should be the territory generally used by the great majority of the subscribers therein rather than a much larger territory, the greater portion of which is seldom used by the majority of subscribers.

Company should collect its revenue for calls between more distant portions of the territory from those who make use of such service rather than those who use only local service.[58]

A plan such as Jackson's would have a significant impact on the amount of communications between different areas or zones of the metropolitan and suburban districts. In fact, the plan was intended to effect a reduction:

We estimate that the proposed zone arrangement, with its toll charge for inter-zone service, will reduce the number of inter-

zone messages by something less than two-thirds of their existing value, which is but a small effect in comparison with the
gain to the majority of subscribers which we believe will be obtained by putting the cost of the longer haul messages on the
subscribers who originate them, and to whom the cost belongs.[59]

Additional insight into Jackson's perception of the impact
of this type of reduction can be obtained from his work in
Chicago:

It is obvious from our examination that the Chicago telephone
service is burdened by an expensive and undesirable parasite of
useless telephone calls which must be removed before the service can be made the most satisfactory and the price of service
can be reduced to a minimum.

The frivolous and useless messages can be largely cut off by
making measured rates of service since each small charge becomes a marked deterrent to the needless use of the telephone.[60]

Jackson seemed to hold a utilitarian view of telephone usage.

The Massachusetts Highway Commission had much the same
orientation as Jackson and also had a vision of the social value
of the proposed changes.

The Commission is also confident that the proposed rates will,
when put into effect, not only lead to great improvements in the
service, but also to substantial increases in the number of subscribers, thus increasing the community value of the telephone
as well as the company's revenue.[61]

In the year after Jackson's plans were implemented, reactions
were varied. NET and the Highway Commission reported favorably about price reductions, extension of service to nonsubscribers,[62] and how smoothly the plan was working.[63]

However, the reaction of the United Improvement Association[64] was at best unfavorable. A report prepared by the association contained several arguments against the new rates.[65]

The new schedule, it was claimed, was substantially an increase
of rates not warranted by the financial needs of the company or
by the expense of the cheaper class of service. When everyone is
talking about a metropolitan movement in the Greater Boston
area, it said, the telephone company alone came forward with a
proposition to disintegrate and divide. Also, the proposed
changes tend to restrict and decrease the use of the telephone.

The tone of the report is best characterized by a quote that
described measured service as the "style of the pick-pocket

who noisily dropped a penny in the blind beggar's tin cup and quietly extracted a dime."[66]

Apparently, for many, communication costs were important within the Greater Boston area. Who were these individuals? In December 1911, the *Boston Post* conducted a telephone survey to determine business and residential users' reactions to the new telephone rates. They reported that the costs were greater for most; the exceptions were usually professional people—notably physicians and householders.[67] Many argued that the new rates resulted in great inconvenience and a loss of business; an enforced practice of economy was making the telephone less efficient as office equipment or as a tool of commerce. Most wanted a return to the wider zones (all sections of Boston and contiguous cities).[68]

The mechanisms used to combat the "higher rates" varied. Some used the phone less often and substituted letters. Others waited until they returned home to the suburbs from their Boston offices before making suburban calls. Many discovered ways to obtain the most service for the lowest cost. A common practice for many Boston firms was to have two telephones—one phone had unlimited metropolitan service and was for calls within that district, the other had measured service for metropolitan and suburban districts and was used for calls into the suburban district. Tolls were charged for measured phones in terms of additional message units; the 3¢ charge for each additional message unit was cheaper than toll charges incurred with unlimited service (outside the unlimited area). Most of the firms which voiced strong opposition were retail and wholesale merchants, as well as certain service firms (e.g., attorneys). Also some of the manufacturing firms who had decentralized to locations in East Boston, Charlestown, Chelsea, and other districts of Greater Boston expressed concern and now felt that access to their plants was curtailed.

Out of similar situations, with constrained calling zones, other urban areas adopted Foreign Exchange Service, initiated by AT&T in 1911.[69] For a relatively small charge, a suburban business could have a phone on the central exchange so city customers and associates could call toll free. In a similar manner, a city business could reach its suburban customers.

The concepts developed and the geographical separations employed in Jackson's zone plan have survived to the present, but some of the inferences from the Jackson traffic study concerning the relationships between urban growth and telephone growth are far more important for this paper. To understand the telephone use in suburban and metropolitan areas, the Jacksons (with the cooperation of NET) conducted a traffic study during 1908 and 1909. Jackson summarized their findings:

Of all the messages originating in the suburban exchanges, only 14 percent are suburban interzone messages, 20 percent go into the metropolitan zone and the remaining 66 percent are strictly local zone messages. Seventy-two percent of the messages originating within the metropolitan zone are local to that zone, and a very large part of the remaining terminate within the limit of 7½ miles. In 1908 the suburban exchanges within the 5 mile radius covered 18.3 percent of the total suburban territory, served 65.6 percent of all the suburban subscribers, and took care of more than three fourths of the Boston and Suburban trunked calls.[70]

They also noted that since the reduction in tolls from 10 ¢ to 5 ¢ (suburban district to Boston),[71] the traffic had increased until nearly 80 percent of the Boston and suburban trunked calls were within that radius.

Jackson recorded the number of main stations (rather than accumulating data on the number of telephones) in each given locale.[72] In 1910, 13.5 percent of the population of the Boston and suburban district resided in the central area and had 30.3 percent of the area's main stations, a density of 14.3 main stations per 100 inhabitants. In the suburban district, within five miles of the center of the metropolitan zone (known as the 5 ¢ zone) were 62.4 percent of the population and 45.9 percent of the main stations, resulting in a density of 4.7 stations per 100 residents. In the outer suburban zones, 24.1 percent of the population lived, and they had 23.8 percent of the main stations for a density of 6.4.[73] Overall the density of the Boston and suburban district was 6.3 main stations per 100 inhabitants. The telephone showed a marked decrease in density in the high population area immediately adjacent to the metropolitan district where most of the interzone traffic originated or terminated. Outside the ring, the phone density increased to about the overall average of the district and seems to correlate with the population of that area. The density was low in an area of lower-middle-class and working-class housing, few of whose residents could afford telephones. The area also contained a significant amount of industry. Telephone density appeared to correlate with the residents' economic resources and with the extent of businesses located in the area.

Table 2 illustrates the distribution of traffic from different types of subscribers. Subscribers to metropolitan and suburban[74] wide unlimited service were mostly businesses located in the central business district. Conversely, most unlimited service suburban-area subscribers were residential users. Together, they generated about 70 percent of the traffic in the entire zone.

The telephone's use by businesses was extensive; 7 percent of the users generated 34 percent of the traffic. Some residential subscribers also used the phone extensively for suburban calls. One might speculate about the relative isolation of farm life and its attendant use of the phone. A telephone enabled the newcomer to the streetcar suburb to maintain contact with relatives and friends scattered about the metropolis and, to a lesser extent, with those remaining in the central area.

Low-income areas had fewer phones than high-income areas or areas with a significant industrial base.

The highest telephone densities were in some of the more prosperous suburbs, specifically Winthrop, Brookline, South Newton, Winchester, and Lincoln, each of which had more than one phone per ten inhabitants.

For interzone traffic (central-suburbs), which was about 25 percent of the total traffic, the amount of traffic to an exchange seems to correlate with the population living in the exchange area. Exceptions occur in instances with relatively high telephone density close to Boston (e.g., Brookline) or in areas of known industrial use or mixed industrial (e.g., Charlestown).[75]

## SOME RELATIONSHIPS

From the data that we have reviewed about Boston's growth and its telephone service growth, what types of relationships are suggested? First, the telephone was not crucial to the initial decentralization of Boston; it had already begun before the telephone arrived. In some ways initial decentralization created a demand for improved intracity communications—a demand by those already decentralized and those unwilling or unable to

**Table 2**
Distribution of Traffic

| Service | Range of Service | % of No. of Main Telephones | % of Main Tel. Traffic | Phone Location (% of Main Tel.) Metro. | Sub. | % of Main Tel. Traffic (location) Metro. | Sub. |
|---|---|---|---|---|---|---|---|
| **Unlimited** | Metropolitan & Suburban | 7 | 34* | 6 | 1 | 27 | 6 |
| | Business | 6 | 30 | 5 | | 25 | 5 |
| | Residential | 1 | 3 | 1 | | 2 | 1 |
| | Suburban | 42 | 35 | | | | |
| | Business | 1 | 2 | | | | |
| | Residential | 41 | 33 | | | | |
| **Measured** | Metropolitan & Suburban | 24 | 21 | 15 | 9 | 15 | 6 |
| | Business | 17 | 16 | 12 | 5 | 12 | 4 |
| | Residential | 7 | 5 | 3 | 4 | 3 | 2 |
| | Suburban | 10 | 4 | | | | |
| | Business | 5 | 2 | | | | |
| | Residential | 5 | 2 | | | | |
| **Coin-Box** | Metropolitan & Suburban | 17 | 6 | 8 | 9 | 3 | 3 |
| | Business | 11 | 4 | 5 | 6 | 2 | 2 |
| | Residential | 6 | 2 | 3 | 3 | 1 | 1 |
| **Total\*\*** | | | | 29 | 71 | 45 | 55 |

Source: Compiled from data in D. C. Jackson and W. B. Jackson, *Report of the Highway Commission.*
*Inconsistency caused by round-off error.
**Does not include PBX subscribers, hence biased in favor of suburban district.

move further from downtown Boston without improved communications. Significant growth in new residential construction did not occur until after telephone exchanges were interconnected. Could the telephone have provided the means by which developers could rapidly obtain supplies and services and increase their housing production? Although demand for houses could not be great without trolley transportation to them and municipal services for those purchasing them, could those homes have been constructed rapidly on such a wide scale without the telephone? Could the downtown district have handled the increased traffic necessary if face-to-face communication had been the only means usable? Would the mail have been fast enough? To what degree did telephone service availability influence the upper middle class to move to the outer extremes of the streetcar line? Was the ability to remain in semicontact with friends and relatives living elsewhere in the metropolis important in their decision? To someone from the city, the outer suburbs (with a large mix of strangers) could sometimes be as isolated as farm life. Friends a couple of towns away, living on a different streetcar line, were hours away from each other. Without cross-radial transportation, journeys of this sort required a trip into the city, and then back out again.

The telephone was used extensively by businesses throughout this time to transfer directions and orders and to extend prevailing business practices. By providing communication to the center with its markets and services, the phone helped small retail-oriented businesses to grow in the suburban area; orders could be placed with wholesalers downtown. While the phone supported this type of decentralization, it allowed the continued centralization and dominance of key urban businesses. Firms doing business with several different suburban manufacturers could remain where they were or move elsewhere without losing clients not in close proximity. Other firms could move their manufacturing facilities to nearby suburbs while retaining headquarters downtown without losing operational control. As Gottman and Abler have noted in Chapters 14 and 15, this would transform the city gradually to a site of business headquarters, government agencies, and firms in the service business sector. With newspaper advertising and a telephone to receive orders, large retail firms could stay in the city until better cross-radial transportation was developed. The telephone extended the range and influence of the city on the "hinterland," whether it was a suburb or frontier town. While facilitating residential decentralization, the telephone allowed many key institutions and businesses to remain centralized and allowed the city to continue its dominance in the nation's life.

Facilitating decentralization of some functions while directly allowing others to centralize[76] was common to the telephone in many ways. As the telephone spread to rural areas, so too did it spread the city's influence, ideas, values, avocations, and beliefs throughout the countryside. Rice, writing in the era, comments:

The rural resident is less and less class distinct from the people of town. It is no "jay" who calls you up by telephone in your store or office and asks you to deliver twenty-five dollars worth of edibles or machinery or legal advice at his place eight miles and forty minutes out; he is of necessity an enlightened gentleman farmer of the twentieth century.[77]

Many factors actively changed the sense of community in suburban areas, but for Rice the telephone was paramount. He comments:

If trolley service has brought the dweller in the country place nearer to the town, the instrument which projects talk has really landed him right in it. Space has been virtually annihilated—the telephone offers "up-to-date rapid transit." So the telephone is entitled, perhaps, to even greater consideration than the trolley as an urbanizing influence.

The tendency is altogether to bring those things which the city especially esteems within easy reach of the whole section so that feelings of local consciousness disappear.

The telephone is in short about the greatest urbanizer on record.[78]

Rice's enthusiastic commentary offers a perspective on changes brought to small suburban towns by the expansion of Boston and their transformation from small, independent, close-knit communities to members of a larger, urban-centered community.

This perspective differs slightly from Warner's view of community life during Boston's initial decentralization. Warner posits the fragmentation of local community life and the lack of a large scale community to deal with metropolitan problems— problems resulting from the centerless nature of suburban development and social class isolation there. If the telephone was as successful in urbanizing rural areas as Rice claimed, it could not overcome other influences in constructing or reconstructing communities based on physical location. The telephone gave access to alternative communities of interest. Most residential subscribers made infrequent calls to the city compared to calls within their town or the immediately surrounding towns—this was the basis of Jackson's recommendations. Rather than building a new metropolitan community, the telephone was utilized to maintain or build smaller communities. In a sense, existing residential class patterns were reinforced because only those of "means" had telephones. Mark Twain's 1880 sketch "A Telephone Conversation," described in Chapter 9, illustrates new adaptations in telephone usage. After a prolonged conversation covering a wide range of topics, Twain's parties agree to meet at 4. Although it is not certain that they mean 4 the same day, one is left with the impression that the two parties will meet later in the day and continue their discussions. The telephone replaced the back fence and so was local in its influence. The frequency of local contacts was amplified primarily because of the habitual use of the telephone.[79] Perhaps the intensification of local contacts helps preserve local patterns of attitude, habit, and behavior, and inhibits cultural leveling;[80] as such it may build and develop a sense of local community. Although viewed as a uniting influence by some, the telephone could not unite Boston's metropolis enough to deal with either its own problems or the local problems of its rapidly developing suburbs.

How did the telephone have a dual impact on urbanization? How did it facilitate centralization of some activities and decentralization of others? In short, because it was a tool for people with diverse interests and values, its impact was bound to be varied.

## NOTES

1. F. J. Kingsbury, "The Tendency of Man to Live in Cities," *Journal of Social Science*, Vol. 33, November 1895, in E. C. Banfield, *The Unheavenly City* (Boston, 1970), p. 22.
2. M. M. Webber, "Order in Diversity: Community Without Propinquity," in *Cities and Space*, edited by L. Wingo (Baltimore, 1964), p. 30.
3. Webber, "Order in Diversity," p. 31.
4. N. Johnson, "Urban Man and the Communications Revolution," *Nation's Cities*, July 1968, p. 9.

5. L. Moses and H. F. Williamson, "The Location of Economic Activity in Cities," *Papers and Proceedings of A.E.R.*, May 1967, p. 216.

6. S. B. Warner, Jr., *Streetcar Suburbs* (New York, 1970), p. 15.

7. Oscar Handlin, *Boston's Immigrants* (Cambridge,Mass., 1959), p. 56.

8. Ibid., p. 75.

9. R. A. Woods and A. J. Kennedy, *The Zone of Emergence*, 2nd ed. (Cambridge, Mass., 1969).

10. S. Thernstrom, *The Other Bostonians* (Cambridge, Mass., 1973), p. 20.

11. Ibid., p. 11.

12. American Telephone and Telegraph Company, *AT&T Annual Report 1910* (New York, 1911), p. 23.

13. F. L. Rhodes, *John J. Carty* (New York, 1932), p. 17.

14. J. C. Goulden, *Monopoly*, rev. ed. (New York, 1970), p. 49.

15. N. R. Danelian, *AT&T: The Story of Industrial Conquest* (New York, 1939), p. 40.

16. *AT&T Annual Report 1910*, p. 24.

17. New England Telephone Company, *Semi-Annual Report for Period Ending March 31, 1884* (Lowell, 1884), p. 3.

18. NET, *Report Ending December 31, 1884* (Lowell, 1885), p. 11.

19. According to the December 1884 Report, the suburban territory included exchanges in Jamaica Plain, Newton, Brookline, Hyde Park, Woburn, Winchester, Somerville, Malden, Chelsea, and Dedham.

20. NET, *1885 Annual Report* (Lowell, 1886), p. 4.

21. Ibid.

22. Ibid., p. 14.

23. AT&T, *Events in Telephone History*, (New York, n.d.). AT&T publication.

24. Formation of AT&T with the primary responsiblity of providing long-distance interconnection.

25. *Committee on Mercantile Affairs, Arguments and Testimony for the Remonstrants at Hearings as to Telephone Rates and Tolls and Other Questions Affecting Telephone Companies*, Boston, 1886.

26. NET, *1887 Annual Report* (Lowell, 1888).

27. The telephone directory listings gave the type of business of a subscriber or indicated that the phone was in his "residence." Since there were few residential users the telephone directory was, in fact, a business directory.

28. NET, *List of Subscribers January 1, 1887, and Business Directory of NET&T - Boston Division* (Boston, 1887).

29. The area of the suburban portion of the Boston District included Arlington, Bedford, Braintree, Brighton, Brookline, Cambridge, Canton, Charlestown, Chelsea, Dedham, Dorchester, East Boston, Everett, Hingham, Holbrook, Hyde Park, Jamaica Plain, Lexington, Lincoln, Malden, Medford, Melrose, Milton, Nantasket, Needham, Newton, Norwood, Quincy, Randolph, Reading, Revere, Sharon, Somerville, Stoneham, Stoughton, Wakefield, Waltham, Watertown, West Roxbury,

Weymouth, Wilmington, Winchester, Winthrop and Woburn. NET, *List of Subscribers April 1887 and Business Directory* (Boston, 1887).
30. The rates quoted above need to be compared to incomes of that time. An office clerk in Boston in 1902 was paid $12 per week, a book-keeper $20, a carpenter $15, and a laborer $10 (Thernstrom, *The Other Bostonians*, p. 300).
31. NET, *1894 Annual Report* (Lowell, 1895).
32. Ibid.
33. Ibid. The initiation of measured service required large additions in telephone plant. If the telephone company perceived this unfulfilled demand earlier, they did not react to it until their period of monopoly was over.
34. NET, *1897 Annual Report* (Lowell, 1898).
35. NET, *1900 Annual Report* (Lowell, 1901).
36. NET, *1904 Annual Report* (Lowell, 1905).
37. President's Task Force on Communication Policy, *Staff Paper 5*, Part I, June 1969, p. 31.
38. R. Gabel, "The Early Competitive Era in Telephone Communication, 1893–1920," *Law and Contemporary Problems*, Spring 1969.
39. Massachusetts Highway Commission, *14th Annual Report of the Masssachusetts Highway Commission for the Fiscal Year Ending November 30, 1906*, (Boston 1907), p. 107.
40. *Boston Post*, December 21, 1906.
41. *Boston Post*, December 18, 1906.
42. *Boston Post*, December 21, 1906.
43. Bureau of the Census, *Special Reports - Telephones and Telegraphs*, 1907.
44. *Boston Post*, December 19, 1906.
45. *Boston Post*, December 20, 1906.
46. *Boston Post*, December 25, 1906; December 27, 1906.
47. *Boston Post*, December 30, 1906. Apparently, the Metro Home Telephone Company never brought the issue before the courts. No record exists in the Suffolk County Superior Court nor was any claim registered against the city by Metro Home. Within four years, NET's competition had disappeared, either by dissolution of the company or by purchase. It is not known which occurred.
48. *Boston Post*, December 22, 1906.
49. *Boston Post*, January 2, 1907.
50. Twenty-nine of the thirty-three were subscribers to the Jamaica Plain exchange. NET, *1911 Telephone Directory, Boston Division Fall Issue* (Boston, 1911). In addition, two were located in Somerville, and one in West Roxbury. Most were small businessmen; seven were in the grocery, fruit, or provision business, three were lawyers, two dentists and two druggists, three real estate, insurance and investments, one jeweler, one retail boots and shoes, two carpenters and builders, one paper-hanger, one surveying instruments store, one stable, one hackman, and

one builder's iron supply house. The remainder were residents who could not be identified.

51. Among them was Edward A. Filene who advocated a single telephone system (*Boston Post*, December 13, 1906).

52. Based on an examination of the 1911 Telephone Directory.

53. Massachusetts Highway Commission, *1907 Annual Report* (Boston, 1908), p. 136. Stone and Webster, Inc., *Report to the Massachusetts Highway Commission Concerning Exam of NET Accounts* (Boston, April 22, 1907). G. Albree, *Report to the Massachusetts Highway Commission on NET* (Boston, 1907).

54. Committee on Gas, Oil, and Electric Light, *Telephone Service and Rates* (Chicago, September 3, 1907). D. C. Jackson, W. A. Crumb, and G. W. Wilder, *Report of the Special Telephone Committee* (Chicago, 1907).

55. D. C. Jackson and W. B. Jackson, *Report to the Massachusetts Highway Commission on Telephone Rates for Boston and Suburban District* (Boston, February 14, 1910).

56. About ⅓ cent per call.

57. Under pressure from suburban subscribers, the Highway Commission allowed measured service within the entire metropolitan and suburban district.

58. Massachusetts Highway Commission, *Recommendations of the Massachusetts Highway Commission to the New England Telephone and Telegraph Company Relative to Rates and Service for the Boston and Suburban District* (Boston, August 23, 1910), pp. 9–10.

59. Jackson and Jackson, *Report to the Massachusetts Highway Commission* p. 49.

60. Jackson, Crumb, and Wilder, *Report of the Special Telephone Committee* p. 71.

61. Massachusetts Highway Commission, *Recommendations to New England Telephone*, p. 15.

62. NET, *1911 Annual Report* (Boston, 1912).

63. Massachusetts Highway Commission, *1911 Annual Report* (Boston, 1912).

64. The relationship between the United Improvement Association and other pressure groups is unknown. The only available information about the association indicated that it was comprised of citizen associations and other improvement associations in Boston neighborhoods and surrounding communities. Many of these associations attended the original petition hearings, and their members were probably among the petition signers.

65. B. C. Lane, *The Telephone in Greater Boston*, United Improvement Association (Boston, 1911).

66. Ibid., p. 11.

67. "Many Protests and Some Praise for New Telephone Rate Schedule," *Boston Post*, December 17, 1911.

68. Note that parts of the City of Boston were within the suburban zone.

69. When Foreign Exchange Service was offered in the Boston area is not known.

70. Jackson and Jackson, *Report to the Massachusetts Highway Commission* p. 41.

71. Changed in 1908 before the traffic study upon recommendation of Jackson.

72. Main stations are the number of telephones in an area, not counting extension phones or PBX telephones but counting PBX access lines to exchanges. With the exception of main stations on multiparty lines, they are a measure of potential simultaneous traffic demand.

Since many businesses had PBX's with extension phones, main station counts tend to be biased in understating the percentage of business phones and the percentage of phones in the downtown area.

73. Jackson and Jackson, *Report to the Highway Commission*,
p. 36.

74. The area referred to is the calling area and not the subscriber's location.

75. Based on 1906 traffic reported in D. C. Jackson's *Report to the Massachusetts Highway Commission Being an Answer to Three Questions Asked by the Commission Growing Out of the Investigation of N.E.T.&T.* (Boston, 1908).

76. Just as the telephone was necessary to build Boston's suburbs, it was also necessary to build Boston's skyscrapers and to allow them to function without an army of messengers. Chapters 1, 6, and 14 in this volume have elaborated on this aspect in greater detail.

77. F. Rice, Jr., "Urbanizing Rural New England," *New England Magazine*, January 1906, pp. 528–531, 540.

78. Ibid., pp. 537–540.

79. M. M. Willey and S. A. Rice, *Communication Agencies and Social Life* (New York, 1933), p. 153.

80. Ibid., p. 154.

# IV

**The Telephone
and Human Interaction**

## Editor's Comment

To talk to others who are unseen and far away is an experience which, before the telephone, occurred only in mythology. Gods, devils, and angels talked from the sky across the world, but not mere mortals. The authors of the next three chapters ask how people behave when given the power of talking at any distance while being deprived of the power to see those with whom they talk.

The previous section on the telephone and the city and an earlier chapter by Suzanne Keller have already posed some of the same questions. Do people form noncontiguous communities when they can, or do they still communicate by telephone primarily with those they also frequently see face to face?

Bertil Thorngren marshals empirical evidence for the latter view; most telephone conversations, he finds, are with people located within easy visiting distance. In short, the telephone does not create a totally new set of relations; it is an additional channel for communicating with friends and coworkers—the same people one sees in person and occasionally sends notes to. It reinforces, Thorngren argues, existing networks of contacts, rather than creating socialized societies of telephone friends.

Alex Reid's paper suggests some reasons for this. His paper comprehensively reviews the literature on the differences between conversations and conferences with or without visual contact. Many laboratory and field studies have been done, in Britain, the USA, and elsewhere, of the differences between face-to-face and telephone conversations. The findings are consistent and sometimes surprising. For many purposes, nonvisual voice communication is just as good—perhaps better—than face-to-face interaction; that is true, for example, where the purpose is exchange of information. However, in other circumstances, such as creating new relationships, people find personal meetings necessary.

But when the telephone rings, one does not know who is summoning. In earlier chapters Briggs, Perry, and Abler quoted turn-of-the-century British commentators on the social dilemmas that this posed. One may protect oneself, these commentators suggested, by having the servant answer. In today's democratic society most of us respond to the summons ourselves, at least at home, but one question dominates our mind until we learn "who is it?" Schegloff, in his paper, examines behavioral patterns in the brief first exchanges of greetings as the caller and respondent try to resolve that anxious question.

Etiquette books and phone company publicity once tried to urge people to answer the phone by an immediate self-identification, e.g., "the Smith residence." But with rare exceptions, which Schegloff analyzes, people declined to do that. They protect their identity until they know who is calling and prefer to be recognized rather than identify themselves. The result—a grammar of greetings that Schegloff records—is stylized, conventional, and different from that which prevails in any other realm of human discourse.

Concealing one's identity and gradually revealing it is, however, not a unique kind of human behavior, but it was not an everyday matter until the ring of the telephone. As Schegloff reminds us, one of the great themes of classic folklore is communication with one whose identity is hidden. When telephoned, we are gripped by dramatic excitement as we wait for the revelation, just as the audience watches with excitement for the true identity of Oedipus or Telephus to be revealed.

The game of guessing who said "hello" makes sense, however, only because so much telephone conversation is among close associates who know each other's voices well. The telephone may bridge distance, but it also reinforces the bonds of personal relationships.

# 17

**Silent Actors:
Communication
Networks
for Development**

Bertil Thorngren

## BACKGROUND

In retrospect, the telephone may have been more crucial to social and economic development during the past 100 years than
many had observed and expected. There now seems to be a
growing consensus that telephony has had long-run consequences too important to be overlooked.

The mechanisms linking telecommunications to social and
economic development, however, are still not clear. Because of
the matter's complexity, there are different lines of reasoning regarding the impact of telecommunications on development with
regard to both the past and future.

Two main lines can be identified:

1. Some see a new communications technology as almost
automatically bound to produce new communication linkages
between previously unconnected places and individuals. According to this view, the existence of new linkages promotes social and economic development, irrespective of other measures
taken.[1]

2. Others point out that without close "fitness" to already existing structures, a new communication network will not long survive, and its development effects will be poor. In short, and
according to the view of most telecommunication administrations, the telephone of the 1880s (as well as the new media of
the 1980s) is supposed to have no independent, active role in
the development processes; other factors are the leading ones.

## THE SCOPE OF THE DISCUSSION

We shall focus here on feasibility problems and the mechanisms linking telecommunications to social and economic development.[2] This includes questions such as:

1. Under what circumstances can new communication networks, linking previously disconnected groups, work and survive long enough to have a lasting impact on social and economic development?

2. Are there unfulfilled needs for communication to which telecommunications can make unique contributions, or can they bridge critical gaps in existing networks?

3. If the answers to the first two questions are positive, what are the consequences for further research, including field trials of new technology?

## A FRAMEWORK FOR DISCUSSION

I begin with a point made by Professor Pool in a recent book. He notes that a few critical linkages might connect groups much larger than the actual communicators; thereby, the potential network available to an individual or an organization is enlarged.[3] My question is: what happens if the crucial link is disconnected due to some outside circumstance? To use the title concept of this paper, will "silent actors" inherent in the network be activated to keep the original network alive, or are the two subsets of communicators to become disconnected forever? Is planned action needed to restore the original network, or do communication networks have the capacity for "self-repairing" so that the original, or even more viable, combinations are created?

Such questions are not impossible to answer, as long as the discussion is confined to small group behavior studied under controllable conditions. These have been fairly well studied, and similarly, in the case of mass communication, statistical distributions can be observed.

When it comes to intermediate cases, however (such as government or business organizations that contain thousands of employees in different locations), the dynamics of the linkage pattern are still mostly unknown. Further analysis of interorganizational networks is an important task in itself, and also be-

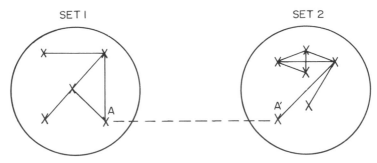

**Figure 1**
If link AA' exists, everyone in the population is linked.

cause the elucidation of organizational networks might help us to understand communications at large. Public and private organizations also have a heavy impact on information flows that connect individuals outside the organizations themselves.

## EMPIRICAL STUDIES

In his contribution, "Comparing the Telephone with Face-to-Face Contact," Alex Reid reviews the substantive British research comparing different media in the laboratory. He points out that telephone contact differs from traditional face-to-face contact in two ways. First, it transcends distance; second, it transmits only audio information. Laboratory experiments permit singling out these factors, or testing them in different combinations.

By contrast, the field studies discussed here depict large-scale networks, where the telephone and face-to-face communication are used side by side, or sequentially. A communication process between two organizations (or regions) may begin in any medium. During the process, however, a switch may take place between media. Often this happens in combination with a switch of the actual communicators as the message is passed upwards, downwards, or sideways in the organizations (or regions) involved. Thus, a communication process normally has a history in which persons not presently participating may still have had a heavy influence on the design of the networks at large, as well as on the choice of media at a specific point in time. Once one has passed the background information, additional narrow media may be more profitably used for a while.

To observe the "silent actors" which affect the viability of the networks, large-scale and repeated field studies are needed. A first attempt was made in Sweden in 1965. A pilot study covering all headquarters employees of a relocating Stockholm company was launched on a before and after basis. Data was collected four months before and eight months after the actual relocation from Stockholm to Eskilstuna, a small city 120 kilometers away. Follow-up studies were done in 1969 and 1974 to observe long-run adaption to the new location. Self-completed diaries were used to cover all communication with outside sources (oral, telephone, and written) during two six-day stretches in each of the observation periods. Also included was the communication between headquarters and fourteen plants in different locations in Sweden. The results indicated a substantial reshuffling of contact from higher to lower echelons but a surprisingly low degree of substitution of face-to-face by telephone contacts despite a travel time to Stockholm of more than two hours in each direction. Thus, the ratio of face-to-face to telephone contacts has been quite stable, both in the short run and in the ten-year period after the relocation.[4]

Some years later, more extensive studies covering hundreds of organizations were launched in Sweden and the United Kingdom. In Sweden, close to a hundred production and service companies of varying scale were covered by another diary study. Data from more than 10,000 telephone calls (longer than two minutes) and 3,000 face-to-face meetings was collected covering about twenty different dimensions for each contact event: e.g., length of meeting, number of participants, degree of advance planning, frequency of earlier contact, scope of information, and degree of feedback. The composition of groups (proportion of new vs. established participants) was of specific interest, as were a number of questions on travel.

To supplement the contact data, other kinds of data were collected regarding the characteristics of the individual participants and their companies. For individual employees, data was collected on education, age, number of years with the firm and in present position, function, and salary level. Structures of employment and production within the firm were studied for a

**Table 1**
The Structure of Contact Events in Four Studies

| % of events | KOMM 68 | | London 69 | | GOV. LOC. 70–72 | KOMM 71 | |
|---|---|---|---|---|---|---|---|
| | Face-to-Face | Telephone | Face-to-Face | Telephone | Face-to-Face | Face-to-Face | Telephone |
| **Time Spent** | | | | | | | |
| 2–10 minutes | 14 | 52 | 19 | 87 | 4 | 18 | 83 |
| 10–30 minutes | 29 | 17 | 29 | 12 | 16 | 21 | 16 |
| 30–60 minutes | 19 | 1 | 19 | 1 | 16 | 13 | 1 |
| 1–2 hours | 19 | 0 | 18 | 0 | 25 | 19 | 0 |
| 2 hours or more | 19 | 0 | 15 | 0 | 39 | 29 | 0 |
| **Number of participants** | | | | | | | |
| 2 | 42 | 98 | 55 | — | 30 | 42 | 95 |
| 3–5 | 46 | 2 | 31 | — | 32 | 39 | 5 |
| 6–10 | 8 | 0 | 9 | — | 16 | 13 | 0 |
| More than 10 | 4 | 0 | 5 | — | 22 | 5 | 0 |
| **Planning time** | | | | | | | |
| None | 25 | 86 | 17 | 83 | 20 | 23 | 86 |
| Same day | 20 | 6 | 13 | 9 | 10 | 15 | 6 |
| Day before | 19 | 5 | 12 | 4 | 6 | 12 | 4 |
| 2–7 days | 23 | 2 | 31 | 2 | 21 | 25 | 3 |
| 1 week or more | 13 | 1 | 27 | 2 | 43 | 25 | 1 |

| | | | | | | | |
|---|---|---|---|---|---|---|---|
| **Contact frequency, minutes** | | | | | | | |
| Never before | 20 | 14 | 25 | 11 | 28 | 24 | 19 |
| One or more occasions | 33 | 31 | 38 | 34 | 39 | 31 | 32 |
| Once a month | 21 | 22 | 13 | 14 | 12 | 21 | 23 |
| Once a week | 18 | 23 | 10 | 23 | 11 | 16 | 19 |
| Daily | 8 | 10 | 14 | 18 | 3 | 8 | 6 |
| **Direction** | | | | | | | |
| Gave information | 27 | 38 | 20 | 29 | 25 | 22 | 46 |
| Received information | 25 | 41 | 15 | 38 | 18 | 21 | 34 |
| Mutual exchange of information | 48 | 21 | 65 | 23 | 56 | 57 | 21 |
| **Range of discussion** | | | | | | | |
| One limited question | 38 | 71 | 57 | 87 | 56 | 47 | 78 |
| Several limited questions | 40 | 21 | 35 | 15 | 36 | 36 | 20 |
| General broad information | 22 | 8 | 8 | 1 | 8 | 17 | 2 |
| **Region** | | | | | | | |
| Greater Stockholm | 85 } | 76 } | | | | 69 | 53 |
| Center of Sweden | | | | | | 16 | 27 |
| South of Sweden | 9 | 22 | | | | 6 | 11 |
| North of Sweden | | | | | | 5 | 9 |
| Foreign country | 6 | 2 | | | | 5 | 1 |
| **Traveling time** | | | | | | | |
| Less than 10 minutes | 32 | | 27 | | — | 13 | |
| 10–30 minutes | 49 | | 48 | | — | 37 | |
| 30 minutes to 1 hour | 10 | | 11 | | — | 21 | |
| 1–2 hours | 4 | | 8 | | — | 11 | |
| More than 2 hours | 5 | | 6 | | — | 18 | |

**Table 1** (continued)

| | % of events | | | | | | | |
|---|---|---|---|---|---|---|---|---|
| | KOMM 68 | | London 69 | | GOV. LOC. 70–72 | | KOMM 71 | |
| | Face-to-Face | Telephone | Face-to-Face | Telephone | Face-to-Face | Telephone | Face-to-Face | Telephone |
| **Transportation** | | | | | | | | |
| Walking | 19 | | 22 | | — | | — | |
| Car | 69 | | 56 | | — | | — | |
| Bus, underground | 12 | | 17 | | — | | — | |
| Train | 3 | | 4 | | — | | — | |
| Aeroplane | 0 | | 1 | | — | | — | |
| Total number of observations | 2,169 | 8,184 | 1,544 | 5,266 | 929 | | 20,243 | 98,743 |

three-year period. This background data included investment, turnover, value added, and R&D efforts, as well as measures of R&D progress. The purpose of this extensive data collection was to establish linkages between the study of communication behavior on the level of organization and other aspects of behavior. (A first report of the results can be found in Thorngren [1970]; another report will follow upon a repeat study to be carried out in 1978, ten years after the primary data collection.)

Parallel data from other countries are already available. Dr. John Goddard of the London School of Economics launched his study of London offices in 1969 using the same diary; it has also been a standard tool for data collection in more recent studies. Some of the results from Goddard's study, "LONDON 69," are presented at the end of this chapter along with comparable data from other studies. (For a full presentation, see Goddard [1971 or 1973].)

Other studies include extensive investigations carried out before the dispersal of government offices in Sweden and the United Kingdom. The Swedish exercise, "KOMM 71," covered *all* employees (20,000 individuals) within thirty-four government offices in Stockholm; the British counterpart, "GOV. LOC. 70-72," also had a very wide coverage. There are minor differences in the two studies. The British study focused on meetings, and data collection extended over a longer time. The Swedish study was carried out in one shot to achieve cross-checking of communication between the agencies. However, as all four studies followed the same format, cross-national, cross-sectoral comparisons are feasible.

The four major studies discussed here, together with related studies in other European countries, represent a data base of more than 500,000 observations of contact events reported in twenty dimensions.

## SOME EMPIRICAL RESULTS

Full presentation of the results cannot be made in this context; only an overview can be offered. The tables below report comparisons, dimension-by-dimension and separately, for telephone and face-to-face contacts. (For an example of more elaborate comparisons, see Collins [1972].)

On a general level, however, the most striking feature of the comparisons is not the obvious differences between face-to-face and telephone contacts (that coincide with the results from laboratory experiments), but the close relationships between different media that form part of a joint multimedia, multiperson network connecting large segments of society over long periods of time.

For example, even the crude data presented here contradict the notion that telephone contacts have a propensity to create new linkages. The percentage of new relations with someone previously uncontacted is markedly higher for face-to-face contacts, whereas telephones are for more regular contacts (monthly, weekly, daily). The proportion of occasional contacts is similar for the two media.

More interesting than such static descriptions of the communication process is the dynamic by which telephone contacts contribute to the variety and viability of the contact networks. In turn, these are enriched by elements of face-to-face meetings as part of the contact chain over time. Once one has the background information from earlier meetings, narrower channels (such as the telephone) become more powerful than they would be in isolation. Space permits only a few illustrations, highlighting the local nature of communication and the role of third parties as silent actors in the process over time.

## COMMUNICATION IS LOCAL

Remembering that the data refer to large organizations (such as central government agencies and companies operating nationally and internationally), perhaps it is surprising to find that they are so heavily dependent on local communication networks. As much as one-third of the travels are less than ten minutes away, and two-thirds are within a thirty-minute radius. Because telephone contacts are basically with the same set of sources, they do not extend as far over space as might be expected. A more detailed analysis of production patterns has demonstrated that the local networks are not explainable by reference to routine contacts alone, but rather to complex, continuously renewed interactions with new sources in the near environment. Personal meetings and telephone contacts are of a mutually reinforcing variety; viability of contact with a self-repairing capacity is difficult to achieve over long distances, at least with present telecommunications.

## COMMUNICATION IS EXTENDED BY THIRD PARTIES OVER TIME

Quite often, the links between two organizations are not the obvious direct ones. Rather, a third (or fourth) party may act as the common denominator carrying the potential for occasional direct contacts. Most of the contacts that occur, as indicated by our data, are local. However, the existence of occasional long-distance communications among the few individuals who link the local subsets may be essential for the viability of the organization. The different bodies might not be aware of each other's existence but could still benefit from the existence of intermittent and normally silent actors. This is a process over time, where switching between different media and groups goes on in a Markov-type chain in which averaging smooths the inherent dynamics. However, the dependence over time might also be seen as the key to the obvious spatial dependence.

## CONCLUSIONS FOR FUTURE FIELD TRIALS

Retrospectively, the data presented here might both delimit and extend our conclusions about the role of telecommunication. The capacity of telephony to transcend distance, create new linkages, and replace face-to-face meetings might have narrower margins than expected from laboratory experiments. On the other hand, the importance of telephony as part of a multimedia flow might be even more critical than expected, keeping other networks such as personal contact viable and extended over time and space. Telephony might even be crucial for the existence of other networks that have a more direct relation to development. Without repeated brief telephone contacts, the personal relationship that is exercised only occasionally might atrophy with time. At the same time, without personal contact as background, telephone networks do not develop.

Looking into future field trials, we need to find a working compromise between two opposing dangers in testing new systems:

1. One danger is defining the relevant network participants too narrowly, thereby excluding not only important user groups but also normally silent actors with whom potential contact is necessary for the long-run survival of the relationship. [5]

2. The opposite danger is beginning with a new communication system that is so weakly connected to the mainstream that the network meets premature death. As already mentioned, empirical field data might be needed to strike a correct balance; research in that direction is going on at an increasing scale at many European universities.

In the meantime, awaiting full elucidation of the complexities of contact networks, alternative delimitations and sensitivity analysis of the effect of different network borderlines might be a necessary precaution lest the silent actors remain silent forever.

## NOTES

1. See, for example, Suzanne Keller's paper in this volume.
2. The concept of development is used here in a wide sense to include not only potential effects on income levels, employment, etc., but also potential effects on participation in decision processes, etc.
3. Figure source: Ithiel de Sola Pool and Wilbur Schramm, eds., *Handbook of Communication* (Chicago: Rand McNally, 1974).
4. See Thorngren (1976).
5. The introduction of loud speaking telephones (as well as the picturephone) are well-known examples of the danger of too narrowly conceived user groups. Use of the new equipment was expected to percolate from the top levels of the organization to the bottom, even though the lower echelons might have more need for free hands to find the papers the boss is requesting. Similarly, picturephones might be more useful in production and R&D than in top management. Processes started at the wrong end will not diffuse.

## REFERENCES

Collins, H."Organizational Communications." Communications Studies Group. London, 1972. Mimeo.

Goddard, J.B. "Communications and Office Location: A Review of Current Research." *Regional Studies* 5 (1971).

_____. "Office Linkages and Location." *Progress in Planning*, Vol. 1, Part 2 (1973).

Pool, I., and Schramm, W. *Handbook of Communications*. Chicago: Rand McNally, 1974.

Reid, A. A. L. "Face to Face Contacts in Government Departments. First Report of the Contact Sheet Survey." Communications Studies Group. London, 1971. Mimeo.

Thorngren, B. "Regional External Economies." *EFI*. Stockholm, 1976.

_____. "How Do Contact Systems Affect Regional Development?" *Environment & Planning* 2 (1970).

_____. "Communication Studies for Government Office Dispersal in Sweden." Office Location and Regional Development, An Foras Forbartha. Dublin, 1973.

# 18 Comparing Telephone With Face-to-Face Contact

## A. A. L. Reid

Telephone conversations differ from face-to-face contact in two ways. First, they transcend distance, but they also transmit only audio information. The advantages of transcending distance are obvious, but to what extent are they outweighed by the disadvantage of being limited to an audio channel? What effect, if any, does the absence of the visual channel have on human conversation?

The answers are clearly relevant to an understanding of the telephone's social impact and also perhaps to the planning of new telecommunications services by suggesting the extent and ways in which to improve the telephone to simulate more closely full audio-visual face-to-face communication. The question of difference between telephone and face-to-face contact is therefore as relevant to the future as it is to the past and seems appropriate for centennial review.

## FIELD DATA

Telephone and face-to-face conversation may be compared in the field or laboratory. In the field, data may be gathered either by direct observation or questionnaires (usually referred to as "contact diaries" or "contact record sheets"). The work of Mintzberg (1970, 1971) and Palmer and Beishon (1970) exemplify the observation method. Mintzberg conducted a week-long study of the chief executives of five medium-to-large organizations (a consulting firm, a school system, a technology firm, a consumer goods manufacturer, and a hospital). Structured as well as unstructured anecdotal data were collected in three records. A chronology record logged activity patterns throughout the working day. A mail record noted for each of 890 pieces of mail processed during the five weeks its purpose, format, sender, the attention it received, and the action taken. A contact record listed, for each of 368 verbal interactions, the purpose, medium (telephone call, scheduled or unscheduled meeting, tour), form of initiation, and location (Mintzberg, 1971).

Palmer and Beishon studied five managers and a supervisor (in electricity, gas, beverages, engineering, electronics, and paper-making industries) for a total of 365 hours. Each participant wore a halter microphone and pocket transmitter; the signals were picked up and recorded on a portable tape recorder carried by the observer, who also recorded a commentary on the second track of the tape (Beishon and Palmer, 1972). The participants' activities (sampled at ten-second intervals) were classified into six behavior categories: (1) observing; (2) interacting with others; (3) reading, writing, dictating; (4) telephoning; (5) thinking; (6) washing, walking, miscellaneous. Palmer and Beishon were particularly interested in the determinants of behavior, i.e., the stimuli for particular episodes. The stimuli were classified as information received, a conversation, reading, information via telephone, a self-initiated action with no overt information or causal source, and a previous activity.

Although such studies provide stimulating insights into how different communication media are used, their samples are too small for general comparisons to be drawn. By contrast, the use of self-completion questionnaires can provide data from relatively large samples of respondents. For example, Goddard (1973) surveyed 705 employees of seventy-two firms in central London. He recorded 1,544 meetings and 5,266 telephone calls on contact record sheets.

Some of the differences between telephone and face-to-face contact are shown in Table 1. We have argued that the telephone differs from face-to-face contact in two ways: the telephone can transcend distance, and a visual channel is absent with the telephone. The differences documented in Table 1 between telephone and face-to-face conversation could therefore be attributable to either or both of these causes.

**Table 1**
Comparative Field Data on Telephone and Face-to-Face Contact

|  | Telephone (%) | F/F (%) |
|---|---|---|
| **Length of Contact** | | |
| 2–10 minutes | 87 | 19 |
| 10–30 minutes | 12 | 29 |
| 30–60 minutes | 1 | 19 |
| 1–2 hours | 0 | 18 |
| More than 2 hours | 0 | 15 |
| **Arrangement of contact** | | |
| Not arranged | 83 | 17 |
| Same day | 9 | 13 |
| Day before | 4 | 12 |
| 2–7 days | 2 | 31 |
| More than week | 2 | 27 |
| **Initiation of contact** | | |
| Myself/another person in firm | 52 | 49 |
| Person outside firm | 48 | 51 |
| **Frequency of contact** | | |
| Daily | 18 | 14 |
| Once a week | 23 | 10 |
| Once a month | 14 | 13 |
| Occasional | 34 | 38 |
| First contact | 11 | 25 |
| **Range of subject matter** | | |
| One specific subject | 84 | 57 |
| Several specific subjects | 15 | 35 |
| Wide range of subjects | 1 | 8 |
| **Concerned with sales or purchases** | | |
| Directly | 36 | 38 |
| Indirectly | 23 | 25 |
| Not at all | 41 | 37 |
| **Main purpose of contact** | | |
| Give order or instruction | 13 | 7 |
| Receive order or instruction | 3 | 1 |
| Give advice | 5 | 6 |
| Receive advice | 9 | 5 |
| Bargaining | 3 | 8 |
| Give information | 11 | 7 |
| Receive information | 26 | 9 |

**Table 1** (continued)

|                                  | Telephone (%) | F/F (%) |
|----------------------------------|:-------------:|:-------:|
| Exchange information             | 20            | 28      |
| General discussion               | 7             | 13      |
| Other                            | 5             | 16      |
| **Total number of contacts**     | 5,266         | 1,554   |
| For meetings only                |               |         |
| Number of people at meeting      |               |         |
| One other                        |               | 61      |
| 2–4 people                       |               | 26      |
| 5–10 people                      |               | 8       |
| More than 10 people              |               | 5       |
| For meetings outside workplace   | Total         | 523     |

Source: Goddard, J. B., *Office Linkages and Location*, Pergamon Press Ltd., 1973.

## INTRODUCTION AND EXAMPLE

To associate cause and effect it is necessary to control for all variables except one. This can most easily be done in experiments in the laboratory. A substantial body of data from small-group laboratory experiments comparing telephone and face-to-face conversations under controlled conditions has accumulated in the last five years. The main purpose here is to review the findings of that work, summarized in Table 2. Much of the work has been at the Communications Studies Group, University College, London. I am grateful for the opportunity to draw upon the manuscript of a forthcoming book by three members of the Communications Studies Group—*The Social Psychology of Telecommunications* by John Short, Ederyn Williams, and Bruce Christie (to be published by John Wiley). The experiments reviewed briefly here are described there in much more detail and in a wider context.

By way of introduction and example, a single experiment will first be described and discussed in detail. The experiment by this author, Brian Champness, and David Prichard was the first full-scale laboratory experiment carried out by the Communications Studies Group and is therefore a suitably simple (and

**Table 2**
Laboratory Experiments

| Reference | No. of participants | Task | Media | Effects on outcome of withdrawing vision |
|---|---|---|---|---|
| **Information Transmission** | | | | |
| Champness & Reid (1970) | 72 | Letter transmission | A, A, F | none |
| **Problem Solving** | | | | |
| Champness & Davies (1971) | 44 | Slow worker on production line | A, F | none |
| Davies (1971a) | 40 | Factory location | A, F | none |
| Davies (1971b) | 80 | Factory location | A, F | none |
| Chapanis et al. (1972) | 40 | Assembly & location | A, F, W, W | none |
| Woodside et al. (1971) | | Resource allocation & crossword | A, A, F | none |
| **Conflict** | | | | |
| Wichman (1970) | 88 | Prisoner's dilemma | A, F, N, V | Reduction in the amount of cooperation. |
| Dorris et al. (1971) | 134 | Negotiation | A, F | More failures to reach agreement, & longer time taken to reach agreement. |

| | | | | |
|---|---|---|---|---|
| Morley & Stephenson (1969) | 40 | Wage negotiation | A, F | Trend toward side with "stronger" case achieving more victories. |
| Morley & Stephenson (1970) | 40 | Wage negotiation | A, F | Side with "stronger" case achieved more victories. |
| Morley (1971) | 122 | Wage negotiation | A, F | none |
| Short (1971a) | 60 | Wage negotiation | A, F | More failures to reach agreement. Side with "stronger" case achieved more victories. |
| Short (1971b) | 96 | Conflict of opinion & conflict of objectives | A, F, T | Side with the "strength of his convictions" did relatively less well. |
| Short (1971c) | 64 | Conflict of objectives | A, F | none |
| Short (1972a) | 120 | Conflict of opinion | A, F, T | More opinion change. |
| Short (1972b) | 144 | Conflict of opinion | A, F, T | Trend toward greater opinion change. Less accurate estimation of the other's opinion. |
| Short (1972c) | 112 | Conflict of opinion | A, F, T, T | More opinion change. |
| Short (1973b) | 96 | Conflict of opinion | A, F, T, T | More opinion change. |

**Table 2** (continued)

| Reference | No. of participants | Task | Media | Effects on outcome of withdrawing vision |
|---|---|---|---|---|
| Young (1974b) | 48 | Conflict of opinion | A, F | Trend toward more opinion change. |
| **Person Perception** | | | | |
| Reid (1970) | 23 | Interview | A, F | No difference in accuracy; lower confidence in judgment. |
| Reid (1970) | 24 | Truth/lies | A, F | No difference in accuracy; lower confidence in judgment. |
| Young (1974a) | 36 | Interview | A, F, T | No difference in accuracy; trend toward lower confidence in judgment. |
| Klemmer & Stocker (1971) | 59 | Discussion | A, T | none |
| Williams (1975a) | 144 | Discussion & assignment of priorities | A, F, T | In free discussion, other & conversation rated less favorably. |
| Williams (1972) | 96 | Discussion & assignment of priorities | A, T | Medium, conversation, and other rated less favorably. |
| Weston & Kristen (1973) | 96 | Discussion | A, F, T | Medium and other rated less favorably. |

| | | | | |
|---|---|---|---|---|
| Young (1975) | 90 | Business game | | Other person rated as showing less understanding. |
| Williams (1973a) | 144 | Departmental managers' negotiation | A, F, T | none |
| Williams (1975c) | 180 | Brainstorming on travel topic | A, F, T | Reduced cooperation and lower rating of the other person. |

Note: The following abbreviations are used for media. A: audio. F: face-to-face. N: no communication. T: television or video-telephone. V: vision only, i.e., face-to-face without speech. W: written, including teletypewriter.

somewhat nostalgic) example. It is reported in Champness and Reid (1970).

In this experiment, thirty-six pairs of male students each engaged in three communications tasks, using three media: face-to-face, acoustic speech path (no-vision), and telephone. In each transaction one participant (the "sender") communicated the contents of a 200-word business letter to another participant (the "receiver"). All possible orders of media were combined with all possible orders of the three letters, in a fully balanced design. Each letter contained four paragraphs, the second and third of which included "key information." In one letter, for instance (reproduced here as Figure 1), the second paragraph stated: "it *may* be possible to fabricate the new pack in plastic." The third paragraph stated: "if the new pack is fabricated in plastic, it *will* be *uneconomic* to adopt a perforated type of opening" (no words were underlined in the actual letters used).

These statements containing the key information were arranged to be either positive or negative (e.g., "possible" or "impossible") and either certain or uncertain (e.g., "*will* be possible" or "*may* be possible"). The two statements in each of the three letters were arranged to balance, as far as possible, the combinations of positive, negative, certain, and uncertain. The other sections of the letters were designed to be similar in both style and content. Any effects due to differences between the letters were controlled for in the experimental design.

The written instructions to sender and receiver asked them to make sure that the key points of the letter had been grasped by the receiver and allowed them as much time as they wished for the task. Full discussion was allowed, but the receiver was not allowed to see the letter or to make notes during the conversation. After the conversation, sender and receiver separately completed questionnaires. The sender's questionnaire contained a six-point rating scale designed to measure how well he felt the receiver had grasped the key points of the letter and (in the case of the telephone condition) a five-point scale measuring the user's opinion of the telephone connection's quality. The receiver completed a similar questionnaire, which asked him to write a summary of the key points of the letter as he understood them. Finally, the receiver completed a second questionnaire consisting of multiple choice incomplete sentences. The receiver was

Dear Mr. Robinson,/

    I have had some discussions on the lines you suggested,/ and am now able to give you an opinion on the feasibility of adopting a perforated type of opening/ for our new pack./ Firstly, although this is clearly only one of many considerations that will have to be borne in mind,/ I am told/ that it may be possible to fabricate the new pack in plastic./ Such a decision would have advantages and disadvantages,/ many of which will have occurred to you./ Perhaps you could let us know your own reactions to this possibility;/ they would be of great value to us in our further investigations./ Secondly, it has been very clearly impressed on me by my colleagues in this department/ that if the new pack is fabricated in plastic,/ it will be uneconomic to adopt a perforated type of opening./ Again, this depends on many factors/ to each of which we have tried to give appropriate weight./ You will have your own views on this subject, but that is the considered opinion here in this office./ Either way, we would be glad of your views on this matter,/ and hope that we may hear from you before too long./

    Yours faithfully,

    J. Brown

**Figure 1**
Sample letter used by Champness and Reid (1970).

asked to complete the sentence using the choices he thought
best summarized the key points of the letter. Each letter contained two key statements; each statement could be either positive or negative and either certain or uncertain. Each letter could
therefore take any of sixteen forms, and the letters were worded
so that each of these sixteen forms was plausible. The multiple
choice questionnaire presented all sixteen variables, from which
the receiver had to select one.

The purpose of the experiment was to measure whether the
absence of the visual channel had any effect on the perceived
and actual efficiency of information transmission. The design
enabled the following measures to be taken: (1) time taken for
the transaction; (2) scores on the rating scales; (3) error point
scores; and (4) accuracy scores.

Errors were scored from the receiver's multiple choice questionnaire. The number of error points was defined as the number of units separating the original statement from the receiver's
choice on the following scale:
positive certain—e.g., "will"
positive uncertain—e.g., "may"
negative uncertain—e.g., "may not"
negative certain—e.g., "will not"
Accuracy scores were based on the total amount of the letter
that the receiver was able to recall on his answer sheet. Each of
the three letters could be divided into nineteen information
points (see Figure 1). Receivers' written summaries for each trial
were typed on separate sheets; they were then allotted code
numbers, so that they could be scored "blind" (i.e., scorers did
not know from which subject or trial each sheet originated).
Three judges independently scored each of the 108 summaries,
using the nineteen information points as their criteria. Thus, if
a subject's summary contained all the information points, he
scored a maximum "accuracy score" of nineteen. A single accuracy score was produced for each transaction by taking the
mean of the three judges' scores.

Before turning to the results of this experiment, we note that
a substantial body of social psychology experiments indicates
that visual cues, such as gaze direction, facial expression, gesture, and posture, are associated in regular and predictable
ways with particular types of face-to-face conversation and with

particular types of communicator. For example, it has been established that gaze behavior is related to the pattern of utterances in conversation, to the personality, sex, and role of the participants, and to the physical distance between them (Exline (1963); Argyle and Dean (1965); Kendon (1967); Argyle, Lalljee, and Cook (1968). Writers such as Argyle (1969) have inferred that these visual cues perform the following important functions in conversation:

1. Mutual attention and responsiveness (to provide evidence that the other person is attending).

2. Channel control (to indicate the way participants should take turns in speaking and listening).

3. Interpersonal attitudes (to indicate attitudes and intentions).

4. Illustrations (to accompany and illustrate what is being said— for example, by gesture).

5. Feedback (to indicate whether the other person understands, believes, or disbelieves, is surprised, agrees or disagrees, is pleased or annoyed).

If the visual channel does perform these important functions then perhaps one would expect communication behavior in a purely audio communication system (such as the telephone) to differ significantly from that in face-to-face contact, either because of the lack of vision and/or the physical separation in the telephone condition.

Somewhat surprisingly, the letter transmission experiment showed no significant effect of medium of communication on any of the three performance measures (time taken, error point scores, and accuracy scores). While accuracy and time taken varied widely with the sender-receiver pair, the average error point score was only half an error point. In terms of error point score, therefore, there was no room for variance by condition. However, later experiments, which cure this fault, still support the finding. There were significant correlations between the three transaction times for a pair, between the three accuracy scores for a pair, and between a pair's transaction time and accuracy score for each medium. These correlations showed that pairs tended to be consistent in the time they took for each transaction and that pairs who took longer were more likely to achieve high accuracy scores.

Like the performance measures, the subjective measures (of

the sender's and receiver's confidence that the key points had been grasped) showed no significant effect due to communication medium. In summary, this experiment suggests that for two-person tasks involving information transmission between persons similar to those who took part, the telephone is just as efficient as the face-to-face meeting. However, this may not hold true for other types of conversation. The task chosen for the letter transmission experiment was deliberately simple. The role of the visual channel may be more critical in complex tasks and in those involving interpersonal emotions. In the latter case, the role of visual cues in transmitting attitudes and in providing feedback (on whether the other person believes or disbelieves, is surprised, pleased, or annoyed) would be expected to be particularly important. The experiments listed in Table 2 cover a wide range of tasks that vary along these dimensions. They are reviewed below in three groups—problem-solving tasks, conflict tasks, and person perception tasks.

## PROBLEM-SOLVING TASKS

Champness and Davies (1971) arranged for twenty-two pairs of civil servants to discuss a human relations problem either face-to-face or by telephone. The problem was that an old, loyal worker was holding up a serial production process in a motor component factory. The participants had twelve minutes to reach agreement; then they separately completed a questionnaire recommending a solution to the problem and rating their satisfaction with the solution. There was no significant effect of communication medium on any of these outcome measures.

This experiment lacked an objective measure of the quality of the solution and hence of the effectiveness of the communication medium. Two further experiments (Davis 1971a; Davis 1971b) used a "concept attainment" task that could be objectively scored. The first experiment used forty participants, the second used eighty (all from the Civil Service). Pairs of participants (communicating with each other face-to-face or by telephone) were told they were factory location planners employed by a company having sixteen factories in different parts of the country. Consultants had found that differences in export figures were due to differential availability of certain transport facilities.

They were given a table showing which of the factories were good exporters and which of eight transport facilities were available to each of the sixteen factories. Participants were asked to discuss the problem and arrive at a mutually agreeable solution as to which transport facility (or combination of transport facilities) was most strongly associated with good export performance. The table can be constructed to achieve the desired degree of difficulty. All answers could be objectively and precisely scored, but only by exhausive comparison of all possible answers; this comparison can be made by the experimenter but not in practice by the participants. Detailed measures of solution quality did not reveal any statistically significant effect of communication medium.

Two problem-solving experiments carried out at Johns Hopkins University using pairs of students communicating via face-to-face, audio, handwriting, and teletypwriter did not find any significant difference between face-to-face and audio in time to completion (although the written modes took almost twice as long as either face-to-face or audio). The first task was to assemble a lightweight wheeled cart; the instructions were given to one participant and the parts for the cart to the other (Parrish et al. 1971). The second problem required the pair to find the name and address of the physician closest to a designated address; a page from the Yellow Pages directory was given to one participant and a marked map to the other. Another task was to identify and sort electrical lamp sockets. Although different media produced widely different "numbers of messages," Chapanis concludes that for this type of task "gestures, facial expressions, and handwriting appear to contribute little extra to pure oral communication" (Chapanis, 1971).

Another series of two-person problem-solving experiments at Bell Northern Research Laboratories compared the three modes of face-to-face, audio, and audio with television graphics display. The first task was "resource allocation," essentially a numerical problem where each participant had partial information. For the second task, one participant was given a crossword puzzle to be solved jointly by the pair. Neither task showed any significant difference in performance between media (Woodside et al., 1971).

In summary, the experiments using problem-solving tasks

show no differences in efficiency between the face-to-face and audio media. In tasks where the outcome could vary, there was no significant difference in the quality of the outcome; in the Chapanis experiments (where there was one possible outcome whose quality was, therefore, fixed) there was no significant difference in the time taken. In practical terms, therefore, the telephone seems an excellent substitute for face-to-face contact in tasks of this type.

The experiments did, however, show some interesting effects of medium on the process of the communication. For example, Champness and Davis (1971) found that in the face-to-face condition there was a highly significant correlation ($p$ less than .001) between the speech lengths of the two members of each pair, whereas no correlation existed in the telephone condition. Davis (1971a, 1971b) found that participants took significantly longer in face-to-face contact than in the telephone contact and consequently exchanged more suggestions. Chapanis et al. (1972) likewise found that participants exchanged more messages in the face-to-face than in the telephone condition. Yet in problem-solving tasks these differences do not affect the practical outcome—perhaps because they are too weak or because the participants can adapt their behavior to counteract the effects. However, it will be seen below that in other tasks this is not the case.

## CONFLICT TASKS

Wichman (1970) reports a prisoner's dilemma bargaining game carried out by pairs of participants under four conditions: no communication; video only (subjects forbidden to talk and white noise played over a loudspeaker); audio only (subjects screened from one another); and face-to-face. The results showed an increasing degree of cooperation through the four conditions; the most substantial change occurred with the addition of the audio channel. Wichman suggests that the face-to-face condition gives participants the feeling that the other can be influenced more, thus making it appear more worthwhile to act cooperatively.

Dorris, Gentry, and Kelley (1971) arranged for participants to carry out a negotiating task (involving a two-minute discussion on how to split a number of points between them) while they stood facing one another with or without a screen intervening. They found that without vision, participants took significantly longer to reach agreement and more often failed to do so.

These isolated experiments suggest that in conflict tasks the medium of communication can affect both outcome and process. They suggest further that withdrawal of vision reduces efficiency of communication, by reducing cooperation and agreement. However, experiments using more complex and realistic bargaining tasks show that the communication medium's effects may be much more difficult to interpret in terms of efficiency.

Morley and Stephenson carried out a pair of experiments (1969, 1970) where student participants acted as management and labor representatives in an industrial wage dispute. The conversations were conducted either face-to-face or by a microphone/loudspeaker system. In the first experiment, the briefing material gave an initially stronger case to the management representative; in the second experiment, the labor representative was given the stronger case. Combining the data from the two experiments, the side with the stronger case was found to be significantly more successful without vision than in the face-to-face discussion. Morley and Stephenson's explanation for this effect is based on two concepts: the concept of "formality" of communication medium and the concept that negotiations involve both "interparty" and "interpersonal" conflict. In terms of formality, it is argued that absence of vision produces a more "formal" discussion; to increase "formality," extra conditions were included in the experiment in which neither side was allowed to interrupt the other. Morley and Stephenson suggest that the more formal the medium, the greater the emphasis on "interparty" communication (i.e., negotiation between two organizations, rather than between two people). The less formal face-to-face negotiation will thus produce more emphasis on the human and reciprocal processes of interpersonal communication; this results in greater generosity and yielding by the side with the stronger case. This explanation implies that the effects of communication medium on outcome are extremely selective— they affect certain types of negotiation but not others. The subtle and elusive nature of the effects was borne out in an experiment by Morley (1971) using 122 participants (compared with the 80 used in the two previous experiments) that produced no effects of medium of any kind. This may be explained by Morley's use of a new and much more voluminous set of briefing material which the participants may not have fully understood in the time available.

The largest single body of research into the effects of communication medium on conflict tasks was carried out by Short at the Communications Studies Group, University College, London, between 1971 and 1973. His first experiment (Short, 1971a) was a replication of Morley's 1971 experiment; it gave one party in the negotiation a much stronger case but allowed the participants a longer time to master the brief. The experiment confirmed the findings of Morley and Stephenson (1969, 1970) that the telephone outcomes were significantly more favorable to those with the stronger case than the results in the face-to-face condition. Significantly more failures to reach agreement occurred in the telephone negotiations than in the face-to-face negotiations ($p$ less than 0.05); six out of the seven such failures occurred in the telephone condition.

The interpretation of Morley and Stephenson's (apparently replicable) result hinges on the concept of "strength of case." To clarify the sources of the effect, Short's second experiment (1971b) distinguished two types of conflict: conflict of objectives (arising from incompatabilities in objective) and conflict of opinion (arising from the different experiences and values of the participants). A task was designed in which one member of each pair faced a conflict of objectives and the other faced a conflict of opinion. To achieve this, the individual facing a conflict of opinion (A) was allowed to choose his order of priorities among a list of nine potential areas of expenditure for a hypothetical company (e.g., employee education or capital investment). An inverse order of priorities was then assigned to the other participant (B), for whom the task became a simple conflict of objectives. The task was to agree on three areas in which expenditures should be cut. Forty-eight pairs of civil servants performed the task by one of three modes: face-to-face, two-way television, and audio. It was found that A, the individual facing the conflict of opinion, was relatively more successful in face-to-face contact; B, the individual facing a conflict of objectives, was relatively more successful in the audio condition. The television condition behaved in similar fashion to the face-to-face condition, suggesting the differences were due to the absence of the visual channel rather than physical separation.

In Short's third experiment (1971c) thirty-two pairs of civil servants met face-to-face and by telephone to agree on three items from a list of fifteen alternatives; each alternative had a separate payoff for each member of the pair (a conflict of objectives). Four indices were used to describe the outcome of the task: (1) A plus B, the combined payoff, represents the effectiveness with which the two sides coordinated to mutual advantage: (2) A minus B represents the extent to which the winner succeeded in imposing on the loser; (3) the winner's payoff; (4) the loser's payoff. No effect of medium was found for any of these four indices; nor was there any effect of medium on the number of agreements reached, since all pairs succeeded in reaching agreement. The conversations were tape-recorded, transcribed, and analyzed in detail; this, too, showed no effect of medium on the pattern of offers during the negotiation.

The fourth experiment in this series (Short, 1972a) investigated the effect of medium on the resolution of a conflict of opinion. Sixty pairs of civil servants completed an initial questionnaire in which they recorded their reaction to each of nineteen mildly controversial statements (e.g., "students should be financed by loans rather than grants," "public transport in large towns should be heavily subsidized out of taxation to discourage the use of the private car"). As in Short (1971b), the degree to which participants believed in the case they were asked to argue was deliberately manipulated as a variable by having each pair carry out two discussions. In the first discussion, they were assigned a statement on which the experimenter knew (from the questionnaires) they had sharply differing views. They were told to discuss this statement, with the objective of reaching a common view on it somewhere along the scale from "strongly agree" to "strongly disagree." In the second discussion, a statement was assigned to them on which the experimenter knew (again from the questionnaires) that they had the same view. However, in this second discussion, one of the pair was instructed to be a "devil's advocate" and argue the opposite point of view. This second condition resembled the situation in Short (1971b) where an insincere individual is arguing against an individual who is sincere in his case. Finally, at the end of each discussion participants were asked to record their private opinions on the statement they had discussed. As in Short (1971b), the sincere individual tended to do better face-to-face than via audio; however, in this experiment the difference was only slight and did not approach statistical significance. This casts further doubt on the already doubtful interpretation of the results of Short 1971a in terms of "sincerity." The design of the experiment made it possible, for the first time, to measure the degree of opinion change occurring via each medium. Analysis of variance showed a significant main effect of medium on opinion change ($p$ less than 0.05). Surprisingly, opinion changed significantly *more* in the audio condition than in the face-to-face condition. The video condition was intermediate and not significantly different from either.

Short's fifth experiment (1972b) was broadly similar, with seventy-two pairs of civil servants attempting to resolve conflicts of opinion via face-to-face, television, or audio contact. Although the trend towards greater individual opinion change after audio was substantial, it did not reach significance. Participants in the audio condition were found to be significantly less able to estimate their partner's opinion on the issue than were partners in either of the other two conditions.

In Short's sixth experiment (1972c), 112 civil servants were initially asked to rank eight topics in order of decreasing importance as national problems (e.g., industrial pollution, mental illness, private car). Randomly assigned pairs then discussed the topics either face-to-face, via television, or via audio, their task being to reach agreement on a ranking. Individual post-discussion rankings were then obtained by questionnaire. Again, opinion changed significantly more after audio discussion than after face-to-face or television discussion.

The task involving the nineteen mildly controversial statements (used in Short 1972a and 1972b) was again used in two further replications. Short (1973), using forty-eight pairs of participants, found significantly more opinion change in the audio than in the face-to-face condition. Young ( 1974b), using forty-eight participants, found a nonsignificant trend in the same direction.

In summary, the clearest and most consistent finding in this series of thirteen experiments using conflict tasks is the unexpected result that audio discussions produce more opinion change than do face-to-face discussions. The results of the relevant experiments are brought together in Table 3. In terms of opinion change, the difference between face-to-face and audio is highly significant ($p$ less than 0.0005); the difference between television and face-to-face is less so ($p$ less than 0.02), and although there were indications of a possible dependence on task, there was no significant difference overall between the two telecommunications conditions.

The finding is rather unexpected; a hypothesis based on the proposition that face-to-face telecommunication is richer and more effective than audio communication would predict greater opinion change face-to-face than via audio. No fully satisfactory explanation has yet been produced, but there may be some truth in the hypothesis that the face-to-face condition accentuates the interpersonal and social aspects of the conversation, thus distracting participants from the task of exchanging and influencing opinions.

**Table 3**

Opinion Change Findings

| | Face-to-face | | Video | | Audio | | Number of participants per condition |
|---|---|---|---|---|---|---|---|
| | Mean | S. D. | Mean | S. D. | Mean | S. D. | |
| Short (1972a) | 100 | (178) | 152 | (223) | 160 | (205) | 40 |
| Short (1972b) | 100 | (134) | 132 | (126) | 133 | (166) | 48 |
| Short (1972c) | 100 | (54) | 107 | (73) | 133 | (57) | 28 |
| Short (1973b) | 100 | (159) | 198 | (238) | 221 | (227) | 24 |
| Young (1974b) | 100 | (157) | — | | 138 | (158) | 24 |

Note: Results are adjusted so that the face-to-face mean = 100 in each experiment.
The results of the 1972c experiment are not strictly comparable to the other results because a different method was used.
Source: This table is reproduced from Short, Williams, and Christie *The Social Psychology of Telecommunications* (to be published by John Wiley).

## PERSON PERCEPTION TASKS

Two early pilot experiments in this series were carried out by
Reid (1970). In the first experiment participants were asked to
act the role of an assessor for the award of a £200 travel scholarship intended to enable a recent graduate of a British university
to travel abroad. Participants (drawn from the principal grade of
the Civil Service) were told that they were to interview a candidate for five minutes and that they would afterwards be asked
to express some opinions about his suitability for the award.
Eleven participants interviewed a confederate candidate face-to-
face, and eleven by telephone. Media were alternated to balance
for trends in the confederate's performance. After each interview the participant judged the candidate on twelve rating
scales. The results showed no significant difference between the
spread of judgments via the two media; nor was there any significant effect of medium on the type of judgments. However,
face-to-face interviewers expressed significantly higher confidence in their judgments than did the telephone interviewers ($p$
less than 0.02). The second pilot experiment was designed to
test whether truth and lies could be more easily distinguished
face-to-face than over the telephone. In each conversation one
participant spoke for five minutes on a specified topic to two
listeners—one seated opposite him in full view, the other on a
telephone in a separate room. The five-minute period was
broken down into ten thirty-second portions; the speaker was
instructed (by means of a visual signal invisible to the listeners)
to tell either truth or lies in each portion. Each listener had to
record whether he felt the preceding portion had consisted of
truth or lies and to state his confidence in making that judgment (very confident/moderately confident/not at all confident).
The two listeners then exchanged places, and the exercise was
repeated using a new topic. This whole exercise was repeated
eight times using a total of twenty-four participants. The experiment did not find any effect of medium on accuracy of perception of truth and lies. Again, the participants tended to express
greater confidence in the face-to-face judgments.

A more thorough experiment using an interview task has
been carried out by Young (1974a). Each of eighteen interviewers had three ten-minute interviews with three different inter-

viewees (the former actually being experienced professional interviewers in the Civil Service, the latter being university students). The three interviews were carried out by three different media (face-to-face, audio, and television). After each interview, the interviewer rated the interviewee on twelve seven-point scales such as "kind," "rational," and "trustworthy," giving a confidence rating for each judgment. At the end of the three interviews, the interviewer ranked all three interviewees on the same scales. Interviewees similarly filled in scales giving their impression of the interviewers and the interviews, ranked the three interviewers, and gave opinions of their own personality on the twelve scales that had been used by the interviewers. The results showed no significant media differences in "accuracy" (similarity between interviewer's judgments and interviewee's self-rating) or in confidence of judgment. There was, however, a trend toward greater confidence in face-to-face judgments and toward greater "accuracy" face-to-face. Although the interviewee's judgments of interviewers and interviews showed no media effects, there was a scatter of significant media differences in the interviewers' ratings of the interviewees. For example, the audio interviewees were considered more "dominant" than the interviewees over the other two media.

A different type of task was used by Klemmer and Stocker (1971). The reasoning was that media effects might be strongest in relatively undefined situations in which interpersonal communication might come to the fore. It was also felt that such effects were likely to be strongest in the case of discussions between strangers. Each participant therefore had two ten-minute conversations (one over a video-telephone, the other via audio), the only constraint being a suggested topic. After each conversation, the participants rated their partners on eight rating scales. The first subexperiment of thirty-five participants showed a single difference in ratings between the media; people met via the video-telephone were considered more submissive than those met via the audio system. However, the second subexperiment of twenty-four participants failed to confirm this difference and found no other effects of communications medium.

A major experiment by Williams (1975a), on somewhat similar lines, used 144 participants (all Civil Servants) each of whom had two fifteen-minute conversations. Each of these conversa-

tions was with a different partner and via a different medium
(two out of face-to-face, telephone, or television). Half the conversations were free discussion; the other half involved the
directed discussion of priorities. In the free discussion conversations, the face-to-face partners were strongly preferred to the
partners in the other two modes. In the mildly competitive
"priorities" conversations, the television partners and conversations were strongly preferred. Williams suggests that the additional interpersonal tension inherent in the more competitive
task may have led participants to prefer a less intimate communications medium; this explanation is consistent with Argyle's
intimacy equilibrium theory (Argyle, 1969). Whereas previous
experiments of this type had asked participants to judge their
partners on a rating scale, this experiment used instead a
"forced choice" method; each participant had to choose, for
example, which of his two conversations was more friendly,
pleasant, and which of two partners was more helpful, trustworthy. This forces the participants to reveal fine differences
in their judgments—differences that may be lost in the tendency
to group personal evaluations around one or two points on a
rating scale.

A similar but more elaborate experiment was conducted by
Williams and Stocker (Williams, 1972). Each of ninety-six participants had one ten-minute conversation with a different person via each of two media (video-telephone and audio). For half
of the participants, the partner was a complete stranger; for the
other half, the partner was a friend. Three tasks were used: the
free discussion and priorities tasks used in Williams (1975a),
and a third "persuasion" task in which participants chose four
from a list of forty-four problems of modern life as the most important. They were then told "to persuade the other person that
your list is correct and his incorrect." After each conversation,
participants expressed their preferences (using in each case
eight scales) between partners, between conversations, and between media. The results showed significant media effects on
five media scales, seven conversation scales, and five partner
scales. In most cases, the video-telephone tended to lead to
more favorable evaluations (more efficient medium, more interesting conversation, more positively evaluated partner).

Weston and Kristen (1973) had sixteen groups of six students each discuss their courses face-to-face, or over audio or television links. Each group had three discussion sessions, always using the same medium. The post-experimental questionnaires included various statements (e.g., "I had quite a bit of trouble knowing how the people at the other end were reacting to the things I said") with which participants had to express the extent of their agreement on rating scales. Of twenty-seven scales, twelve showed a significant effect of medium ($p$ less than 0.05); in all cases, television was rated more fovorably than audio. Participants were also asked to judge how many members of the group had agreed with them and had understood their remarks. Those at the other end of an audio link were rated significantly lower in both these respects than the members of the television or face-to-face groups.

Young (1975) found a similar result in an experiment where ninety participants took part in three-person mixed media business discussions. When subjects were asked how well partners understood their views, in regard to agreeing or disagreeing with them, they rated those at the other end of audio links as understanding less. However, in a conflict experiment already described (Young 1974b), actual understanding was calculated by comparing A's rating of B on a series of adjectival scales with B's rating, on the same scales, of how he felt A saw him. Although accuracy measured in this way was better than chance, there was no significant media effect. This reflects the earlier findings that the face-to-face condition may increase participants' confidence in their judgments, without increasing the actual accuracy of the judgments.

The final pair of experiments, by Williams (1973a, 1975b), were designed to test the hypothesis that by separating the group into subgroups, telecommunications can affect the lines along which coalitions form within a group discussion. In the first experiment, forty-eight three-person groups were formed; two members of the group were face-to-face, and the third was in a separate room (connected by either audio or television link). The participants posed as departmental managers in a small firm where a budget cut had to be made in one department. Their task was to agree which department would suffer the cut. In addition to these data on the decisions, participants completed forced-choice rating scales to evaluate the partners, conversations, and media. The results showed no effect of medium on the decision (it had been hypothesized that the remote participant in the three-person group would be more likely to suffer the budgetary cut). Nor were there any strong effects of medium on the rating scales. Williams suggests that the obvious media manipulation may have led the participants to "lean over backwards" in favor of the remote member of the group, thus negating any media effects.

The second experiment in this group (Williams, 1975b) was designed to overcome the obvious isolation of the remote third member. Forty-five groups of four office workers took part, fifteen groups communicating face-to-face, fifteen via television with two people at each end, and fifteen via audio with the same arrangement. The participants were asked to generate ideas about traveling in Britain through a "brainstorming" discussion. One of the participants, designated as secretary of the group, noted each idea that was proposed and seconded. In addition, anyone who wished could have their name noted as a dissenter from each idea. The results showed that for both telecommunications conditions participants at the same node supported each other significantly more than they supported either of the individuals at the other node. The patterns of dissent were not significantly different from chance in the face-to-face and television conditions; in the audio condition, however, the dissenter was significantly more likely to come from the opposite side ($p$ less than 0.001). The participants' evaluations of the other members of the group also differed from chance in the audio condition; members at the other end of an audio link were rated significantly lower on "intelligence" ($p$ less than 0.02) and "sincerity" ($p$ less than 0.01).

In summary, this group of ten experiments on person perception suggests that medium of communication does affect evaluative ratings of the conversation and the conversation partner, with marked differences between face-to-face and audio systems but much less difference between face-to-face and television. It seems that by separating groups into subgroups, telecommunications can affect the lines along which coalitions form. There are also indications that participants' confidence in their evaluative judgments is higher in the face-to-face than in the audio condition, despite the lack of experimental evidence that face-to-face contact actually increases the participants' accuracy of judgment.

## CONCLUSIONS

This paper began by asking what effect, if any, the withdrawal of the visual channel has on human communication. The answer that emerges from these experiments goes some way toward explaining the extraordinary success of the telephone over the last 100 years. In information transmission and problem-solving conversations, the withdrawal of vision has no measurable effect of any kind on the outcome of the conversation. In conflict and person perception conversations, however, the medium does affect the outcome. Even here the practical importance of the differences remains open to question. For example, in real life the telephone is likely to be used in conjunction with face-to-face meetings and interpersonal judgments are made in the face-to-face meetings. Most of the effects of medium have been subtle, small, and elusive. Consistent results (for example, the effect on opinion change in conflict tasks and the effect on evaluations in person perception tasks) have been achieved only by using ingenious experimental designs and highly sensitive measures. Moreover, these differences do not show the telephone inferior to face-to-face contact in any simple sense. For example, if the objective of negotiation is to change the opinion of the other person, then the telephone could be preferred as producing more opinion change. And while the face-to-face condition produced more favorable and confident evaluations, there is no evidence that these evaluations were any more accurate.

The experiments described here provide a convincing demonstration of the telephone's effectiveness. The results suggest that in many cases the benefit of adding a facial display to the telephone will be very small indeed. It also shows that Bell's means of conveying speech captures to a remarkable extent the whole process of human conversation.

## REFERENCES

Argyle, M. (1969). *Social Interaction* (London: Methuen).

Argyle, M., and Dean, J. (1965). "Eye-contact, distance and affiliation," *Sociometry* 28, pp. 289–304.

Argyle, M., Lalljee, M., and Cook, M. (1968). "The effects of visibility on interaction in a dyad," *Human Relations* 21, pp. 3–17.

Beishon, R. J., and Palmer, A. W. (1972). "Studying Managerial Behavior," *International Studies of Management and Organisation*, Volume 2, no. 1 (Spring 1972), edited by Rosemary Stewart.

Champness, B. G., and Reid, A. A. L. (1970). *The efficiency of information transmission: A preliminary comparison between face-to-face meetings and the telephone*, Communications Studies Group paper number P/70240/CH.

Champness, B. G., and Davies, M. F. (1971). *The Maier pilot experiment*, Communications Studies Group paper number E/71030/CH.

Chapanis, A. (1971). "Prelude to 2001: Explorations in human communication," *American Psychologist* 26, pp. 949–961.

Chapanis, A., Ochsman, R., Parrish, R., and Weeks, G. (1972). "Studies in interactive communication: I. The effects of four communication modes on the behaviour of teams during cooperative problem solving," *Hum. Fac.* 14, pp. 487–509.

Davies, M. F. (1971a). *Cooperative problem solving: An exploratory study*, Communications Studies Group paper number E/71159/DV.

—————— (1971b). *Cooperative problem solving: A follow-up study*, Communications Studies Group paper number E/71252/DV.

Dorris, J. W., Gentry, G. C., and Kelly, H. H. (1972). *The effects on bargaining of problem difficulty, mode of interaction, and initial orientations*, University of Massachusetts.

Exline, R. V. (1963). "Explorations in the process of person perception: Visual interaction in relation to competition, sex, and need for affiliation," *Journal of Personality and Social Psychology* 31, pp. 1–20.

Goddard, J. B. (1973). *Office Linkages and Location* (Pergamon Press Ltd.).

Kendon (1967). "Some functions of gaze direction in social interaction," *Acta Psychologica* 26, pp. 22–63.

Klemmer, E. T., and Stocker, L. P. (1971). "Picturephone versus speakerphone for conversation between strangers," unpublished proprietary data.

Mintzberg, H. (1970). "Structured observation as a method to study managerial work," *Journal of Management Studies,* February 1970.

_____ (1971). "Managerial work: Analysis from observation," *Management Science* 18, no.2, (October 1971).

Morley, I. E., and Stephenson, G. M. (1969). "Interpersonal and interparty exchange: A laboratory simulation of an industrial negotiation at the plant level," *British Journal of Psychology* 60, pp. 543–545.

_____ (1970). "Formality in experimental negotiations: A validation study," *British Journal of Psychology* 61, p. 383.

Morley, I. E. (1971). "Formality in experimental negotiations," presented at British Psychological Society, Social Section Conference, Durham.

Palmer, A. W., and Beishon, R. J. (1970). "How the day goes," *Personnel Management,* April 1970.

Parrish, R. N., Ochsman, R. B., and Chapanis, A. (1971). *Studies in man-machine communication: The effects of four communication modes on the efficiency of teams problem-solving,* Department of Psychology, Johns Hopkins University, Baltimore.

Reid, A. A. L. (1970). *Electronic person-person communications,* Communications Studies Group paper number P/70244/RD.

Short, J. A. (1971a). *Bargaining and negotiation—An exploratory study,* Communications Studies Group paper number E/71065/SH.

_____ (1971b). *Conflicts of interest and conflicts of opinion in an experimental bargaining game conducted over three media,* Communications Studies Group paper number E/71245/SH.

_____ (1971c). *Cooperation and competition in an experimental bargaining game conducted over two media,* Communications Studies Group paper number E/71160/SH.

_____ (1972a). *Conflicts of opinion and medium of communication,* Communications Studies Group paper number E/72001/SH.

_____ (1972b). *Medium of communication and consensus,* Communications Studies Group paper number E/72110/SH.

_____ (1972c). *Medium of communication, opinion change and solution of a problem of priorities,* Communications Studies Group paper number E/72245/SH.

_____ (1973). *The effects of medium of communication on persuasion, bargaining and perceptions of the other,* Communications Studies Group paper number E/73100/SH.

Weston, J. R., and Kristen, C. (1973). *Teleconferencing: A comparison of attitudes, uncertainty and interpersonal atmospheres in mediated and face-to-face group interaction*, Department of Communications, Canada.

Wichman, H. (1970). "Effects of isolation and communication on co-operation in a two-person game," *Journal of Personality and Social Psychology* 16, pp. 114–120.

Williams, E. (1972). *Factors influencing the effect of medium of communication upon preferences for media, conversations and persons*, Communications Studies Group paper number E/72227/WL.

_____ (1973a). *Coalition formation in three-person groups: communication via telecommunications media*, Communications Studies Group paper number E/73037/WL.

_____ (1975a). "Medium or message: Communications medium as a determinant of interpersonal evaluation," *Sociometry* 38, pp. 119–130.

_____ (1975b). "Coalition formation over telecommunications media," *European Journal of Social Psychology*

Woodside, C. M., Cavers, J. K., and Buck, I. K. (1971). *Evaluation of a video addition to the telephone for engineering conversations*, unpublished.

Young, I. (1974a). *Telecommunicated interviews: An exploratory study*, Communications Studies Group paper number E/74165/YN.

_____ (1974b). *Understanding the other person in mediated interactions*, Communications Studies Group paper number E/74266/YN.

_____ (1975). *A three-party mixed-media business game: A programme report on results to date*, Communications Studies Group paper number E/75189/YN.

Note: Copies of the Communications Studies Group papers cited here may be obtained from Post Office Telecommunications THQ/TSS6, 88 Hills Road, Cambridge CB2 1PE, England.

# 19

## Identification and Recognition in Interactional Openings

### Emanuel A. Schegloff

What people do on the telephone is talk. These conversations are natural materials for investigators of social interaction, not necessarily because of any special interest in the telephone but because they are instances of conversational interaction.

Materials drawn from telephone conversation can have special interest. That participants lack visual access to one another helps obviate arguments about the possibility of successfully studying conversation and its sequential organization without examining gesture, facial expression, etc. Telephone conversation is naturally studied in this manner and shows few systematic differences from conversation in other settings and media. The gross similarity of telephone and other talk has contributed to our confidence that much can be learned from it about the organization of conversational interaction.

For the last several years, my colleagues and I have been investigating the sequential organization of social interaction. We work with audio and video tapes of natural interaction and transcripts of those tapes to detect and describe the orderly phenomena comprising conversation and interaction.

Our work has yielded descriptions of several of the systematic

This paper has been substantially shortened to meet space limitations. A fuller version may be obtained from the Research Program on Communications Policy, MIT, Publication Number RPCP/76-4.

An earlier version of some portions of this paper was presented at the Conference on the Pragmatics of Conversation, Institute for Advanced Study, Princeton, New Jersey, April 1974.

organizations composing conversation: the organization of turn-taking,[1] the organization of repair,[2] and less systematically as yet, the organization of sequences.[3] A number of reports have dealt with elements of the overall structural organization of the unit "a single conversation," which operates on openings, closings, and on some aspects of what transpires in between.[4]

It is in the overall structural organization of a conversation—in its opening and closing—that the distinctive characteristics of various "types" of conversation may most prominently appear. In the opening of a conversation the "type" of conversation can be proferred, displayed, accepted, rejected, or modified.[5]

The openings of telephone conversations generally do have a distinctive shape. We regularly find a sequence not often found in "face-to-face" conversation—a sequence where the parties identify and/or recognize one another. Even where no such sequence occurs, the issue (identification/recognition) is worked through. This paper is about those sequences and that issue.

Identification is central to "gatekeeping," that is, to control over which of a set of potential cointeractants actually do talk to each, and how. One basic rule appears to be that such persons may (or may be required to) enter into interaction as have done so before. Necessary qualification, refinement, and supplementation aside, I am noting in a slightly different way what others have noted before:[6] "acquaintanceship" is one major basis for the undertaking of an interaction.

If access to interaction is organized and partially restricted, and if acquaintanceship is one basis for its occurrence, then recognition by one person of another will be important because recognition is central to the possibility of a "social relationship" such as acquaintanceship. Identification, and its recognition sub-type, can therefore be expected to be subject to systematic and potentially elaborate organization.

In human social interaction, identification and/or recognition of others is largely accomplished by visual appearance, as our few descriptions of these phenomena make amply clear.[7] When personal recognition of "other" occurs, and especially when it is potentially reciprocal, a subtle or elaborate display of its accomplishment is made and constitutes a "social" rather than "cognitive" event; it is, therefore, an event in interaction.[8]

Regularly, and especially because of their accomplishment by visual inspection in the "prebeginning" of interaction, these events resist study and appreciation. One biologist reports that for social insects "recognition . . . seems outwardly a casual matter, usually no more than a pause and sweep of the antennae over the other's body."[9] If we read "eyes" for "antennae," much the same could be said of humans. A body of materials where identification and recognition of potential coparticipants in interaction is routinely problematic, and where the solution is carried through in such a manner as to make it more accessible to empirical inspection, it is of considerable potential value to students of the organization of social interaction.

Telephone conversations supply such a body of materials. In them recognition is often relevant but cannot be accomplished visually before, or as a condition for, the beginning of the interaction. Recognition has a sequential locus in the talk, occupying or informing a sequence of conversational turns, and is thereby accessible to research approaches concerned with these units.

Our data base consists of about 450 openings; the parties vary on the standardly relevant parameters—age, sex, region, social class, etc. Here, as elsewhere in our studies of the sequential organization of conversation, they are not relevant to the matters under discussion.

## THE LOCUS OF THE IDENTIFICATION ISSUE AND ITS FORMS

For reasons that will become apparent, the sequential focus on the work of identification and recognition in the conversations under discussion is in the second turn, i.e., the caller's first turn. Those turns are overwhelmingly constructed from a very small set of component types. Nine types may be listed, with exemplary displays; some occur infrequently and/or largely in combination with others. The Appendix contains a glossary of symbols used in the transcripts. The arrows locate the phenomenon for which the segment is cited.

1. Greeting terms.

    A:   H'llo:?
→ B:   hH*i*:,
(TG, #1)                                       (1)

    M:   Hello
→ J:   Hello
(MDE, #91)                                     (2)

    C:   He*llo::*,
→ A:   Good *mor*ning.
(NB, #112)                                     (3)

    B:   Hello:,
→ R:   Howdy.
(ID, #277)                                       (4)

2. Answerer's, presumed answerer's, or intended answerer's name or address term (in varying combinations, or first name, title + last name,[10] nickname, etc.) in one of a range of interrogative or quasi-interrogative intonation contours.

    C:   Hell*o:*
→ M:   Miz Parsons?
(JG, #73a)                                      (5)

    I:   Hello,
→ N:   *I*rene?
(ID, #244)                                       (6)

    L:   Hello here.
→ S:   Colonel Lehroff?
(CDHQ, #353a)                                  (7)

    C:   Hello:
→ R:   ·hh Mother?
(JG, #41c)                                       (8)

    M:   Hello:
→ E:   Gina?
(MDE, Supp.)                                     (9)

    M:   Hello?
→ J:   Marcia?
(MDE, #99)                                     (10)

3. Answerer's, presumed answerer's, or intended answerer's name or address term (in varying combinations of name compo-

nents) in one of a range of assertive, exclamatory, or terminal intonation contours.

|   |   |   |
|---|---|---|
| C: | Hello? | |
| → M: | Charlie. | |
| (CF, #155) | | (11) |

|   |   |   |
|---|---|---|
| T: | Hello::, | |
| → E: | Uh Tiny. | |
| (CDHQ, #306) | | (12) |

|   |   |   |
|---|---|---|
| P: | Hello? | |
| → L: | Phil! | |
| (CDHQ, #299) | | (13) |

|   |   |   |
|---|---|---|
| M: | Hello | |
| → G: | Mommy, | |
| (MDE, #98) | | (14) |

4. Question or noticing concerning answerer's state.

|   |   |   |
|---|---|---|
| I: | Hello:, | |
| → A: | Did I waken you dear, | |
| (ID, #235) | | (15) |

|   |   |   |
|---|---|---|
| A: | Hello, | |
| → B: | Hi. // C'n you talk? | |
| (DS, #184) | | (16) |

|   |   |   |
|---|---|---|
| F: | Hello:, | |
| → S: | Hello. You're home. | |
| (RK, #190) | | (17) |

|   |   |   |
|---|---|---|
| F: | Hello:, | |
| → R: | Franklin are you watching? | |
| (RK, #189) | | (18) |

5. "First topic" or "reason for the call."

|   |   |   |
|---|---|---|
| F: | Hello: | |
| → R: | Whewillyoubedone. | |
| (JG, #75a) | | (19) |

|   |   |   |
|---|---|---|
| F: | (...)-o. | |
| → C: | Yeah I'm jus leaving. | |
| (JG, #55) | | (20) |

|   |   |
|---|---|
| M1: | ((Hello)) |
| → M2: | What's goin' on out there, I understand y'got a robbery, |

(WGN, #2)                                                             (21)

   L:   H'llo:,
→ C:   Hi, 'r my kids there?
(LL, #8)                                                              (22)

6. Request to speak to another ("switchboard" request).

   A:   Hello
→ B:   Is Jessie there?
(NB, #118)                                                           (23)

   M:   Hello:,
→ C:   May I speak to Bonnie,
(ID, #289)                                                           (24)

7. Self-identification.
   B:   H'llo?
→ D:   Hi Bonnie. This is Dave.
(ID, #234a)                                                          (25)

   R:   Hello,
→ M:   Hey:: R:i:ck, thisiz Mark iz Bill in?
(#198)                                                               (26)

   M:   Hello?=
→ C:   =Hello it's me.
(MDE, Supp.)                                                         (27)

8. Question re identity of answerer.

   L:   Hello:,
→ M:   H'llo, is this Kitty?
(LL, #27)                                                            (28)

   M:   Yhello,
→ L:   H'llo who's this,
(LL, #23)                                                            (29)

9. A joke or joke version of one of the above (e.g., mimicked intonation, intentionally incorrect identification, intentionally funny accent, etc.).

   Ba:   Hello?
→ B:    Hello?,
   Ba:   Hello?
→ B:    Hello?
   Ba:   Hi Bonnie.
   B:   Hi he heheheheh ·hh

       Ba:          heheheh
(ID, #287a)                                                      (30)

       L:    H'llo::,
→  M:    H'llo:: ((intended intonation echo))
              (1.0)
       L:    H'llo?=
→  M:    =H'llo? ((intended intonation echo))
       L:    Oh hi.
(LL, #9)                                                         (31)

       C:    Hello?
→  G:    Grrreetins. ((gutteral "r"))
(CF, #160)                                                       (32)

       C:    Hello?
→  G:    Helloooooo,
(CF, #160)                                                       (33)

       C:    Hello?
→  G:    Is this the Communist Party Headquarters?
(CF, #147)                                                       (34)

       M:    Hello?
→  G:    Hi= This is your daughter chewing on beets.
(MDE, #93)                                                       (35)

   Nearly all second turns are composed of these component
types, either singly, as generally presented above, or in combinations of various sorts. If one omits requests to speak to another (see collection 6 above) as a single component or one of
several (usually the other is a greeting that precedes), the overwhelming majority of second turns after "hello" are composed
of collection 1 (greetings), collection 2 (other's name interrogative), collection 3 (other's name declarative), or a combination of
collections 1 + 2 or 1 + 3. The turn types constructed with these
nine components and the combinations of them initiate a range
of different types of sequences: greeting sequences, request
sequences, request for confirmation sequences, question/answer
sequences, apology sequences (post "Did I wake you?", for example), and others. In each of them, however, the identifica-
tion/recognition issue is addressed.
   All the data segments displayed above have "hello" (however
inflected) as their initial turn. Elsewhere,[11] I have examined the

other major type of initial turn—self-identification. At a phone whose callers are not expectably recognizables and are not expectably oriented to answerers as recognizables, answerers' first turns are routinely designed to afford categorical confirmation that the caller reached what he intended, typically by self-identification (e.g., "American Airlines"). Such self-identification projects a type of identification for caller (e.g., "customer"), and aspects of the type of initiated conversation getting underway (e.g., "business"). For a phone whose potential answerers are recognizables and where answerers can orient to potential callers who are recognizables, answerers' first turns regularly supply a voice sample—"hello" is its conventional vehicle—as materials from which confirmation of reaching the intended locus may be achieved, but they supply no overt self-identification. The confirmation may be achieved by recognition, and the caller's first turn is the place where such recognition, or trouble with it, may be displayed.

It is by reference to this placement that the turn-types constructed from the components listed above address the identification/recognition issue. Even the request to speak to another (the "switchboard request"), which seemingly claims the nonrelevance of identification or recognition of current recipient,[12] displays a recognition of recipient as "not the intended recipient" or an inability to recognize answerer as intended recipient. The vast majority of second turns address the identification/recognition of answerer directly and initiate an identification/recognition issue for caller.

I will initially focus on the latter issue. I will group together those second turn components that specifically initiate an identification/recognition *sequence* (types 2, 7, and 8) when constituting the sole or final component of the turn and those that are informed by that issue while not overtly addressing a sequence to it (all the other components).[13] To simplify presentation, I will consider from the first group mainly the turn-type composed of interrogative name (type 2 above), sometimes preceded by a greeting, and from the second group the turn-type composed of a greeting alone or greeting plus name in "assertative" intonation.

## RECIPROCAL RECOGNITION EN PASSANT AND ITS FAILURES

In human interactions, greetings are generally the first exchange of a conversation. They are also the end of a phase of incipient interaction (referred to earlier as "prebeginnings"). Routinely, the actual exchange of greeting terms follows a set of other activities: glances, eye aversions, pace changes, body, head, and arm maneuvers, and others.[14] One important component of the prebeginning is identification of other(s) in the scene. Among the outcomes of the prebeginning can be no interaction, a passing exchange of greetings, greetings followed by further talk, or some talk not begun with greetings, etc.

If some talk is undertaken after identification, the first turns regularly display understandings of the outcome of the prebeginning phase. For example, a greeting (e.g., "hi") in first turn can display a claim of recognition by the speaker of its recipient and can make reciprocal recognition relevant if it has not already occurred nearly simultaneously.[15] An "excuse me" in first turn can display an identification of its recipient by the speaker as a "stranger," as well as displaying, for example, that something other than a full or casual conversation is being initially projected—perhaps a single sequence, very likely of a "service" type.

On the telephone, visual access is denied, and typically there is no prebeginning; but by the caller's first turn, the answerer's first turn has occurred, along with its voice and manner. In his first turn a caller's use of only a greeting term, a greeting term plus an address term "terminally intoned," or other of the earlier-listed components in this class constitutes a claim that the caller has recognized the answerer from the answerer's first turn. And it invites reciprocal recognition from the single, typically small turn it constitutes. In doing so, it initiates an effort to have the identifications (in such cases, the recognitions) accomplished *en passant*, while doing an otherwise relevant part of the opening (a greeting exchange), and without building a special sequence to accomplish that work.

An initial greeting in second turn[16] has two aspects at least. First, it is the first part of a basic sequential unit we call an "adjacency pair," whose simplest form is a sequence of two turns, by different speakers, adjacently placed, typologically related such that the occurrence of some particular type of the first part strongly constrains what occurs in the next turn to be one of a restricted set of second parts.[17] In the adjacency pair initiated by a first greeting, its recipient properly responds with a second greeting or greeting return. Second, it is a claim to have recognized the answerer and a claim to have the answerer recognize the caller. These two aspects of the caller's initial "hi" are intertwined. A first greeting having been done, a second greeting is what should relevantly occupy the next turn. But as the first greeting displays recognition, so will a second greeting; it will thus do more than complete the greeting exchange, it will stand as a claim that the answerer has reciprocally recognized the caller.

Answerers regularly follow callers' initial greetings with return greetings, thereby accomplishing both an exchange of greetings and an exchange of recognitions.

|          | A:   | H'llo:? |
|----------|------|---------|
| → | B: | hHi:, |
| → | A: | Hi:? |

(TG, #1)                                                                                 (36)

|          | M:   | Hello |
|----------|------|-------|
| → | J: | Hello |
| → | M: | Hi |

(MDE, #91)                                                                              (37)

|          | A:   | Hello::, |
|----------|------|----------|
| → | B: | HI:::. |
| → | A: | Oh: hi:: 'ow are you Agne::s, |

(NB, #114)                                                                              (38)

The callers' "recipient-designed" use of such a turn type in Turn 2 (the caller's first turn) is regularly successful. It is employed with recipients that callers suppose will recognize them from a small voice sample. Recipients of such turns are aided in accomplishing the recognition by information supplied by the form of the turn: the caller has rights and grounds for supposing that he can be so recognized. Such information can considerably restrict the set of candidate recognizables—they search to find who the caller is.

Recipients display their reciprocal recognition by doing a

sequentially appropriate second part for the type of sequence initiated by the caller. This sequence is (in the cases here under consideration) not overly directed to identification and can be occupied with some other opening-relevant job, such as greeting.

If the answerers/recipients do not recognize callers from the initial "Hi" or other sequence start, several courses are available to them. They may withold the return greeting in order not to claim a recognition they have not achieved.

    C:   Hello?
→ G:   Hello.
→      (1.5)
(CF, #130)                        (39)

    C:   Hello?
→ Y:   Hello Charles.
→      (0.2)
(CF, #145)                        (40)

    L:   Hello,
→ B:   Hi Linda,
        (0.1)
(ID, #212a)                      (41)

The caller's first turn is followed by a gap of silence that affords the caller an opportunity to back down from the turn-type claiming that he be recognized from the voice sample alone.

Several alternative courses may follow. The caller may back down from the constraint placed on the next turn by supplying additional resources for recognition, thus weakening the claim of recognizability. Two of the sequences presented above follow this course:

    C:   Hello?
    Y:   Hello Charles.
→      (0.2)
→ Y:   This is Yolk.
    C:   Oh hello Yolk.
    Y:   How are you heh heh
    C:   Alr(hh)ight hah hah It's hh very funny to hear (hh)
        from you.
(CF, #145)                      (42)

```
    L:    Hello,
    B:    Hi Linda,
→         (0.1)
→   B:    's Bonnie.=
    L:    =Yeh I know=I've been trying to call you . . .
```
(ID, #212a)                                                         (43)

In both segments, the upgrading of resources by the caller as
a way of backing down from the strength of the initial claim to
recognizability (i.e., from voice sample alone) is sufficient to allow the achievement of recognition, which is displayed in the
next turn. In both cases also, the snag in the sequence is further
dealt with. In (43), it is dealt with by a claim that the full self-
identification was unnecessary, the recognition having been already achieved (indeed, only a slight gap developed before B's
upgrade). In (42), C, having recognized the caller and displaying
by the "oh" (which marks both success, and success "just
now") that he had not recognized before, produces the return
greeting, but uses "the big hello" that is used with "long time
no see" recipients or "unexpected" callers. One turn later, the
recipient explicitly comments on the unexpectedness of this caller, finding therein warrant and diagnosis for having failed to
recognize from the voice sample alone.

The third segment we are examining (39), where a gap follows
the Turn 2 greeting, is resolved in a different manner. Here, despite the long gap, the opportunities for the caller to upgrade
the recognitional resources are not taken.[18] Finally, the recipient
breaks the gap with "who's this."

"Who's this" makes explicit and embodies in a sequentially
consequential turn C's failure to recognize. It does not simply
declare the failure of recognition, as in another segment in the
corpus, "I can't place you"; rather, it is a form of question we
call a "next turn repair initiator."[19] Such repair initiators are directed to some trouble in a prior turn which the speaker of the
prior turn has not repaired elsewhere in that turn or in the
"transition space" immediately following it. Generally, such repair initiators afford the prior speaker (the speaker of the trouble source) another opportunity in the following turn to repair
that trouble. If that is done, the speaker of the repair initiator
may in the turn after *that* do whatever turn-type was made sequentially appropriate by the turn containing the trouble. Thus,

if the trouble-source turn was a first part of an adjacency pair, its second part may follow the next turn repair initiator and the solicited repair.

In the segment under examination here, the Turn 2 "Hello" is offered as the resource from which recognition should be achieved. The gap of 1.5 seconds has displayed the incipient failure of recognition and provided an opportunity for the caller to repair the turn; for example, the caller can upgrade the resources from which the recognition might be made. She does not do so. "Who's this" locates the trouble source (the insufficiency of the resources for achieving recognition) and provides another opportunity in the following turn for G to repair the trouble by giving her name. Had the sequence developed that way, and the name been a sufficient resource to allow recognition, then the still relevant second part of the greeting pair might have been produced. The sequence would then have gone:

C:   Hello?
G:   Hello.
      (1.5)
C:   Who's this.
G:   ((Gloria))
C:   ((Oh hi, Gloria.))

In such a sequence, the caller would have been marked as the speaker of the trouble source, the difficulty having been with the resources supplied for recognition, as in (42) and (43).

The sequence, however, did not develop this way.

C:   Hello?
G:   Hello.
      (1.5)
C:   Who's this.
G:   Who *is* this.=This is your (0.2) friendly goddess,
G:   OHhh, hhh, can I ask for a wish
(CF, #130)                                                        (44)

G somewhat turns the tables on C by affording *him* yet another opportunity to accomplish the recognition from less than a full self-identification, making *his failure to recognize*, rather than *her failure to give her name*, the trouble source. She does this by availing herself of a device available to recipients of questions, the "joke-first answer."[20] Her joke-first answer preserves the sequentially appropriate type of turn for the question it follows—a self-identification. But instead of self-identifying by name, she does a joke self-identification which supplies clues for recognition (e.g., that she is a friend, that she is female, etc.) as well as a further voice sample. C does thereupon recognize her, displaying his recognition with the "success marker" described earlier.

The data segments examined show differences in who moves to fix the snag displayed by the gap after the second turn, with derived interactional consequences for the issue, who is "at fault": the caller for not supplying sufficient recognition resources or the recipient for failing to recognize.

The segments examined all began with a greeting or a greeting and address term in the second turn; but as was noted at the beginning of this discussion, this Turn 2 turn-type was selected for convenience from the set of Turn 2 turn-types which do not overtly address the identification issue. We have found that the identification theme underlies these sequences; their success is also a success of reciprocal recognition, and their failure not that of greetings but of recognition.

This holds equally true for the other not overtly identificational turn-types which are used in the caller's first turn. If recognition—or at least some identification—is not achievable from that turn, then the identification issue is raised in the ensuing turn. Regularly, no such trouble arises because the use of a non-identificational turn-type at Turn 2 is recipient-designed. When it is not, or when it fails despite its recipient design, then identification trouble becomes overt. Even the call to a bank during a robbery answered by the robber cited earlier (21) shows this sequence:

```
     M1:   ((Hello))
     M2:   What's goin' on out there, I understand y'got a
           robbery,
→          (0.8)
→    M1:   Uh yes, who's this speaking, please?
     M2:   WGN
     M1:   WGN?
     M2:   Yessir,
           (0.7)
```

M1:    Well this is the robber, (0.2) or the so- so called
       robber, I guess,

(WGN, #2)                                                      (45)

Nearly every turn-type in the second turn that appears to evade
the identification/recognition issue is vulnerable to its immediate appearance. This issue appears as "who's this," as a gap
understood as displaying the need for self-identification by the
caller, or as a gap that only gets resolved by self-identification.
Thus, it seems that the identification/recognition issue is generi-
cally relevant at second turn, whatever the overt composition of
the utterance placed there. This is true whether the turns overtly
address the matter or not. We have been examining the set of
cases in which they do not, but we will shortly turn to the other
major class of turn-types, those overtly addressed to the identi-
fication/recognition issue at Turn 2. But first, an additional
matter concerning the class under discussion requires brief
treatment: deception.

We have noted that the nonovertly identificational second
turn is recipient-designed, selected for use with a recipient who
the caller supposes will recognize him from it; the caller thereby
displays a claim on the answerer and on their relationship. The
recipient's failure to recognize may reflect on the state of the relationship; as we have seen, some sequential maneuvering can
place the blame on one or another party.

Another possibility open to the nonrecognizing answerer—
other than withholding a next turn—is deception. Here the answerer returns the greeting (if that is what is required by the
turn-type at Turn 2) although no recognition has been accomplished. It is likely that many answerers' return greetings are
deceptions when produced, but they are never "caught" because the caller's next turn allows the answerer to identify the
previously not quite recognized voice—a resource answerers
may rely upon in choosing this tack. Such deceptive uses may
thus routinely escape notice both by callers and by analysts.
Sometimes, however, they do become visible and are "caught."
One instance will suffice:

A:    Hello
B:    Hi:
A:    Hi: (0.3) Oh *Hi* Robin

(EN, #183)                                                     (46)

We noted earlier the use of some "oh's" to mark success and to mark success "just now." In (46) above, the "oh" displays the point where recognition is achieved, and the regreeting is shown to be "honest" by affixing the caller's name to it. We (and the caller) are allowed to see that A's first "Hi" in Turn 3 was deceptive; it claimed recognition, even though it had not been achieved.

## OPENINGS WITH IDENTIFICATION SEQUENCES

The caller's first turns which initiate a special identification sequence fall into two classes: those directed to self-identification, and those that appear occupied with identification of the answerer.

On the whole, self-identification is not prevalent in the caller's first turn,[21] but examination of openings where it does occur reveals the following:

1. Many of the instances have a caller's first turn in the *form* of a self-identification, which nonetheless operates in the manner of the turn-types discussed in the previous section. For example:

```
        M:   Hello?=
→   G:   =Hello it's me.
        M:   Hi.
(MDE, Supp.)                                                (47)
```

```
        P:   Hallo?=
→   C:   =Hi it's only me.
        P:   Hallo there, you,
(CG, #182a)                                                 (48)
```

In such openings, though the form of caller's first turn is self-identificatory, it works largely through voice recognition (supplemented by use of the "It's me" form, which may be used specially by nuclear family members).

2. Another subset of the instances have a self-identification in the caller's first turn followed in that turn by another turn component, regularly the first part of an adjacency pair. The latter turn component then sequentially constrains the next turn; in that case, the self-identification does not occupy its own se-

quence. Two main types of component follow the self-identification: switchboard requests and "How are you" questions. Thus:

S:   Hello:.
→ P:   Pt. ·hh H:i. This is Penny Rankin from Lincoln I'm a
       friend of Pat's. can I speak t'her at all?
S:   She:re.
(RF, #180)                                              (49)

"Switchboard requests" are regularly followed by an identification question (e.g., "who's this," "who is calling," etc.) from the answerer, and cases such as the above appear to be anticipations of this question. Such anticipation is especially in point when the requested party is not usually associated with that number, in which case the unprefaced switchboard request is vulnerable to being heard as a wrong number. This is the case in (49) above and is clearly shown in (50):

I:   Hello:,
→ JM:   Hello. i- This is Jan's mother.
I:   Oh yes.
JM:   Is Jan there by any chance?
(ID, #233)                                              (50)

Here the switchboard request is started in Turn 2 ("i-" being the start of "is Jan there . . . ") but is cut off in favor of self-identification first. It is also characteristic (though not invariable) that Turn 2 pre-switchboard self-identifications are not made by first name (hence for recognition). They are made by first name and last name, sometimes by (or supplemented by) recognitional descriptions, and sometimes use the frame that shows that the self-identification is not intended to solicit recognition: "My name is."

Another component type that follows a Turn 2 self-identification is the "How are you" question.

R:   Hello.
→ L:   Hi Rob. This is Laurie. How's everything.
R:   ((sniff)) Pretty good. How 'bout you.
L:   Jus' fine. The reason I called was ta ask . . .
(LM, #199)                                              (51)

Such cases share with others, where Turn 2 self-identification is

not followed by another turn component, the feature of project-
ing an abbreviated opening and a quick move to first topic or
reason for the call. The following is an instance of such abbre-
viation, where the Turn 2 self-identification is not followed by
another component:

     C:   Hello.
→  G:   Charlie?=Gene.
     C:   Oh, Hi.=
→  G:   =The whole weekend I forgot to tell you, I have this
          book, . . .
(CF, #164)                                                              (52)

These cases, then, appear to be related to (19)–(25) where the
first topic is initiated in the caller's first turn. Here, the risks of
nonrecognition entailed by that procedure are avoided by self-
identification in the caller's first turn, at the cost of one turn
(the first topic being initiated in the caller's second turn) but
avoiding a fully expanded opening section.[22]

     Self-identification by name occurs mainly in the caller's sec-
ond turn. When the opening develops in that way, the caller's
first turn is occupied with the major turn-types yet to be discussed here: an address term for the answerer (or intended or
presumed answerer), in the interrogative or quasi-interrogative
intonation, alone or preceded by a greeting term. Because the
use of Title + Last Name in that position frequently displays
that recognition is not a relevant outcome of the sequence, I will
not deal with cases of that sort in what follows.

     Sometimes the use of an interrogative name in the caller's
first turn displays uncertainty in the caller's recognition of the
answerer from the initial "hello," and sometimes the caller's
recognition of the answerer is incorrect. For example:

     M:   Hello?
→  E:   Tina?
     M:   This is Martha.
→  E:   Well if I had said "Martha" you would've said "This is
          Tina."
     M:   Oh, Esther! // hih hih
     E:   (yah) hih hih heh heh hah
     E:   Hi:*
     M:   hih heh* hah

→ M:    ·hhh HI:: I didn' recognize your voi:ce. Either.
(MDE, Supp.)                                                          (53)

However, in many cases of this form of Turn 2, the caller's
recognition of the answerer seems certain enough (consider
again (52) above, in which interrogative name is used, though it
appears that the caller does not doubt the identity of the answerer). In any case, the form has other sequential uses and
consequences as well.

One sequential consequence of the "confirmation-request"
form of this turn-type may be appreciated by contrast with a
turn-type identical in all respects but intonation: the "assertative" name with or without a preceding greeting. We have seen
that the latter form constrains its recipient to greet in return,
which displays reciprocal recognition (whether achieved or not).
The alternative was to ask the caller for a self-identification displaying potential failure and disappointing an expectation the
caller claimed a right to have.[23]

The "interrogative name" form of second turn adds to these
possibilities "confirmation" and "disconfirmation + correction"
(the former overwhelmingly occurs) as turn-types for next turn;
both possibilities allow avoidance of identification/recognition
of the caller at that turn position. It is then a more flexible instrument than the same components in "assertative" or "declarative" intonation; it allows—but does not require—deferral
of recognition of a caller.

Most importantly, an interrogative name in the caller's first
turn operates as a "presequence." The most accessible instance
of "presequences" is the pre-invitation: questions such as "Are
you doing anything?" "What are you doing?" etc., in the opening are understood to preface an invitation. If the answer to the
pre-invitation is no, the next turn will have the invitation; other
answer types may result in no invitation or in a report of what
the invitation would have been: "I was gonna say let's go to the
movies." A range of sequence types can take "pre-"s; there are
pre-requests, pre-announcements,[24] etc., as well as the "generalized pre-"—the summons—described elsewhere.[25] The Turn 2
interrogative name takes a form identical to some summonses,
but it is more properly understood as a pre-self-identification.

One use of some pre-sequences, such as pre-invitations, is

the avoidance of rejection. A prospective inviter can guard against being rejected by using a pre-invitation first; if the answer to the pre-invitation projects rejection, the invitation may be withheld.

Another use of some pre-sequences, such as pre-requests, is the avoidance of relatively less preferred first parts of adjacency pairs. For some projected outcomes, alternative sequential routes are possible, of which one may be structurally preferred to another. My late colleague Harvey Sacks argued that offers were structurally preferred to requests as a way of accomplishing transfers. Where such preferences between alternative sequence types—and therefore between alternative first parts of adjacency pairs—operate, a pre-sequence can elicit from its recipient the preferred first part. A pre-request can thus elicit an offer next, obviating the need for actually making the request (sometimes, the offer may be of something other than the projected request).

Using an address term in an interrogative or quasi-interrogative intonation in the caller's first turn operates in the second of these ways. Although generally, in person reference, name is the preferred form of identification for recognition,[26] all the evidence we have reviewed suggests that *for achieving recognition from coparticipant*, self-identification by name is less preferred than recognition by "inspection." The heavy use of non-identification-relevant turn-types discussed in the preceding section and the substantial absence of self-identification from the second turn—even when identification is directly addressed in a sequential locus where identification/recognition are focused and where the answerer's initial, strong interest is in "who is calling"—all point to the relative dispreference for self-identification as a route for achieving recognition. How does an interrogative name in the caller's first turn serve as a pre-self-identification which can potentially avoid self-identification, and why is it needed?

I have frequently referred to the intonation of the turn-type under discussion as "quasi-interrogative." When displaying a serious doubt about the identity of the answerer, callers sometimes employ a fully inflected interrogative intonation (often following a slight gap of silence after the first "hello"). But in the occurrences I am terming "quasi-interrogative," a less inflected intonation is used, rather like what has elsewhere been termed a "try marker."[27]

In the organization of reference to third persons in conversation, recognitional reference (of which name is the basic type) is the preferred reference form "if possible." The latter constraint concerns the speaker's supposition that the current recipient knows the one being referred to and that the recipient can be expected to suppose that the speaker so supposed. This is a specification in the domain of reference to persons of the general recipient design preference: don't tell the recipient what you ought to suppose he already knows, use it.

This principle builds in a preference to oversuppose and undertell, but even with oversupposition (or because of it) a speaker regularly employs the name with a slight upward (or "quasi-interrogative") intonation, marking the reference as a "try"; it is this intonation which characterizes the caller's Turn 2 pre-self-identification use of the answerer's name.

When recognition of self by other is at issue, as it is at Turn 2, the speaker's supposition concerns whether the recipient knows him and by what resource that recognition can be secured; the relevant recognitional resource may be voice sample. But here the preference to oversuppose and undertell may be especially guarded against oversupposition, if not countered by an inclination to undersuppose and overtell, avoiding the interactional consequence of presumptuous and embarrassingly disappointed claims, and the technical organizational consequence of dispreferred sequence expansion. When the supposition that the recipient can recognize by voice sample is doubted, then the try marker may be employed, qualifying not the recipient's knowing who he (recipient) is, and not (in these cases) the speaker's knowing who the recipient is, but qualifying the supposition that the recipient will know, from the voice sample which that turn supplies, who the speaker (i.e., caller) is. The reciprocity of recognition is nicely caught in the use of form which also can display the possible inadequacy of voice sample for the caller's recognition of the answerer.

The try-marked address term in the caller's first turn can then work as a pre-self-identification by (1) providing a voice sample; (2) displaying a doubt that the recipient can recognize the caller from it; (3) providing a next turn where the recipient can display recognition if it is achieved; (4) providing an option in the next turn which will not exhibit failure of recognition if recognition does not occur; and thereby (5) allowing the caller to supply—or project the possibility of supplying in his second turn—self-identification by name from which the recipient can achieve recognition (if recognition is not achieved from second turn and displayed in the turn following). The pre-self-identification thus provides the possibility of success without recourse to the less preferred route of self-identification, while retaining the possibility of the less preferred route should the pre-sequence not avoid it.

The pre-self-identification can have a number of outcomes. Its greatest success, achieved in many cases, is "evidenced recognition" in the next turn. From the supplied voice sample and sometimes from other clues put into the turn (such as wholly or partially self-identifying address terms, e.g., "Mommy"), the answerer achieves a recognition and displays it in the next turn in a way that obviates the possibility of deception. The basic form of evidence is inclusion in the recognition-exhibiting turn of caller's name, usually as an address term, occasionally as a "try."

```
        I:    Hello:,
   →    B:    H'llo Ilse?
   →    I:    Yes. Be:tty.
   (ID, #231)                                                    (54)

        F:    Hello?
        M:    Hello,
   →    M:    H'llo, Donna?
   →    F:    Oh. yeah, Hi Jim,
   (JH, #86)                                                     (55)

        I:    Hello:,
   →    D:    Hello mo:m?
   →    I:    Debbie?
   (ID, #296)                                                    (56)
```

No self-identification by name is then in point, and the opening continues.

A closely related (but somewhat weaker) class of outcomes, which adds another substantial proportion of cases, is that of "unevidenced recognition claims." Again, the device to display

a claim of recognition is a greeting term. It may occupy the turn
(after the pre-self-identification) alone, or it may be preceded by
the "oh" previously described as a marker of "success" and
"success just now," which upgrades its strength as a claim of
recognition.

<pre>
      A:   Hello?
 →  B:   Shar'n?
 →  A:   Hi!
(RB, #185)                                                    (57)
</pre>

<pre>
      C:   Hello.
 →  M:   Hello, Charlie?
 →  C:   Oh, hi.
(CF, #153)                                                   (58)
</pre>

It may also be upgraded by adding other turn components after
it, especially first parts of adjacency pairs (the characteristic one
in this sequential environment is "How are you") which set
constraints on the next turn to be a fitted second part, and
thereby immediately advance the opening past identification.

<pre>
      C:   Hello?
 →  M:   Hello, Charlie?
 →  C:   Oh hi. // How are you.
      M:   How *are* you.
      M:   Hey listen . . .
(CF, #146)                                                   (59)
</pre>

The greeting as recognition claim may be upgraded production-
ally, by raised amplitude, pitch, or duration.

<pre>
      S:   Hello::?
 →  J:   H'llo, Sima?
 →  S:   hhhHI!
(TAC, #122)                                                  (60)
</pre>

The import of these unevidenced recognition claims can be
equivocal; they are regularly taken by callers to display recognition. An exchange of recognitions is thereby completed, no self-
identification by caller is necessary, and the opening proceeds
to other components. Sometimes (especially when the recogni-
tion-marking greeting is not upgraded), the caller proceeds to a
self-identification in his second turn anyway, perhaps sensitive
to the deception potential of an unevidenced recognition claim,

and the inclination to undersuppose or at least not press over-supposition too far.[28]

```
   B:    ·hhh Hello,
→  Ba:   Hi Bonnie,
→  B:    Hi.=
→  Ba:   =It's Barbie.=
   B:    =Hi.
```
(ID, #275a)                                                          (61)

```
   J:    Hello,
→  B:    Hello Jim?
→  J:    Hi-,
→  B:    Hi, it's Bonnie.
→  J:    Yeah I know
```
(ID, #246)                                                           (62)

Thus the "upgraded" recognition claims appear to operate as the evidenced recognitions in allowing deletion of self-identification by the caller, and so shade into that class of turns after the pre-sequence.

The major class of next turns after the pre-self-identification is even more equivocal. The prototype turn component here is "yes" or "yeah," but the range of its intonational shadings is vast; its sequential import seems at least partially tied to them. In many cases, "yeah" or "yes" in the next turn is treated by the caller as evidencing the pre-sequence's failure to achieve the answerer's recognition, and the projected self-identification is then produced in the caller's second turn. For example:

```
   L:    H'llo:
→  P:    Laura?
→  L:    Yeah,
→  P:    This is Pam.
   L:    Hi.
```
(LL, #13)                                                            (63)

```
   M:    Hello.
→  C:    Hello, Mary?
→  M:    Yes?
→  C:    Hi. This is Bernie Hunter.
   M:    Oh hello. How are you.
```
(CF, #177)                                                           (64)

However, even the apparently polar intonational values—a
clearly interrogative "Yes?" or an emphatically assertive
"yeah!"—which might appear to display confirmation-of-
answerer's-identity-but-no-recognition-of-caller on the one
hand, and confirmation plus recognition on the other, do not
unequivocally elicit regular sequels. There are Turn 4 self-identifications after enthusiastic "yeah"s which might be taken to
exhibit recognition:

|     | H: | Hello? |
| → | G: | Henry? |
| → | H: | Yeah! |
| → | G: | Yeah. It's Gary. Is Neil there? |

(LL, #33)                                                           (65)

|     | E: | Hello, |
| → | G: | Eddy, |
| → | E: | Ye:h. |
| → | G: | Guy Huston. |

(NB, #109a)                                                        (66)

The caller may not self-identify by name in the fourth turn after
a pre-self-identification, even when the prior turn has been a
fully interrogative "Yes?" For example:[29]

|     | L: | H'llo, |
| → | E: | Laura? |
| → |   | (0.5) |
| → | L: | Yeah? |
| → | E: | Hi, |
|     |   | (0.5) |
|     | L: | Hi. // Erin? |
|     | E: | Didju- |
|     | E: | Yeah. |

(LL, #17)                                                          (67)

|     | H: | H'llo:? |
| → | R: | Harriet? |
| → | H: | Yeah? |
| → | R: | Hi! |
|     | H: | Hi:. |

(RB, #186)                                                        (68)

```
    C:    Hello.
→ J:    Hello, Charlie?
→ C:    Yeah?
→ J:    Did I wake you up?
    C:    No. It's alright.
```
(CF, #171)                                                                    (69)

It is not only after an interrogative "yeah?" that the callers
may fail to deliver (in their second turn) what apparently had
been prefigured in their first. In twenty-five of approximately
sixty calls where callers used interrogative or quasi-interrogative
address terms in Turn 2 or its equivalent, no self-identification
by name appeared in Turn 4 or its equivalent. Instead, these
second caller turns contained all the various turn components
earlier listed as components of Turn 2, excepting the one actually used in the preceding Turn 2 of that call. Thus, roughly half
of the twenty-five were composed of greetings and/or some version of "How are you," three are "Did I wake you," three are
switchboard requests, five start first topic, and one is a mock
self-identification:

```
    A:    Hello:,
→ B:    Hello M::A?
→ A:    Ye:AH!=
→ B:    =It's m:e.
```
(RF, #179)                                                                    (70)

Those components behave sequentially in Turn 4 as in Turn
2: they invite recognition from less than name self-identification, but (1) they do not require it at Turn 3, which lowers the
degree of their claim; (2) they supply a second voice sample,
upgrading the resources for recognition which were provided;
and (3) more than in Turn 2, the turn's composition provides
additional clues for recognition (for example, in the five first
topic starts).

The withholding of self-identification by name from Turn 4 as
well as Turn 2, when Turn 2 has prefigured it, supplies additional evidence of the relative dispreference for that recognition
resource, and the persistence of the effort to secure recognition
from inspection. It shows a second way in which the pre-self-
identification contributes to potential avoidance of self-identification. Not only does it elicit a "safer" (i.e., from "who's this")

position for possible recognition at Turn 3, but if recognition does not occur there the other less-than-self-identification turn-types may be tried at Turn 4 and receive recognition at Turn 5. The persistence toward recognition without self-identification at Turn 4 can inform the answerer that the caller has reason to suppose that such recognition is possible, even if not at Turn 2. In only two cases of the twenty-five where Turn 4 employs a non-self-identificatory turn-type does "who's this" (to which it is vulnerable) occur. And in one of them, the "who's this" follows a switchboard request in Turn 4, a turn-type regularly followed by "who's this." In the other twenty-three cases, recognition is secured without self-identification. The pre-self-identification thus has considerable success in allowing avoidance of self-identification. However, most openings in which a pre-self-identification is employed have it done at Turn 4.

## SPECIAL CASES

It is striking, that when a large number of openings are examined, some run off quite straightforwardly in a very nearly (if not totally) standardized way, while others look and sound idiosyncratic—almost virtuoso performances. But it is worth remembering that the "special" cases are variants engendered by a systematic sequential organization, adapted and fitted by the parties to some particular circumstances. The organization's standard product outlines are discernible through variations of the particular case; also, the standard-looking cases may be nonetheless special to the parties for their local circumstances. I shall close by examining these two themes in turn.

Consider the following segment:

M:    He*llo::*,
A:    Hello Margie?
M:    *Ye*::*//s,
A:    ·hhh We do pai:nting, a:ntiquing,
M:    I(hh) is that *ri:*ght.
(A):  eh!hh//hhh:::::
M:    hnh hnh hnh
A:    nhh hnh hnh! ·hh
M:    ·hh
A:    -keep people's pa'r too:ls,

M:     Y(hhh)! hnh//hnh
A:     *I'm* sorry about that//*that*///*I* din' see that-
(NB, #119)                                                  (71)

A is calling to apologize for keeping overlong some power tools
borrowed from M.[30] It is an interactionally delicate task,
brought off with considerable skill by building it into a joke
constructed in the form of a list. The fact that the joke list is a
vehicle for an apology does not become evident until the end,
when laughter is already in progress; it is a virtuoso performance in a potentially embarrassing situation. Among the many
interests this segment has, its bearing on openings and on iden-
tification/recognition is not prominent; yet what is being
brought off interactionally in the segment depends deeply on
the organization of identification/recognition in telephone openings, and the outlines of one of that organization's standard
sequences is apparent in it.

The segment is based on the sequence type that was described in the preceding section. In Turn 2 the caller uses a pre-
self-identification, though M and A stand in a relationship
which could well have a Turn 2 greeting sufficient to secure recognition. The Turn 2 pre-self-identification projects a self-identification in Turn 4, and Turn 4 is built in the form of a self-
identification. But the turn, and the form, are used to package a
mock self-identification (like "This is your friendly goddess").
The mock self-identification is a joke, but its appreciation as
such depends upon M's recognition of the caller and what is involved in that caller's self-description in this way. This is allowed to be in hand from the beginning of Turn 4 by the voice
sample in Turn 2; it may not have worked to begin with "We
do painting . . . " in Turn 2; also, the mock self-identification
takes a "business" form, and would (if real) be placed in a
fourth turn preceded by an interrogative address term. With all
the special circumstances involved here, the shape of a standard
sequence is visible and depended upon for the special interactional job being done.

On the other hand, what appear to be standard identification/
recognition sequences can have quite idiosyncratic and special
status. Since those for whom recognition is relevant have talked
before, any next opening—and recognition sequence—can have
a prior history of such sequences informing it.

By reference to such a history, a standard-appearing opening can be for the parties quite special. Thus:

A:   Hello,
B:   Mr. Lodge,
A:   Yes,
B:   Mr. Ford.
A:   Yes.
B:   Y'know where Mr. Williams is?
A:   What?
B:   hhhahhhah
A:   Do I know where who?
B:   Leo is.
A:   No.
B:   Oh. Okay.
(HS, #207)                                                    (72)

Here, the standard-appearing sequence is a joke; the parties are "on a first-name basis"; the joke, indeed, interferes with A's recognition of the friend being referred to.

For parties to be "on a first-name basis" can take doing. There is likely to be a historical development, in which the parties may first use Turn 2 interrogative names with first name plus last name self-identifications, then drop the last name, then perhaps one starts displaying evidenced recognition after the Turn 2 interrogative name, finally a greeting alone, perhaps in a distinctive intonation, would be sufficient. The first occurrence of any of these will look like a standard sequence; to its parties it may be a minor event of sorts, a small *rite de passage* between phases of a relationship. In any case, the development is one through a series of standard, organizationally produced sequence types, as is the reverse direction, when a caller may supply more recognition sources after a long hiatus between conversations than had otherwise been the practice with some particular recipient. The "practice" between a pair of persons can come to be a signature, and a special form may be used by a caller for a particular recipient.

In any particular case, such idiosyncratic particulars may be operative, but they are made operative as local adaptations of an independent organizational format and work the way they do by virtue of it. Particular cases can, therefore, be examined for their local, interactional, biographical, ethnographic, or other idiosyncratic interest. The same materials can be inspected so as to extract from their local particularities the formal organization into which their particularities are infused. For students of interaction, the organizations through which the work of social life gets accomplished occupy the center of attention, and whatever of their materials can be extracted and related to such organizations should be. For those whose lives are being led in interaction, those organizations are always filled out by the locally relevant details, the organizations by reference to which that detail is relevant receding into an unnoticed background.

Whatever a telephone conversation is concerned with, however bureaucratic or intimate, routine or unusual, earthshaking or trivial, its parties will initially have identification/recognition as a job. The contingencies of its organization thus have a pervasive relevance—a relevance inherited from less specialized settings of interaction and adapted to a technological innovation which makes it more prominent. As a result, what was associated in the Western mythic past with heroes and elders such as Odysseus and Isaac—recognition when identity is partially masked—has become democratized. Writ incomparably smaller, it has become anyone's everyday test.

## APPENDIX
## THE TRANSCRIPT SYMBOLS*

| // | V: | Th' guy says tuh me-·hh my *son* //  didid. | The double obliques indicate the point at which a current speaker's talk is overlapped by the talk of another. |
| | M: | Wuhjeh do:. | |
| | V: | I // left my garbage pail in iz // hallway. | A multiple-overlapped utterance is followed, in serial order, by the talk which overlaps it. Thus, C's "*Vi*:c," occurs simultaneously with his "hallway." |
| | C: | *Vi*:c, | |
| | C: | Victuh, | |

*Devised by Gail Jefferson; adapted here from Sacks, Schegloff, and Jefferson, "Turn-taking," pp. 731–734.

| | | | |
|---|---|---|---|
| (0.) | V: | . . . dih soopuh ul clean it up, (0.3) | Numbers in parentheses indicate elapsed time in tenths of seconds. The device is used between utterances of adjacent speakers, between two separable parts of a single speaker's talk, and between parts of a single speaker's internally organized utterance. |
| | ( ): | hhehh | |
| | V: | No kidding. | |
| | M: | Yeh *there's* nothin the:re? (0.5) | |
| | M: | Quit hassling. | |
| | V: | She's with somebody y'know ·hh ennuh, (0.7) she says *Wo:w* . . . | |
| ? ? ! , · | V: | Becuss the soo*puh* dint pudda *b*u:lb on dih sekkin flaw en its burnt ou:t? | Punctuation markers are not used as grammatical symbols, but for intonation. Thus, a Question may be constructed with "comma" or "period" intonation, and "question-intonation" may occur in association with objects which are not questions. |
| | V: | A do:g? enna cat is diffrent. | |
| | R: | Wuhjeh do:. | |
| :: | V: | So dih gu:y sez ·hh | Colon(s) indicate that the prior syllable is prolonged. Multiple colons indicate a more prolonged syllable, as in the second instance, in which V's "Wow" covers five syllables in M's overlapped utterance. |
| | M: | *Yeh* it's all in the *chair* all th//at junk is in the chair.)= | |
| | V: | Wo::::::::W)= | |
| | ↖: | =*I* didn' know tha:t? | |
| *italics* | V: | I sez y'know *why*, becawss *look*. | Italics indicate various forms of stressing, and may involve pitch and/or volume. |
| — | V: | He said— yihknow, I get— I get sick behind it. | The dash indicates a "cutoff" of the prior word or sound. |

| | | | |
|---|---|---|---|
| (hh) | M: | I'd a' cracked *up* 'f duh friggin (gla- i) (h)f y'- kno(h)w it) sm(h)a(h) heh heh | The [(h)] within parentheses and within a word or sound indicates explosive aspiration, e.g., laughter, breathlessness, etc. |
| UC | V: | En it *d*int fall OUT! | Upper case indicates increased volume. |
| ( ) | M: | I'd a' cracked *up* 'f duh friggin (gla-i(h)f y'- kno(h)w it) sm(h)a(h) heh heh | Single parentheses indicate transcribers are not sure about the words contained therein. Pairs of parentheses, as in the |
| | M: | Jim wasn' home, //°(when y'wen over there) | third instance, offer not mere- ly two possible hearings, but address the equivocality of |
| | V: | I'll be (right wit- chu.) (back inna minnit.) | each. Empty parentheses indi- cate that no "hearing" was achieved. On occasion, non- |
| | ( ): | Tch! ( ) | sense syllables are provided, in an attempt to capture some- |
| | R: | (Y'cattuh moo?) | thing of the produced sounds. The speaker designation col- umn is treated similarly; sin- gle parentheses indicating doubt about speaker, pairs in- dicating equivocal possibili- ties, and empties indicating no achieved identification of speaker. |
| (( )) | M: | ((whispered)) (Now they're gonna, *h*ack it.) | Materials in double parenthe- ses or double brackets indicate features of the audio materials |
| | M: | ((RAZZBERRY)) | other than actual verbaliza- tion, or verbalizations which |
| | M: | ((cough)) | are not transcribed. Occasion- |
| | V: | ((dumb slob voice)) Well we *use*tuh do dis, en we *use*- | ally an attempt is made to transcribe a cough (which might appear as "eh- *k*hookh!") or a razzberry |

J:      *They*'re fulla          (which might appear as
        sh::it.                  "pthrrrp!").


## NOTES

1. Cf. H. Sacks, E. A. Schegloff, and G. Jefferson, "A Simplest System-atics for the Organization of Turn-Taking for Conversation," *Language* 50, 4, 1974. Henceforth, "Turn-Taking."
2. Cf. E. A. Schegloff, G. Jefferson, and H. Sacks, "The Preference for Self-Correction in the Organization of Repair in Conversation," *Language* 53, 2, 1977. Henceforth, "Self-Correction."
3. E. Schegloff and H. Sacks, "Opening Up Closings," *Semiotica* VIII, 4, 1973. Henceforth, "Closings." Also, "Turn-taking," pp. 716–718, 728–729.
4. E. Schegloff, "Sequencing in Conversational Openings," *American Anthropologist* LXX, 6, 1968; Chapters 2 and 3 in H. Sacks, E. A. Schegloff, and G. Jefferson, *Studies in the Sequential Organization of Conversation* (forthcoming), henceforth *Studies*; H. Sacks, "Everyone Has to Lie," in Blount and Sanchez (1974); "Closings"; and G. Jefferson, "A Case of Precision Timing in Ordinary Conversation: Over-lapped Tag-Positioned Address Terms in Closing Sequences," *Semiotica* IX, 1, 1973.
5. Cf. *Studies*, Chapter 3.
6. E. Goffman, *Behavior in Public Places* (1963), Chapter 7.
7. For example, ibid.; and A. Kendon and A. Ferber "A Description of Some Human Greetings," in Michael and Crook (1973).
8. Cf. Goffman (1963) pp. 112–113.
9. E. O. Wilson, *The Insect Societies*, 1971, p. 272.
10. Title + Last Name can be a distinct type, as will be seen below.
11. In *Studies*, Chapter 3.
12. A cause for considerable resentment on the part of those who fre-quently find themselves in this position; for example, wives answering the phone when husbands' colleagues, often met at social occasions, are calling.
13. There *are* sequentially relevant differences between the forms in each group, but they do not bear on the present discussion.
14. The best description based on recorded (in this case, filmed) data I know of being Kendon's (op. cit.) account of greetings in the setting of a party.
15. Goffman (1963) notes urban/rural differences in this regard, rural folk offering greetings to strangers as well. It seems likely, however, that some aspect(s) of the first turn(s) will display some discrimination between recognizables and others (e.g., in the form of greeting used, the adding of an address term to it, or some less familiar variation).
16. I have argued elsewhere (*Studies*, Chapter 3) that the first turn "Hel-lo" is, sequentially, not fundamentally a greeting.

17. Cf. "Closings," pp. 295–298; "Turn-Taking," pp. 716–718. Question-answer, request-grant/rejection, and the like are instances of types.
18. Cf. "Turn-Taking," p. 715.
19. Cf. "Self-Correction." The most familiar next turn repair initiators are the various one-word questions ("Huh?", "What?", "Who?"), whole or partial repeats of the trouble source in prior turn, "pro-repeats" (e.g., "He did?") of prior turn, and others. "Who's this" does not otherwise appear as a repair initiator (except as an expanded variant of "who?"); it is, then, a next turn repair initiator specialized for use in openings for the identification/recognition issue and is rarely found outside the first several turns.
20. Her "conversion" of C into source of the trouble is marked, as well, by her repeat of his question at the beginning of her turn. As noted in footnote 19 and in the paper cited there, repeats of all or part of prior turn are one form of next turn repair initiator, the repeat marking the trouble source. Here it is not fully exploited as a repair initiator, no room being left after it for its recipient to do a repair.
21. This refers to the type of call under consideration here, in which the first turn is "Hello." In calls whose first turn is a self-identification, self-identification in the second turn is much more frequent. It is because of the different sequential consequences of "Hello" and self-identification in first turn, directly reflected in the second turn, that these constitute different types.
22. Gail Jefferson (personal communication) has proposed that an address term without greeting in second turn may operate similarly to foreshorten the opening. Initial examination of my materials lends some support but leaves the issue open, to my mind.
23. Another possibility should be mentioned here for the turn after a greeting, intermediate between reciprocal recognition and "who's this," and that is an "uncertainty marked" reciprocal recognition, a guess. For example:

    M:    Hello.
    P:    Hehlo.
→       (1.2)
→ M:   Pe:t?
(JG, #43)

    M1:   Hello.
    M2:   Hello. ((intonation echo))
→       (1.0)
→ M1:  This Sid?
(LL, #32)

24. A. Terasaki, "Announcement Sequences," ms.
25. Schegloff (1968), and *Studies*, Chapter 2.
26. H. Sacks and E. A. Schegloff, "Two Preferences in the Organization of Reference to Persons in Conversation and Their Interaction," in Psathas (forthcoming). Henceforth, "Two Preferences."
27. Ibid.

28. Indeed, deception may be clear to the caller if the answerer has not used the correct, recipient-designed greeting—i.e., the greeting in characteristic intonation which "that one" always uses "to me." The issue of recipient-designed opening components, although important, cannot be entered into here.

29. This segment is striking on a number of counts. L's interrogative "yeah" seems clearly non-recognitional, following as it does a gap of 0.5 sec. Still, E does not self-identify after it. Nor does she supplement the greeting with self-identification when a gap develops after it. Finally, we find in L's third turn another evidence of deceptive recognition display, in a different locus from that discussed before, but similarly placed sequentially—after an invitation to recognize from less than self-identification. Her guess—"Erin?"—after the greeting shows the latter to have claimed a recognition that had not been fully achieved.

30. I merely allude here to a rich and elaborate analysis of this segment by my late colleague, Harvey Sacks, included in various sets of lectures of his, hopefully to be published at some time in the future.

## REFERENCES

Blount, Ben, and Mary Sanchez (eds.), *Ritual, Reality, and Innovation in Language Use* (New York: Seminary Press, 1974).

Goffman, Erving, *Behavior in Public Places* (Glencoe, Ill.: The Free Press—MacMillan, 1963).

Jefferson, Gail, "A Case of Precision Timing in Ordinary Conversation: Overlapped Tag-Positioned Address Terms in Closing Sequences," *Semiotica* IX, 1, 1973.

Kendon, Adam, and Andrew Ferber, "A Discription of Some Human Greetings," in R. P. Michael and J. H. Crook (1973).

Michael, R. P., and J. H. Crook (eds.), *Comparative Ecology and the Behavior of Primates* (London: Academic Press, 1973).

Pomerantz, Anita, "Second Assessments," Ph. D. dissertation, School of Social Sciences, University of California, Irvine, 1974.

Psathas, George (ed.), *Papers in the Boston University Conference on Ethnomethadology* (Boston: Goodyear, forthcoming).

Sacks, Harvey, "Everyone Has to Lie," in Blount and Sanchez (1974).

Sacks, Harvey, and Emanuel A. Schegloff, "Two Preferences in the Organization of Reference to Persons in Conversation and Their Interaction," in *Psathas* (forthcoming).

Sacks, Harvey, Emanuel A. Schegloff, and Gail Jefferson, "A Simplest Systematics for the Organization of Turn-Taking for Conversation," *Language* 50, 4, 1974.

Sacks, Harvey, Emanuel A. Schegloff, and Gail Jefferson, *Studies in the Sequential Organization of Conversation* (New York: Academic Press, forthcoming).

Schegloff, Emanuel A., "The First Five Seconds: The Order of Conversational Openings," Ph. D. Dissertation, Department of Sociology, University of California, Berkeley, 1967.

Schegloff, Emanuel A., "Sequencing in Conversational Openings," *American Anthropologist*, LXX, 6, 1968.

Schegloff, Emanuel A., and Harvey Sacks, "Opening Up Closings," *Semiotica* VIII, 4, 1973.

Schegloff, Emanuel A., Gail Jefferson, and Harvey Sacks, "The Preference for Self-Correction in the Organization of Repair in Conversation," *Language* 53, 2, 1977.

Terasaki, Alene, "Pre-Announcement Sequences in Conversation," Social Science Working Paper # 99, School of Social Science, University of California, Irvine, 1976.

Wilson, Edward O., *The Insect Societies* (Cambridge, Mass.: Harvard University Press, 1971).

# V

**Social Uses of the Telephone**

# Editor's Comment

The last two chapters of this book concern novel applications of the telephone: hot lines and education by telephone. They are, perhaps, foretastes of the many new ways in which the telecommunications network is entering daily life.

In Chapters 1 and 2, Aronson and Briggs revealed that the telephone was not always considered a device primarily for pairwise conversations. Initially it seemed equally plausible that the telephone would be used as radio has since come to be used. By the 1940s and 1950s, however, the telephone became so identified with point-to-point voice service that we are only now beginning to relearn its other possibilities. The growth of data transmission for computers has been the most prominent new use of the network.

Uses halfway between mass broadcasting and strictly pairwise conversations are now being investigated in earnest. Teleconferencing is growing rapidly; survey researchers conduct most of their polls these days on the telephone; politicians organize telephone canvassing; credit checks are made by telephone automatically; and cash registers are tied to central accounting over telephone lines. There are increasing numbers of recorded messages, from dial-a-joke and dial-a-prayer to the time and weather.

The two uses of the phone discussed in this section fall into this marginal zone between mass communication and conversation. Hot lines for people in trouble are in one respect simply conversations, but in another sense the existence of an organized counseling institution, with a staff ready to carry on the conversations, gives hot lines something of the character of a medium. Education by telephone is a step still further into the gray area between mass communication and point-to-point. Teleteaching is a kind of teleconferencing wherein, typically, the instructor addresses a number of people at once.

The growth in the telephone network's technical flexibility, particularly with electronic switching, allows complicated programs of user interaction with the system. In the second century of the telephone the network will apparently become the vehicle and mechanism for an ever-expanding part of mankind's activity.

# 20

## The Use of the Telephone in Counseling and Crisis Intervention

### David Lester

In the last decade, there has been a tremendous increase in the use of the telephone for counseling people. This development has two main sources. First, the suicide prevention movement, following the opening of the suicide prevention center in Los Angeles in the 1950s, adopted the telephone as the primary instrument because of its accessibility. Any individual in crisis could usually get to a telephone to call for help. The telephone offers a number of other advantages over traditional modes of counseling, particularly the relative anonymity afforded the distressed individual.

The second stimulus to telephone counseling came from the development of poison information centers. Here, the telephone has the assets of immediacy and accessibility. If a person accidentally (or intentionally) ingests some chemical, immediate counseling about antidotes and treatment can be obtained. The telephone serves to transmit information quickly to people.

These two models of immediate counseling, twenty-four hours a day from a trained staff, have been applied to numbers of other areas. The number of centers now operating is easily over 1,000.

### THE USE OF THE TELEPHONE IN COMMUNITY SERVICES

It is useful to review, briefly, some uses to which the telephone has been put as an instrument of counseling and advising.

This paper contains elaborated and updated ideas discussed by the author in previous publications (Lester, 1974a, 1974b; Lester and Brockopp, 1973).

1. *Suicide prevention*. The telephone has been the major treatment medium of suicide prevention centers. In many centers, counseling by telephone only is offered.

2. *Crisis intervention*. Many suicide prevention centers soon found they were being asked to help in all kinds of crises, not just suicidal crises, so some centers changed their orientations toward more general crisis intervention.

3. *Teen hotlines*. Telephone counseling services soon became directed toward particular groups of the population, most commonly teenagers. The teen hotline functions similarly to a crisis intervention center, except that the problems it handles differ. Many teen hotlines do not attempt to provide twenty-four-hour service but are open for counseling late afternoons and evenings.

4. *Services for the elderly*. Another population selected for special concern is the elderly. Boston's Rescue, Inc., runs a service for senior citizens where a call is placed every day to each member. This protects the members in case of illness or emergency. If the call is unanswered, a volunteer visits the person's home. The calls are made by senior citizens, so social contacts are initiated and renewed in the process of maintaining the service. In Boston the service is free. New York City's service, financed by fees from participating senior citizens, differs from the Boston service because it restricts the socializing aspects. New York City limits calls to roughly a minute and serves a mainly protective function in case of illness.

5. *Services for individuals with particular needs and problems*. For example, Boston's Rescue, Inc., started a special telephone counseling service for homosexuals, with supporting clinic service. Services now exist for abortion counseling, victims of rape, parents who have a history of abusing their children, and so on.

6. *Services of a more general nature*. Buffalo's Suicide Prevention and Crisis Service opened a "problem in living" service to encourage people to call with any kind of problem.

7. *Drug hotlines*. These provide information about drugs and their effects, plus counseling to those involved with drugs. They not only provide general counseling but also help individuals currently on "bad trips" or in states of acute panic.

8. *Poison centers*. Poison control centers provide immediate counseling on treatment procedures. Although these centers

were generally originated by pediatricians to aid in treating children who ingest poisons, the focus has shifted. It has become increasingly obvious that in many cases of "accidental" poisoning, self-destructive tendencies are at work. As a result, the suicide prevention centers and drug hotline services must work closely with the poison control centers to facilitate treatment.

9. *Rumor control centers.* These services were primarily motivated by the riots of the 1960s and the need to quiet the rumors accompanying such social upheavals. Now they have extended their information-giving service to other areas of community concern.

10. *Community problems.* A recent development has been cooperation between radio stations and community groups. In Call For Action, originated by WMCA in New York City, trained counselors tried to help listeners with specific problems: garbage removal, rat and pest control, low-standard housing, voter registration, consumer fraud, traffic safety, pollution, taxes, etc. Stations have occasionally focused upon specific problems: WMCA in New York focused on housing, WWDC in Washington, D. C., on garbage removal.

11. *Counseling.* Last year, Dr. Lloyd Moglen, a psychiatrist, started a program on radio station KQED in San Francisco, where listeners call in with problems. Dr. Moglen counsels them while the conversations are broadcast. His program differs from most radio call-in programs because his aim is counseling.

12. *Minimal services.* Finally, there is a growing number of minimal services which are nonetheless related to the above: the Dial-A-Prayer service, Wake-Up Services, etc.

The telephone plays a central and important part in a large range of services and has proven particularly suitable to the goals of these services. If there is a community need, a relatively cheap telephone counseling service can be set up in a short period of time. The service provided, once advertised, is available to everyone, because most people have a telephone or easy access to one. When the community no longer needs the service, it can be easily dismantled.

The proliferation of such services raises some serious issues. Is it better to have many separate services, or are they better localized in one agency? Can quality control be assured when many unlicensed, uninspected services exist? One central agency, with trained and supervised staff, would provide better quality counseling. Yet would victims of rape, for example, call a general counseling service? Doesn't the provision of a special service, directed toward those victims and manned by sympathetic counselors, who are themselves perhaps victims of rape, facilitate use of the service?

One possible solution is to have separate telephone numbers and individual advertising campaigns, but situating the lines in the same agency. The Suicide Prevention and Crisis Service in Buffalo, at one point, had its counselors answer four different services, each with its own telephone number and advertising program (a suicide prevention service, teen hotline, problems in living service, and drug hotline). In this case, however, it proved difficult to have counselors switch from service to service quickly, turning from a seriously depressed elderly citizen considering suicide to a shy teenager who did not know how to ask out the girl who sits near him in class. Perhaps it is most sensible to coordinate services (and if possible, locate them together) but have separate groups of counselors for each service.

A second issue regarding the proliferation of telephone counseling services is whether any counselor can counsel any caller. Must a counselor be homosexual to counsel homosexuals, a rape victim or female to counsel rape victims, a teenager to counsel teenagers? Or can any competent, trained counselor handle any client and any problem? There are no pertinent empirical data here and opinions differ. Usually, however, when community needs arise, specific interest groups initiate the service, which results in like counseling like.

## THE TELEPHONE IN PSYCHOTHERAPY

The telephone, in increasing ways, is used by qualified professionals engaged in individual face-to-face psychotherapy. For example, Robertiello (1972) reported two cases of psychoanalysis in which a patient who was temporarily unable to visit the psychoanalysts's office (due to travel and illness) continued the sessions by telephone. He reported that the telephone made no difference in one case (where much of the psychoanalysis consisted of discussion of the patient's dreams) and actually helped in the second. In the latter case, transference had been so disruptive that the patient could not stand being in the same room with the analyst. Her emotions interfered with integration of insights into her ego. The telephone sessions enabled her to experience the emotions and also reflect upon the transference.

Beebe (1968) reported on use of the telephone to begin the schizophrenic patient's integration into his family. He regarded the first goal to be the patient's return to involvement in difficult family situations, rather than temporary isolation from them. Physically returning him to his family may be too stressful, but the telephone often provides the right amount of distance. Calling permits contact without closeness, and curbs the fantasies that take over when there is no exchange. In return, the family feels involved and permitted to help undo whatever they have done to the patient.

Beebe reported a case of an acute schizophrenic psychosis in a sailor forced into the service by his parents, who wanted to get rid of him. Early in training he became anxious, confused, and felt he had sinned. At the height of his confusion, he believed he had killed his mother. The call home was a great relief to him, and he became quite lucid and free of psychosis.

Owens (1970), a dentist, demonstrated the effectiveness of inducing hypnosis by telephone, when he called a number of his previous hypnosis patients and used a standard induction procedure to cause a mild state of hypnosis. In all cases, he was successful. He also had success with two patients he had not previously hypnotized. Owens' intent was to explore whether hypnotic states could be induced by telephone, but because his patients reported feelings of relaxation and reduction in dental pain after hypnotic induction, the procedure may have some utility.

The telephone has been used to follow up discharged alcoholics (Catanzaro and Green, 1970), to speed up consultation between patient and counselor, and to permit case supervision for counselors by their supervisors (Wolf et al., 1969). Chiles (1974) used telephone contact to reinforce behavior modification procedures with patients. Each day a call lasting a few minutes is made so the patients can report briefly on aspects of their behavior, such as eating, consumption of alcohol, hostile behavior with relatives, etc. Telephone contact, by maintaining the behavior modification program and reinforcing the patient's self-image, facilitates continuing the behavior modification regimen.

Miller (1973) surveyed a number of psychiatrists and found out that 97 percent used the telephone for handling emergencies, 45 percent used it as an adjunct to face-to-face psychotherapy, and 19 percent used it as the primary mode of treatment. The psychiatrists differed in how easy they found the telephone to be as a mode of communication. It is important, Miller noted, for psychotherapists to know their own reaction to the mode of counseling, the reaction of patients, and the suitability of particular problems to the mode. Miller also noted that generally depression was most difficult to treat by the telephone, while anxiety was comparatively easy.

## THE UNIQUE CHARACTERISTICS OF TELEPHONE COUNSELING

As experience with telephone counseling and psychotherapy has grown, it has become apparent that such counseling has unique characteristics not shared by other modes of counseling.

### Client Control
When a client walks into a counselor's office, the counselor has most of the power. There may be a receptionist to receive the client, and once past that the client faces a counselor who usually sits behind a desk. Perhaps a difference in status is reflected in the counselor's dress, or in the difference between the luxury of the counselor's office and the client's home. The client cannot remain anonymous; even if giving a false name, the client can be recognized again and is often required to give personal information while the counselor, of course, is not. Further, it is difficult to terminate the contact because the client must stand up and leave the office, allowing the counselor time to intercede and discourage the client's departure. As Williams and Douds (1973) have stated, it is very easy for a face-to-face counseling contact to provoke anxiety and humiliation for a client.

With telephone contact, in contrast, the client has much more control, because he can remain anonymous. He need give no information about himself, and he remains unseen. (Even if a counselor could obtain permission from the telephone company to trace the call, this process can take over an hour and a public telephone could have been used.) The client can terminate the contact quite easily by hanging up. (This method of abrupt termination is often the most immediate, effective comment on a counselor's performance.)

This equalization of power was arranged in face-to-face psychotherapy by Nathan et al. (1968). Nathan arranged for client and psychotherapist to sit in different rooms, communicating only by closed circuit television. To see and hear the counselor, the client must repeatedly press a button (at a high rate of 120 times per minute) to maintain maximum volume and clarity of picture. To blot out the counselor the client decreases the rate of button pressing, causing the television picture and sound to fade; the counselor has similar control.

This equalization of control often produces anxiety in the counselor, but it has a facilitative effect for the client. The client who is anxious, feels threatened, or is reluctant to walk into a counselor's office, may be willing to call the counselor. On the telephone, the client maintains a feeling of freedom and a sense that he cannot be hurt or victimized.

This equalization of power is useful for clients in crisis calling a counselor for the first time and for patients in psychotherapy. MacKinnon and Michaels (1970) reported a phobic housewife who revealed disturbing thoughts about her psychotherapist when a snowstorm that prevented her office visit forced her to call him by telephone. She subsequently sought a telephone session when difficult material emerged again.

### Client Anonymity
The client can remain anonymous when talking to a counselor via the telephone, and the possibility of anonymity encourages greater self-revelation and openness. Anonymity minimizes the feeling of possible ridicule, abuse, censure, or hurt because of the counselor's evaluations. The effect of anonymity is an example of the common sociological observation that it is easier to discuss problems with strangers than with acquaintances.

### Positive Transference
The counselor can also remain anonymous. In face-to-face counseling, any client fantasy about the counselor is checked against

reality, and it is likely that the counselor may not live up to the fantasy. A distressed individual may not be able to tolerate this shattering of illusions. Those illusions may give him enough security to make the contact for counseling. The telephone counselor may also fail to live up to the client's fantasy, but since the client is presented with only a voice, the counselor will be more similar to the client's ideal than in a face-to-face contact. If the client can make of the counselor what he *will*, he may also be able to make of the counselor what he *needs* (Williams, 1971). This may enable the client to move out of his distressed state.

Of course, there are dangers in allowing a client to dwell too long in a world of fantasy. Situations can easily occur where the development of fantasy works to the disadvantage of the client (and of the counselor). The client must then be forced to face reality. However, a skilled counselor can use positive transference to help the client move to a stronger psychological state and then subsequently move to a realistic acceptance of what is happening in the counseling process.

## Reduced Dependency

The counselor's anonymity has an additional advantage for clinics employing a number of counselors. In these clinics, counselors usually use a first or assumed name only. The counselors are told to discourage clients' dependency upon particular counselors. Thus, the client become dependent upon the *clinic*, not the *counselor*: if a counselor leaves, temporarily or permanently, the client is less upset than when dependency has been directed to one counselor. This helps suicidal clients, who often respond to a therapists's vacation or absence by attempting suicide.

## Accessibility

Most people have low-cost access to a telephone, and this is critical for clients in crisis (especially in suicidal or homicidal crisis), the elderly, and the infirm. Many are bedridden or too weak or senile to visit a counselor for face-to-face psychotherapy. For such people, the telephone is often the only source of counseling.

MacKinnon and Michaels (1970) noted this advantage with cases from a private psychotherapy service. Psychotherapy was maintained, in one case, with a female client twice weekly while she went to Nevada to obtain a divorce. Another female client was treated for three months while bedridden on her obstetrician's orders.

## Immediacy

Because there is an immediacy to the telephone, many clinics maintain twenty-four-hour counseling services so that distressed individuals can locate a counselor quickly. Psychotherapists refer patients to such services at night, on the weekend, and while they are on vacation. It is helpful if the psychotherapist formulates a treatment plan with the agency for patients in these situations.

Miller (1973) has noted five properties of the telephone for use in counseling that overlap with those discussed above. Its "spatial property" breaks down barriers between the counselor's office and client, corresponding to the accessibility noted above. It permits a more distant relationship than in face-to-face counseling, but one which is quite intimate because the patient's voice is close to the counselor's ear and vice versa.

Its "temporal property" means that the counselor may be called at any time (although, of course, he may bar complete accessibility by having an unlisted number or by disconnecting his telephone). The patient is not limited to counseling sessions for contact with the counselor.

A third property is that it is single-channeled, carrying audio communication only. This allows greater freedom for client fantasy.

Its fourth property is being a "machine," a concrete and rather impersonal object to relate to and through. However, recent studies indicate that patients do not find that it is excessively impersonal for a computer to administer psychiatric intake interviews. In fact, many patients preferred to have a computer, rather than a human, interview them (Greist et al., 1973). So the telephone's being a "machine" may not necessarily make the patient uncomfortable. The fifth property of the telephone noted by Miller is that it is dyadic. Most conversations are between two people and telephone contacts are more often dyadic than are face-to-face contacts.

Miller explored the effects of these properties on the telephone's characteristic uses. The spatial property appeals to patients with oral and dependency needs who can reassure themselves that support is at hand. Ambivalent patients (such as some schizophrenics) may use the telephone to maintain distance and control in the therapeutic relationship. Hostile patients may be able to express their emotions because they feel safer doing so at a distance. The spatial property often makes the counselor feel that he has less control over the counseling relationship and is more vulnerable to unreasonable demands on his time. (Miller noted that in some circumstances it might be appropriate to charge patients for telephone contacts.)

The temporal property appeals to impulsive patients who cannot tolerate anxiety. The psychotherapist often experiences anger with these patients, and Miller suggested that he set firm limits on how much use of the telephone he will accept.

The single-channel property appeals to patients who want anonymity to protect themselves from the psychotherapist. They do not distrust the psychotherapist; they may be merely exploring the possibility of psychotherapy, or they may find it less embarrassing, anxiety provoking, or shameful to discuss problems over the telephone. The lack of visual cues may distress psychotherapists who utilize nonverbal communication. It may impede effective patient evaluation or induce misleading fantasies on the part of the psychotherapist.

The mechanical property of the telephone may appeal to obsessive neurotics and schizophrenics. On the other hand, Miller felt that counselors dislike its impersonal quality. He felt that the dyadic property appealed to those wishing to exclude others from communication with a psychotherapist, but face-to-face psychotherapy is no different in this respect.

Miller noted that the psychotherapist can make active use of the telephone. He can use it to support insecure and unstable patients between regular psychotherapy sessions. He can instruct impulsive patients to call whenever they feel that they might act upon their impulses, and he can instruct patients who block in psychotherapy to call when they recall a blocked thought. He can utilize the telephone for patients who have difficulty talking about particular issues face-to-face. He can also use the telephone to contact significant others to bring them into the treatment process or to evaluate the patient more accurately.

## PROBLEMS ASSOCIATED WITH TELEPHONE COUNSELING

Along with the advantages of telephone counseling, several disadvantages and dangers exist. Although it may be useful for the client to be able to fantasize about the counselor, the reverse is never useful. There are, however, eventual dangers in the client's fantasies about the counselor. At some point during the counseling process, these fantasies must be brought out, examined, and adjusted to reality.

Brockopp (1970) noted that counselors can easily slip into conversation with clients rather than remain in a psychotherapeutic mode. The telephone is strongly associated with conversation, and the counselor may revert to old habits when using it. The telephone also allows greater distance between client and counselor (such as anonymity) but encourages intimacy. The counselor may be relaxed in a comfortable chair, and the client's voice is close to his ear. This intimacy can facilitate or induce a conversational mode.

This tendency toward conversation is a problem for counselors in clinics that maintain twenty-four-hour counseling services. Occasionally a counselor will work alone at night. (It is a poor practice to sleep between calls, for the counselor may resent being awakened by a client in crisis.) A counselor who is awake and alone can come to welcome calls from clients, since such calls help to pass the time. Under these conditions a counselor may seek to prolong conversations with clients because he has nothing else to do, and counseling will often degenerate into "conversation."

If "conversation" develops, the psychotherapeutic process is minimized, distorted, or eliminated, the counselor's objectivity is reduced, confrontation is less likely, the client's anxiety may be reduced to such an extent that he no longer feels a need to work on his problem, and it develops the false assumption that psychotherapy is taking place.

## PROBLEMS WITH TELEPHONE COUNSELING SERVICES

Not all telephone counseling problems can be blamed on the medium. Many agencies providing telephone counseling use nonprofessionals: college students, preprofessionals, housewives, persons who have personally been through crises. It is the personnel rather than the telephone that cause problems. Lamb (1970) has discussed some typical problems nonprofessionals have in counseling, such as the common fantasy of omnipotence and its variants: "But all I'm doing is listening!", "If I talk about it, it may happen" (the "power of positive thinking" error), and "But he's manipulating me" (the "who's in charge here?" error).

Although volunteers and nonprofessionals have long been used in mental health (Gruver, 1971), the growth of telephone counseling services has increased their use. A counseling service needs sixty to eighty volunteers to maintain a seven-day twenty-four-hour service in a major city. Usually, most of these counselors are nonprofessionals who have received at most twenty-four hours of training and perhaps an hour of supervision weekly.

This use of nonprofessionals has raised the issue of whether they perform worse, the same, or better than professionals. McGee and Jennings (1973) argue that nonprofessional counselors have higher levels of empathy than professionals, but McColskey (1973) argues that if we believe clinical training has any value, it is absurd to believe untrained people perform better than trained people. With rigorous selection, adequate training, and good supervision, some nonprofessionals can do a good job with most clients, but can they handle all kinds of crises and be trusted to behave professionally? The first answer is clearly no, since most services find that they must employ professionals as twenty-four-hour back-up consultants. The second answer is also probably no.

The American Association of Suicidology recently debated the ethics of recording calls without the clients' knowledge; the majority of centers considered it unethical. At a center where I worked, we recorded calls for supervision of the counselor and for research purposes. I have heard a call where the counselor fell asleep while a client was talking and one where a counselor began a call by laughing at a client who said she felt like killing herself. (Our counselors recorded their own calls, and surprisingly neither of those erased their performances.)

Most centers argue that their counselors do not make these errors, yet most centers do not monitor their counselors by day or night. The regular staff have no idea of what their counselors are up to. Further, nonprofessional telephone counselors, lacking a concept of professional behavior, may easily become emotionally and intimately involved with clients. Sometimes when counselors leave a service, they contact clients who formerly called them at the service, in order to maintain a relationship with the client. The mental health professions have enough problems today with therapists' unethical behavior. The problems with nonprofessionals are much greater.

As a result of such problems, I have advocated that nonprofessional telephone counselors be closely supervised, more closely than other groups of mental health workers (Lester, 1973). Further, I have advocated replacing the hordes of part-time volunteer nonprofessionals by a few full-time, well-paid, highly trained paraprofessionals whose performance can be accurately monitored.

Since telephone counseling services are limited in the service they can provide, it is important to recognize that telephone counseling by itself is not sufficient to provide assistance. Hoff (1973) has discussed the importance of adequate follow-up, including subsequent medical and psychiatric help personally or by telephone and contact with the significant others of the client. Richard and McGee (1973) described the development of an outreach team that has the training and mobility to make home visits. Such a service is a most useful addition to a telephone crisis intervention service.

Crisis intervention can also be traumatic for the counselor. He may well need support, advice, and the opportunity to share responsibility for the client.

## PROBLEM CALLERS

Telephone counseling services attract clients who present certain problems less common in face-to-face counseling. The most noteworthy example of "problem callers" is the obscene caller.

Telephone counseling services receive calls from males who wish to talk to a female while masturbating. Particular problems are also raised by the silent caller (who calls but refuses to speak), and the nuisance or prank caller. The management of such problems was discussed by Lester and Brockopp (1973).

I have discussed a caller experienced by most telephone counseling services—the chronic caller (Lester, 1971). These clients call regularly—some call five times a day and spend as much as thirty-five hours a week talking to counselors. Since typical services employ large numbers of counselors, a chronic caller may talk to a different counselor each time. It is difficult for each counselor to report to other counselors on problems and progress with the caller and difficult for full-time staff to design an effective treatment plan. Further, it is difficult to enforce a treatment plan, once it is formulated, if counselors are not supervised. (An excellent and respected counselor at one center refused to limit calls from a chronic caller in the way that the treatment plan recommended. He said to do so was inhumane.)

Clients often gratify counselors' needs that are not necessarily relevant to their function. For example, one female caller to the teen hotline in Buffalo became a chronic caller partly because the male counselors at the center liked talking to this attractive-sounding girl. The sexual gratification for both client and counselors was apparent. The professional staff tried limiting her calls to one counselor and inviting her to the center so that she and the counselors could meet and remove fantasy from the involvement. Several months later, however, the problem had not been solved and the girl was still a chronic caller.

I have focused upon chronic callers because the telephone counseling service itself creates this problem, and the psychological condition of the chronic caller may deteriorate because of dependency upon the service. Perhaps the dependencies were more appropriately distributed prior to involvement with the center, and they are certainly not usefully distributed *after* the development of chronic dependency upon the center. Centers often justify continued involvement with chronic callers by hoping that telephone contact reduces the client's chance of hospitalization in a psychiatric facility. The center sees itself as helping the client to continue to exist in the community, but there is usually no supporting evidence for this. Innovations in any field, perhaps especially in mental health, often create new problems that we deal with by creating additional services. Providing easily available mental health services may reinforce behavior that is not advantageous to mental health; it may reinforce obsessive preoccupation with our psychological moods and behavioral symptoms, encourage lack of responsibility for unhappy relatives and friends who have problems in living, and label people as "psychiatrically disturbed" and thereby facilitate their introduction into the career of psychiatric patient (Scheff, 1966). Whether mental health services increase the level of mental health or happiness in the community is an open question.

The "chronic caller," therefore, is the kind of problem that should make telephone counseling services examine the rationale for their continued existence and the effectiveness of their treatment programs.

## CONTINUED EXISTENCE OF AGENCIES

Brockopp (1973) noted that often agencies become less concerned with their function than with their continued existence. They draw their funds from many sources: state and local government, colleges, churches, hospitals, and voluntary associations. They lose sight of the client and focus on procuring furniture and a larger budget for the next fiscal year. Brockopp urged that all innovative agencies should be set up with the understanding that they will be disbanded in five years, transfering successful experimental programs to other agencies. In this way, the agency would focus on its function rather than its continued existence.

This point is an important consideration for all kinds of agencies, but telephone counseling services merit special attention. Telephone counseling services established in response to temporary community needs face identity crises as community needs change. Should the service continue or disband? Baizerman (1975) discussed this issue for teen hotlines which now handle problems different from those they were forced to deal with. Ten years ago, teenage crises concerned drug highs, runaways, a place to bed down for the night, arrest, military draft counseling, etc. These were "real crises" to the counselors. Today, calls concern loneliness, family conflict, and dating problems. The "crisis" has gone out of the crisis call. Teen counselors now wonder why they are counseling, whom they should serve, what they should do. The services seem to have lost their purpose. Some argue that the services should close their doors and disconnect their telephones; others are searching for new purposes.

## THE RESEARCH STIMULUS OF TELEPHONE COUNSELING SERVICES

The development of telephone counseling services has stimulated a good deal of research in a variety of areas.

The suicide prevention movement in the 1950s saw the need for a decision model for counselors to estimate the probability that a client would kill himself. Simple scales were developed to predict suicidal risk (Lester, 1970). There are a number of other behaviors for which simple prediction scales would be useful. For example, perhaps we could predict whether an individual was likely to assault or murder others. Without the stimulus of counseling agencies to deal specifically with assaultive behavior, however, the construction of such predictive scales has been slow.

A number of reports have appeared concerning the selection of telephone counselors (Lester and Williams, 1971; Tapp and Spanier, 1973) and the particular personality traits that characterize such volunteers. A good deal of work has also been done on evaluating the effectiveness of telephone counseling services and their counselors (Lester, 1972; McDonough, 1975).

Telephone services provide a convenient setting for research on the effectiveness of counseling, but the formulation of objective criteria determining whether clients have been helped has proved difficult. How can we measure psychological improvement? In telephone counseling services, we can often find objective but limited criteria—for example, whether the client accepted the suggested referral for a face-to-face psychotherapy session (Slaikeu et al., 1973; Buchta et al., 1973).

Because telephone calls can be recorded, it is easy to simulate calls with an actor playing a patient with a particular problem. The call is recorded and later examined to see whether the telephone counselor functioned adequately (e.g., Bleach and Claiborn, 1974). Most telephone counseling services use a crisis counseling model, so it is easy to listen to calls and rate counselors for technical effectiveness in following guidelines for handling crises (Fowler and McGee, 1973). Judges can also rate the empathy and genuineness that telephone counselors are supposed to show (Carothers and Inslee, 1974).

It is probably true that telephone counseling services have been more aware of the importance of evaluating their effectiveness than have other mental health agencies.

## CONCLUSION

The telephone provides an important tool for the counselor in helping his clients. Because of the telephone's qualities, some clients use it exclusively, and many other clients use it at some point in their counseling. It poses problems for the counselor, but adequate training and experience should enable the counselor to employ the telephone effectively.

Telephone counseling services have been an important influence in the treatment of psychological problems. The services fulfilled community needs and have stimulated much discussion about the role and purpose of mental health agencies. The services have also stimulated a good deal of research on the selection, training, and evaluation of counselors. In many respects, therefore, telephone counseling services have had a welcome catalytic effect on the thinking of mental health professionals.

## REFERENCES

Baizerman, L. 1975. Changing crises. *Crisis intervention* 6 (2), 45–51.

Beebe, J. 1968. Allowing the patient to call home. *Psychotherapy* 5, 18–20.

Bleach, G., and Claiborn, W. 1974. Initial evaluation of hot-line telephone crisis centers. *Community Mental Health Journal* 10, 387–394.

Brockopp, G. 1970. The telephone call. *Crisis Intervention* 2, 73–75.

_____. 1973. An emergency telephone service. In D. Lester and G. Brockopp (eds.), *Crisis intervention and counseling by telephone*. Springfield, Ill.: Thomas, pp. 9–23.

Buchta, R., Wetzel, R., Reich, T., Butler, F., and Fuller, D. 1973. The effect of direct contact with referred crisis center clients on outcome success rates. *Journal of Community Psychology* 1, 395–396.

Carothers, J., and Inslee, L. 1974. Level of empathic understanding offered by volunteer telephone services. *Journal of Counseling Psychology* 21, 274–276.

Catanzaro, R., and Green, W. 1970. WATS Telephone therapy. *American Journal of Psychiatry* 126, 1024–1030.

Chiles, J. 1974. A practical therapeutic use of the telephone. *American Journal of Psychiatry* 131, 1030–1031.

Fowler, D., and McGee, R. 1973. Assessing the performance of telephone crisis workers. In D. Lester and G. Brockopp (eds.), *Crisis intervention and counseling by telephone*. Springfield, Ill.: Thomas, pp. 287–297.

Greist, J., Gustafson, D., Strauss, F., Rowse, G., Laughren, T., and Chiles, J. 1973. A computer interview for suicide risk prediction. American Association of Suicidology, Houston.

Gruver, G. 1971. College students as therapeutic agents. *Psychological Bulletin* 76, 111–127.

Hoff, L. 1973. Beyond the telephone contact. In D. Lester and G. Brockopp (eds.), *Crisis intervention and counseling by telephone*. Springfield, Ill.: Thomas, pp. 132–148.

Lamb, C. Telephone therapy. 1970. *Voices* 5 (4), 42–46.

Lester, D. 1970. Attempts to predict suicidal risk using psychological tests. *Psychological Bulletin* 74, 1–17.

_____. 1971. The chronic caller to a crisis hotline. *Crisis Intervention* 3, 62–65.

_____. 1972. The evaluation of telephone counseling services. *Crisis Intervention* 4, 53–60.

_____. 1973. Psychologists/community crisis services. *Newsletter, Division 31*, 5 (2), 3.

_____. 1974a. The unique qualities of telephone therapy. *Psychotherapy* 11, 219–221.

_____. 1974b. Recent trends in telephone counseling. *Crisis Intervention* 5 (2), 8–15. (b)

Lester, D., and Brockopp, G. (eds.). 1973. *Crisis intervention and counseling by telephone*. Springfield, Ill.: Thomas.

Lester, D., and Williams, T. 1971. The volunteer in suicide prevention. *Crisis Intervention* 3, 87–91.

MacKinnon, R., and Michels, R. 1970. The role of the telephone in the psychiatric interview. *Psychiatry* 33, 82–93.

McColskey, A. 1973. The use of the professional in telephone counseling. In D. Lester and G. Brockopp (eds.), *Crisis intervention and counseling by telephone*. Springfield, Ill.: Thomas, pp. 238–251.

McDonough, J. 1975. The evaluation of hotlines and crisis phone centers. *Crisis Intervention* 6 (2), 2–19.

McGee, R., and Jennings, B. 1973. Ascending to "lower levels." In D. Lester and G. Brockopp (eds.), *Crisis intervention and counseling by telephone*. Springfield, Ill.: Thomas, pp. 223–237.

Miller, J. 1973. The telephone in outpatient psychotherapy. *American Journal of Psychotherapy* 27, 15–26.

Nathan, P., Smith, S., and Rossi, A. 1968. Experimental analysis of a brief psychotherapeutic encounter. *American Journal of Orthopsychiatry* 38, 482–492.

Owens, H. 1970. Hypnosis by phone. *American Journal of Clinical Hypnosis* 13, 57–60.

Richard, W., and McGee, R. 1973. Care team. In D. Lester and G. Brockopp (eds.), *Crisis intervention and counseling by telephone*. Springfield, Ill.: Thomas, pp. 149–154.

Robertiello, R. 1973. Telephone sessions. *Psychoanalytic Review* 59, 633–634.

Scheff, T. 1966. *Being mentally ill*. Chicago: Aldine.

Slaikeu, K., Lester, D., and Tulkin, S. 1973. Show versus no-show. *Journal of Consulting and Clinical Psychology* 40, 481–486.

Tapp, J., and Spanier, D. 1973. Personal characteristics of volunteer phone counselors. *Journal of Consulting and Clinical Psychology* 41, 245–250.

Williams, T. 1971. Telephone therapy. *Crisis Intervention* 3, 39–42.

Williams, T., and Douds, J. 1973. The unique contribution of telephone therapy. In D. Lester and G. Brockopp (eds.), *Crisis intervention and counseling by telephone*. Springfield, Ill.: Thomas, pp. 80–88.

Wolf, A., Schwartz, E., McCarty, G., and Goldberg, I. 1969. Training in psychoanalysis in groups without face-to-face contact. *American Journal of Psychotherapy* 23, 488–494.

# 21

## Telephone and Instructional Communications

### Paladugu V. Rao

This paper's purpose is to survey the applications of the telephone and its peripherals to instructional communications. The survey is not intended to be exhaustive; rather, only applications that have unique characteristics will be reviewed.

## EARLY APPLICATIONS

The first reported instructional application of the telephone was in Iowa in 1939. Dr. Winterstein, Director of Special Education for the State of Iowa, initiated a project for homebound and hospitalized students. With the aid of AT&T and the local phone company, Dr. Winterstein had intercom equipment installed in student homes. All the class lectures were transmitted to the homebound students simultaneously as they were presented to the classroom students. Soon this idea spread across the state. Within two years, the project had benefited more than 1,000 students at a monthly cost of $15.00 per student station. Student participants in the project commented that learning by telephone was just as good as learning in a classroom.[1]

The first college-level application seems to have been undertaken by the University of Illinois' College of Dentistry, at its Chicago Medical campus in November 1947. Six lectures were simultaneously transmitted to thirty dentists in Scranton, Pennsylvania, and to fifty dentists in a classroom on the Chicago campus. Each lecture was two hours long and allowed enough time for interchange of comments between the local and the remote group; the lectures were supplemented by slides at both ends. This project was hailed as a "novel extension service in education," but the short report about its application does not give details concerning the type of equipment used.[2]

In the 1950s tele-lectures (the term assigned to instructional communications by telephone) began attracting attention from American educators, and applications on a limited scale began appearing at all levels of education from graduate to elementary. The amplified telephone, a telephone hooked into an amplifier, made possible the use of tele-lectures with large groups. It enabled students to talk with outstanding people in various fields and became common at some schools. Many applications in the 1950s, however, turned out to be one-shot projects and were not implemented on a continuous basis. No continuing projects are reported in the literature.

The tendency to one-shot projects continued into the early 1960s. In 1963, the psychology students at the University of Omaha used an amplified telephone to have a discussion with the author of their textbook, the outstanding psychologist Dr. Neil Miller of Yale University. Similarly, students at La Crosse State College in Wisconsin heard Dr. Max W. Carbin speak on "The Nuclear Reactor: Its Functions and Purpose" from the University of Wisconsin's Madison campus. Dr. Coleman of Pittsburgh's Carnegie Institute of Technology taught simultaneously to groups gathered at the university campuses of Oklahoma, Syracuse, Wisconsin, Omaha, and Washington. The Fund for Advancement of Education sponsored a project in which Dr. Moses Hadas of Columbia University delivered eighteen lectures on classical literature to southern colleges.[3]

Perhaps one of the best organized and implemented applications of the amplified telephone for instruction was on the campus of Stephens College, Columbia, Missouri, where Dr. Alfred Novak of the college conceived and organized an in-service training seminar to improve teaching abilities for the science faculties of a group of colleges. Drury College of Springfield, Missouri, Kansas Wesleyan University of Salina, Kansas, Langston University of Oklahoma, LeMoyne College of Memphis, Tennessee, Morehouse College of Atlanta, Georgia, and Wilberforce University of Ohio also participated in this project. Dr. Novak acted as coordinator and moderator of the lectures, while great lecturers spoke from their respective locations. Though the seminar was primarily intended to benefit the faculties of participating colleges, faculty members of other nearby colleges were also invited to take part. There were thirteen thirty to forty-five minute lectures, followed by a question-answer period designed so that each partciating college had a chance to ask questions. The guest lecturers included three Nobel Prize winners: Dr. George Beadle of the University of Chicago, Dr. Herman Muller of Indiana University, and Dr. Peter M. Medawar of the British National Institute for Medical Research. These three distinguished scholars spoke to the seminar audience, spread geographically over the eastern half of the United States, from their own offices.

At the end of the project, the participants were asked to evaluate the content and mode of instruction; these evaluations were very favorable. "Without the telephone facilities provided in this experiment," commented one participant, "it is highly unlikely that any student or teacher would have an opportunity to discuss person-to-person the major ideas of such a distinguished company."[4]

## HIGH SCHOOLS AND TELETEACHING

During the 1960s, another type of telephone-based instruction (commonly referred to as teleteaching) gained popularity among elementary and secondary schools. Developed to meet the needs of homebound and hospitalized students, this mode of instruction was used by New York and California school systems on a continuous basis. "The teleteaching program," according to one California teacher, "is an on-going educational plan devised to meet the needs of children who are exempted from school by illness or accident. The program aims to continue the child's educational plan during his rehabilitative state and prepare him for re-entry into his usual school program."[5] The equipment used in the California school systems allows the teacher to sit at a central console and dial any student on the roster by inserting a plastic badge containing a student's telephone number; this minimizes the teacher's effort to establish communication links with different students. This equipment also allows group communication as well as individualized communication, but if needed, the teacher can make the conversation private. Thus, students can discuss their personal problems with the teacher and obtain guidance.

The New York Board of Education for the City of New York conducted an interesting experiment to determine the telephone's usefulness in instructional communications. In this experiment, the subjects were physically handicapped homebound adolescents divided into two groups: control and experimental. Each day both groups had a radio broadcast of fifteen minutes on social preparation, covering such topics as how to use free time, interests, hobbies, and other social activities. After each broadcast, the experimental group had an opportunity to discuss the broadcast with the speaker over the telephone, but no such opportunity was provided for the control group. At the end of this experiment, the experimental group showed significantly more positive orientation toward social interest; however, it should be noted that while homebound students were motivated by the telephone instruction toward social interest, it was not reflected in behavioral change. [6]

The Wayne County school district in rural Utah used the amplified telephone to help their students overcome significant vocabulary deficiencies, among other cultural disadvantages, compared to their urban counterparts. A telephone hookup was established with a companion New York City school, enabling the rural students to receive part of their instruction from the City. This method of teaching did improve vocabulary among rural students. [7]

The Western States Small Schools Project (WSSSP), a multistate cooperative, endeavors to identify and define ways of strengthening the educational programs in small schools that cannot be consolidated. In this application many schools used the telephone to bring in instructional programs unavailable locally. For example, using the amplified telephone with overhead transparencies, basic art was taught from a central location to students in eleven rural schools of the system located in Nevada, Oregon, Idaho, and Utah. [8]

## UNIVERSITIES AND TELETEACHING

Until 1966, teleteaching was limited to a lecture-discussion mode, and oral presentation could not be complemented by written material as in a traditional classroom environment. In 1966, an interesting experiment was conducted at Cornell University by wedding telephone and visual technologies. Cornell's College of Engineering used this combination to teach a course in physical metallurgy to a group of research and development specialists in Towanda, Pennsylvania, fifty-five miles away. The system transmitted oral lectures and handwritten material over telephone lines for long-distance illustrated lectures. In this system the instructor used an electronic pen to draw diagrams that are transmitted over the telephone line and displayed on a TV monitor screen in Towanda.[9] This system, however, seemed to have two drawbacks: it was very expensive, and it did not have the capacity for two-way graphic communication. The graphic transmission was always from the instructor to students.

## THE VERB SYSTEM

Perhaps the most important accessory the telephone has acquired to improve its effectiveness in instructional communications is the electrowriter. Electrowriters became available in the early part of the 1960s, and their instructional applications began in the later part of that decade. The Victor Comptometer Company of Chicago, Illinois, is generally credited with developing and marketing electrowriters. The electrowriter system marketed by the company for instructional applications is popularly referred to as VERB (Victor Electrowriter Remote Blackboard).

The VERB system may be bought in several configurations, depending on the intended applications, but the basic system consists of a transmitter, a receiver, two data phones, and two telephone lines; the system components are compact and portable. Both transmitter and receiver have 17½ square inches of writing area, and both units accept plain paper or custom-designed electrowriter printed forms. In this system, a ball-point pen moves over the paper and varying tones are generated to represent the movement of the pen. The tones are transmitted, and a pen at the receiving end reproduces the movement of the original, thus creating an identical image on the receiver screen. This image is then projected onto a larger screen with an attached overhead projector.

The VERB system also has a transceiver which performs the functions of both transmitter and receiver. In addition, a variety of accessories is available to improve the effectiveness and usefulness of VERB. The system's modular design makes it possible to add these accessories as needed to the basic system.

The invention of the electrowriter made possible the use of the telephone in instructional communications on a continuing basis. Within the last five years, many university extension divisions have installed VERB systems to carry out their educational activities in almost every field of knowledge. A study at West Virginia University found that achievement by the extension classes taught by VERB was equal to or significantly greater than that of the students in the on-campus classes. The same study also revealed that the success of teleteaching was greater when the professor limited continuous lecturing to twenty to twenty-five minutes, used A-V techniques as a supplement, and made two or three personal visits to the class.[10]

The Division of University Extension at the University of Illinois has been using the VERB system to teach extension classes since 1966, and has also used the system to teach graduate and undergraduate courses in engineering, mathematics, agriculture, education, and library science. It was also used occasionally on campus to bring in outstanding faculty members from other campuses. A survey conducted by the Extension Division indicated that the VERB students in general performed as well as, or better than, their on-campus counterparts.[11]

A very recent development that appears to enhance teleteaching is the Remote Blackboard. This device developed by Bell Labs, is being used at the University of Illinois Urbana-Champaign Campus on an experimental basis. It consists of a blackboard panel connected by telephone lines to a television monitor. Whatever an instructor writes on the panel is instantly transmitted and reproduced on the connected television monitor; the instructor can erase the entire panel by pressing a button, or selected portions by using the regular blackboard eraser. This device seems to have the potential of becoming a widely-used accessory in many teleteaching applications.

At the college level, teleteaching seems to be a particularly suitable medium for teaching newly developing fields like information science. Kent State University offered a graduate course on information retrieval in 1971 using the amplified telephone. During this course, students heard tele-lectures from persons associated with such outstanding information retrieval programs as Project Intrex at MIT and Project MEDLARS at the National Library of Medicine. These tele-discussions enhance student interest and contribute a better understanding of a very complex subject. [12]

## TELETEACHING AND OTHER MEDIA

A variety of media (such as overhead transparencies, audio cassettes, slides, programed instructional materials, etc.) have been used to supplement teleteaching. [13] In home applications, teleteaching has been used to supplement instruction offered through other media such as radio, television, and computer. In the 1967–68 school year, the Catholic schools in Brooklyn developed an experimental instructional system consisting of television, telephone, and computer to teach basic computer concepts.

For the experiment, eight half-hour video tapes were made concerning various aspects of computers. Multiple choice tests on the same subject were constructed and stored on a magnetic disk, the computer was programed to select these tests randomly and play them back to any Touch-Tone telephone through its audio response unit, and students were instructed how to dial and respond to the computer using the Touch-Tone telephone.

During the experiment, the telecasts were made to seven groups. After each telecast, four of the seven groups used the telephone to take a multiple choice test from the computer on the topic just covered by the telecast. The questions from the computer came in a normal human voice, and the students responded by pressing an appropriate telephone button. After each test, the computer evaluated the responses and suggested appropriate additional reading materials through its voice response unit.

In the final test, the four groups that used the telephone for additional instruction from the computer did significantly better than the other three groups. Though this experiment was later discontinued because of high costs, it did prove that teleteaching can be successfully integrated with other instructional media.[14]

It is also appropriate to note that many of today's Computer Assisted Instruction (CAI) systems depend on telephone networks for data transmission to remote locations. One of the most widely used CAI systems today is the PLATO system at the University of Illinois. This system offers many features including computer-generated graphics, audio responses, touch panel, and slide projections. The prime reason that this system has gained such wide acceptance, however, is its ability to establish an instant communication link between a user and a consultant whenever such a link is needed. The exchange between users and consultants on the PLATO system is a form of teleteaching supplementing the CAI.

## COSTS OF TELETEACHING SYSTEMS

Costs seem to vary from application to application, depending upon equipment used and rates charged by the phone companies. It is difficult to come up with cost figures without a detailed system study for the intended application. Published literature in this regard is not helpful, as it provides cost figures without sufficient background information for intelligent evaluation. The Stephens College project mentioned earlier reports the following costs for the equipment used there:

| Equipment | Cost |
|---|---|
| Amplified telephone equipment lease | $40 per college month |
| One-time installation charges | $20 per college |
| Speakers headphones | $5–$15 per set |
| 45-minute long-distance calls within U.S. | $120–$125 on the average |
| 45-minute trans-Atlantic call to London | $450 |

The above information may be used as a basic guideline in planning a teleteaching system. Since the dominant cost is long-distance charges, however, one should be aware of changes in rates and the structure of rates in the past decade. In any case, the cost will depend critically on the location of points to be linked and the time of day.

Cost figuring for the VERB system is even more complicated, as it depends on the number of units bought, leased, or first leased and then bought. The problem is further complicated by the type of auxiliary equipment bought or leased for use with the basic VERB system. On the average a VERB receiver and transmitter rent for $50 per month each. A VERB transceiver rents for $100 per month; accessories added to the basic system may cost as much as $110 per month in rental charges. Data phone and transmission line charges must be figured on the basis of specific application.

Teleteaching costs are considerably lower than costs for instructional radio and television. Teleteaching does not require substantial investment in equipment, studios, and soundproof rooms as do radio and television, and the simple and easy to operate telephone equipment needs no special operating personnel. Since the telephone companies assume responsibility for maintaining the telephone equipment and network, the user needs no special maintenance crew either. All these cost advantages put teleteaching within the means of many educational institutions.

## EDUCATIONAL TELEPHONE NETWORKS (ETN)

There seem to be three types of educational telephone networks in operation today:
1. Campus networks, popularly referred to as Dial Access Information Retrieval Systems.
2. Multicampus networks.
3. Statewide networks.

1. *Campus networks.* A campus network's purpose is to make information on a variety of subjects available to students and faculty through campus telephones. Information is stored on audio tapes made available to the users in a schedule or random mode. In a scheduled mode, users listen to certain types of tapes at certain times; in a random mode, any user may have access to any available tape by dialing an appropriate number. Through campus networks, instruction can be extended from the classroom to dorms and other places. One of the best exam-

ples of a campus educational telephone network is at Bellevue
Community College in Washington State. Through this network, Bellevue residents may gain access to programed lessons,
stereo music, tapes of current events, etc., from any telephone
in the city. [15]

2. *Multicampus networks*. These primarily offer teleteaching
programs from one campus to students located at several different campuses or sites; they also provide instructional programs
to adult groups, business, and industry. Many states now have
multicampus networks. The use of this type of network on a
continuous basis by higher educational institutions, in states
like California, Illinois, and Virginia, indicates that it is a valid
instructional tool. [16]

3. *Statewide networks*. Wisconsin appears to be the only state
that now has a statewide educational telephone network. The
Wisconsin telephone network links courthouses, extension offices, and the University of Wisconsin campuses and centers.
Through this network, instruction is provided to 200 specific locations in more than 120 communities scattered all over the
state. All lectures are transmitted using the state FM radio network, and the telephone is used for feedback and discussion.
Each listening station on this network consists of a radio loudspeaker and a telephone hand set; students turn on the loudspeaker at the time of the program and respond to or ask
questions through the telephone. At any point a student may
lift the telephone to ask a question, and the instructor's response comes through the radio speaker so everyone on the network can hear it. A variety of courses have been taught on this
network since 1965. [17]

## EFFECTIVENESS OF TELETEACHING

There have been relatively few studies on the effectiveness of
teleteaching as compared to other modes of teaching. A 1968
study in New York State was concerned with determining
whether the telephone could be used to spread instruction over
a wide area; sixty-nine teachers in fourteen school districts who
were taught diagnosis by telephone were tested against fifteen
others who received instruction in the conventional manner.
The conclusion was that the students taught by telephone did as

well as the others.[18] In a study comparing the effectiveness of class lectures and telelectures, Blackwood and Trent found no significant difference in the amount of learning.[19] Douglass, in a study assessing the effectiveness of telewriting, found that while telewriting can save much time and money, its effectiveness depends heavily on user attitudes, teacher preparation, and adequate service and facilities.[20] Hoyt and Frye compared the effectiveness of six classes taught remotely by amplified telephone with that of identical on-campus classes. When judged on the basis of academic achievement, the telephone classes and on-campus classes were equally successful.[21] The Pellett study found that an Educational Telephone Network (ETN) can be an effective medium for communicating cognitive knowledge for extension in-service training.[22] Though somewhat limited in scope, the research done on the effectiveness of teleteaching indicates that teleteaching is an economical and effective educational tool.

## ADVANTAGES OF TELETEACHING SYSTEMS

Although teleteaching lacks face-to-face instruction, it offers many advantages. Among the reported advantages are the following:
1. Teleteaching provides convenient access to the outstanding resource people in various fields.
2. It accommodates live or taped lectures; hence it simplifies scheduling. In the VERB system both audio and handwritten material may be prerecorded and played back as needed.
3. Teleteaching acts as a high motivational tool.
4. It can be used as a break-in medium for student teachers as there is less chance of getting nervous in teaching a remote class.
5. The needs of hospitalized and homebound students can be met.
6. It permits flexibility in planning and scheduling. One instructor taught a class gathered in Boston from a phone booth on the New York Taconic State Parkway.
7. The live mode of teleteaching allows immediate feedback to the students.
8. Faculty sharing by schools is possible. The University of

Illinois and the University of Wisconsin share their faculties in engineering through the VERB system. This feature also enables schools to offer more complete educational programs.

9. It is perhaps the only way to provide quality instruction to students in remote rural areas.

10. Classes for a limited number of students can be conducted at different locations simultaneously. In other modes of instruction these small classes might not be economical.

11. It is less expensive compared to other instructional media with similar applications. Cost per one hour of telelecture proved to be less than the cost of producing one hour of video tape.

12. Teleteaching has one great advantage over other communications media such as TV, radio, and tape recorder, because it permits two-way communication.

13. The teleteaching method proved to be very ego satisfying for high school students since they were able to talk to outstanding people.

14. It can be integrated into a multimedia program.

15. The use of telephones provides a universal access point to the instructors as well as to the students—just about any telephone can be used.

16. It eliminates travel costs and saves time for the instructors. Instructors no longer have to travel for hours to give a one-hour lecture, nor are the extension classes cancelled by bad weather.

17. Teleteaching is the closest thing yet to live lectures.

18. Studies indicate that students learn as much or more in teleteaching as in a face-to-face lecture.

19. When needed, communication between student and teacher can be personalized.

20. Teleteaching equipment may also be used for administrative applications when not used for instructional applications.

21. Teleteaching can be used as a part of computer-assisted instruction and as a part of a Dial Access Information Retrieval System.

22. No special training is needed to use teleteaching equipment.

23. Teleteaching uses a readily available telephone network.

24. Costs of teleteaching can be shared by a number of schools.

25. Sound levels and projected image sizes can be controlled to accommodate the needs of students and the learning conditions.

A review of available literature on the telephone's instructional applications reveals that the educational enterprise in North America has yet to exploit the telephone's full potential as an instructional communications device. With the advent of computer and audiovisual technologies, however, the use of telecommunication facilities by schools and colleges has been constantly increasing. It is consequently quite reasonable to assume that in the near future the telephone will be used extensively for a variety of instructional applications.

## NOTES

1. "To School by Telephone," *Scholastic Teacher*, April 5, 1950, p. 25.
2. *School and Society*, November 22, 1947, p. 390.
3. Puzzuoli, David A., *A Study of Teaching University Extension Classes by Telelecture* (Morgantown, W. Va.: West Virginia University, 1970), EEIC Report No. ED 042 961.
4. Madden, Charles F., "Amplified Telephone as a Teaching Medium," in Howard E. Bosley and Harold E. Wigren (eds.), *Television and Related Media in Teacher Education: Some Exemplary Practices* (Baltimore: Multistate Educational Project, 1967), ERIC Report No. ED 018 978.
5. Steele, Molly Anne, "Teleteaching: A New Form of Home and Hospital Instruction," *Audiovisual Instruction* 14 (November 1969), p. 80.
6. Lolis, Kathleen, *Evaluation of a Method of School-to-Home Telephone Instruction of Physically Handicapped Homebound Adolescents* (New York: Board of Education for the City of New York, 1968), ERIC Report No. ED 025 090.
7. Merrell, Bussell G., *Some Approaches to Meeting Cultural Deprivation in Students Entering Small Rural Schools* (Salt Lake City: Western States Small Schools Project for Utah, 1968), ERIC Report No. ED 026 181.
8. Clarke, Michael J., et al., *Art-by-Telephone: Design and Evaluation* (Las Vegas: Clark County School District, 1970), ERIC Report No. ED 044 222.
9. "Teaching by Telephone," *Audiovisual Instruction* 12 (September 1967), p. 683.
10. Puzzuoli, David A., "Extension Classes by Telelecture."
11. *The UNIVEX-NET*, a brochure on the University of Illinois VERB network.
12. Heliger, Edward M., *Teaching Information Retrieval Using Telediscussion Techniques* (Kent, Ohio: Kent State University, 1972), ERIC Report No. ED 058 905.
13. Byrd, Phyllis F., "Dial System Initiated in Virginia," *Adult Leadership* 21 (October 1972), pp. 122-123.
14. David, Austin, "ITV, Telephone, and Computer as an Instructional

System: A Feasibility Study," *AV Communication Review* 21 (Winter 1973), pp. 453–466.

15. Bolvin, Boyd M., "Using Technology to Serve Learning Needs of the Community," *New Directions for Community Colleges* 3 (Spring 1975), pp. 33–38.

16. *Los Angeles Community College District: A Plan for Development of an Educational Telephone Network (ETN) to Extend Access to Educational Programs* (Los Angeles: Los Angeles Community College District, 1973), ERIC Report No, ED 085 057.

17. Parker, Lorne A., "Educational Telephone Network and Subsidiary Communications Authorization: Educational Media for Continuing Education in Wisconsin," *Educational Technology* 14 (February 1974), pp. 34–36.

18. *Graduate Instruction via Telephone* (Homer, N.Y.: Supplementary Education Center, 1968), ERIC Report No. ED 032 767.

19. Blackwood, Helen, and Trent, Curtis, *A Comparison of the Effectiveness of Face-to-Face and Remote Teaching in Communicating Educational Information to Adults* (Manhattan, Kansas: Kansas State University, 1968), ERIC Report No. ED 028 324.

20. Douglass, Stephen A., *Telewriter, A Survey of Attitudes, Information and Implications* (Columbia, Missouri: University of Missouri, 1969), ERIC Report No. ED 038 606.

21. Hoyt, Donald P., and Frye, David W. M., *The Effectiveness of Telecommunications as an Educational Delivery System: Final Report* (Manhattan, Kansas: Kansas State University, 1972), ERIC Report No. ED 070 313.

22. Pellett, Vernon L., *The Comparative Effectiveness of the Educational Telephone Network and Face-to-Face Lectures for University Extension In-Service Training* (Ann Arbor, Michigan: University Microfilms, 1970).

# List of Contributors

Sidney H. Aronson is Professor of Sociology at Brooklyn College and the Graduate Center of the City University of New York and is Chairman of the Department of Sociology at Brooklyn College. He specializes in applying sociological perspectives to history.

Asa Briggs, Professor of History and Vice-Chancellor of the University of Sussex, 1966–1976, is now Provost of Worcester College, Oxford. His books include *A History of Broadcasting in the United Kingdom*, the fourth volume of which, *Sound and Vision*, is in press. In 1975 he received the first Marconi Medal for pioneering work in communications studies, in association with Dr. James Killian. He is a British life peer.

Charles R. Perry is an Instructor in History at the University of the South, Sewanee, Tennessee. He is currently completing his Harvard dissertation, "The British Post Office, 1836–1914: A Study in Nationalization and Administrative Expansion."

Jacques Attali is a Director of the Institut de Recherche Universitaire and a Professor of Public Economy at the Ecole Nationale des Ponts et Chaussées.

Yves Stourdze is Assistant Professor at the University of Paris and a researcher at the Institut de Recherche et d'Information Socio-Economique, University of Paris.

Colin Cherry has been Professor of Telecommunication at Imperial College, London, since 1958. His interests have largely been concerned with sociological and psychological aspects of communication technology.

Ithiel de Sola Pool is Professor of Political Science at M.I.T. and director of the Research Program on Communications Policy there. His books include *Talking Back: Citizen Feedback and Cable Technology* and *Handbook of Communication*. His coauthors are M.I.T students.

John Robinson Pierce served at the Bell Telephone Laboratories from 1936 until 1971; he was in charge of the research and communications sciences division there. Both the Echo program and Telstar resulted from satellite work he had initiated. He is currently at the California Institute of Technology.

Henry M. Boettinger is Director of Corporate Planning Research for AT&T. He is also a visiting fellow at Oxford and has written books and articles on management, history, and technology. He is the author of *The Telephone Book*, a social history of telephony, soon to be published.

John Brooks, historian and journalist, is the author of *Once in Golconda*, *The Go-Go Years*, and *Telephone: the First Hundred Years*, among other books. He is president of the Authors Guild of America.

Martin Mayer has written articles and books about a number of subjects, from advertising (*Madison Avenue, USA*) to education (*The Schools*) to finance (*The Bankers*). His most recent book is a more general discussion of current trends, entitled *Today and Tomorrow in America*.

Alan Wurtzel is Assistant Professor and Television Coordinator in the School of Journalism, University of Georgia. He has published numerous articles dealing with mass communication and its societal effects and is currently working on a major research grant from the American Broadcasting Company to examine the effects of television on children.

Colin Turner is Assistant Professor of Media Studies at Queens College of the City University of New York and is studying the cultural significance of media as tools of adaptation.

Brenda Maddox is an American author who lives in London and often writes about communications technology and policies. Her publications include "Communications: the Next Revolution" (London: The Economist, 1968) and *Beyond Babel: New Directions in Communications.* She is a frequent contributor to *The Economist* and *New Scientist.*

Suzanne Keller, Professor of Sociology, holds a joint appointment in the Department of Sociology and the School of Architecture at Princeton University. Her interests include the design of new communities and planning for the future. Among her works are *Beyond the Ruling Class, The Urban Neighborhood,* and *Sociology, A Text* (with D. Light).

Jean Gottmann is Professor of Geography in the University of Oxford and Fellow of Hertford College. He holds a Chair at the Ecole des Hautes Etudes en Sciences Sociales, Paris. Recurrently member of the Institute for Advanced Study, Princeton (1942–1965). Author of *Megalopolis* and other works.

Ronald Abler is Associate Professor of Geography at Pennsylvania State University, where he has been teaching since 1967. His major research interests are postal and telephone history and the influence of communications media on settlement patterns, especially cities.

J. Alan Moyer is a telecommunication systems engineer and a Ph.D. candidate in Urban Affairs at Boston College.

Bertil Thorngren is Associate Professor at the Stockholm School of Economics. He has been advisor to a number of organizations including the Royal Commission on the Dispersal of Government Offices and the OECD on the topic of communications and regional development.

Alex Reid is Head of Long Range Studies at the Telecommunications Headquarters of the British Post Office. He was formerly Director of the Communications Studies Group at University College, London, where he took his Ph.D. in research on the substitution of telecommunications for face-to-face contact.

Emanuel A. Schegloff teaches in the Department of Sociology, University of California, Los Angeles. His major research interest is in the organization of social interaction. A collection of papers in this area (jointly with Harvey Sacks and Gail Jefferson) will soon be published under the title *Studies in the Sequential Organization of Conversation.*

David Lester, M.A. Cambridge University, Ph. D. Brandeis University. Currently Professor of Psychology at Richard Stockton State College, Pomona, N. J.. Formerly Director of Research at the Suicide Prevention and Crisis Service in Buffalo, New York.

Paladugu Rao is Head of the Information Systems Department at Eastern Illinois University, Charleston, Illinois. He is the author of publications in the area of instructional communications and data processing.

# Index